Transmission Expansion Planning: The Network Challenges of the Energy Transition

Sara Lumbreras • Hamdi Abdi • Andrés Ramos
Editors

Transmission Expansion Planning: The Network Challenges of the Energy Transition

Springer

Editors
Sara Lumbreras
Institute for Research in Technology (IIT),
ICAI School of Engineering
Comillas Pontifical University
Madrid, Spain

Hamdi Abdi
Engineering Faculty,
Electrical Engineering Department
Razi University
Kermanshah, Iran

Andrés Ramos
Institute for Research in Technology (IIT),
ICAI School of Engineering
Comillas Pontifical University
Madrid, Spain

ISBN 978-3-030-49427-8 ISBN 978-3-030-49428-5 (eBook)
https://doi.org/10.1007/978-3-030-49428-5

This Springer imprint is published by the registered company Springer Nature Switzerland AG.
The registered company address is: Gewerbestrasse 11, 6330 Cham, Switzerland

Preface

The transmission network is a key element in the power system and is currently challenged by the deep transformation experienced by the energy sector. Namely the increase in renewable energy penetration and the integration of markets creates a need for new corridors and reinforcements to the existing grid.

Besides, the Transmission Expansion Planning (TEP) problem is highly non-linear and complex, as the physics of power flows and the idiosyncrasies of technological choices deeply influence network dynamics. This demands detailed modeling that should be balanced with manageable computation times. For this, a wide array of solution methods has been proposed in the literature.

This book reviews the recent developments in the transmission planning context and puts them in context, providing a much-needed comprehensive perspective for both practitioners and scholars. In particular, the book covers the following:

- An introduction to the problem, which includes a stylized formulation and a summary of the main modeling choices and solution methods applied in the literature, is presented in Chap. 1, "The Key Role of the Transmission Network."
- A particularly interesting development of the past few years has been the introduction of non-classical methods to solve complex optimization problems, which offers some advantages but risks inappropriate implementation. Chapter 2, "Metaheuristics for TEP," presents a critical review of these methods and provides guidelines for their application.
- During the past years, there has been a growing need for solving large interconnected systems, as well as systems with increasingly wide sources of uncertainty. The particularities of large-scale TEP in practical contexts are presented in Chap. 3, "TEP at a National Level."
- In a large-scale context, it can be extremely useful to apply reduction techniques to maintain the most relevant information about a case study while disregarding nonessential information. Chapter 4, "Reduction Techniques for TEP," deals with this matter extensively.
- The need for transmission over longer distances favors the idea of a super-grid, a wide-area transmission network that would make it possible to trade large

volumes of power across thousands of kilometers. This super-grid can be an overlay to the mainland transmission network or be built offshore, serving the functions of integrating offshore wind power as well as communicating different countries across the sea. Chapter 5, "The Emerging Offshore Grid," presents the main characteristics of this super-grid and how it is crucially impacted by regulatory considerations.

- Long-distance transmission of power can be made more efficient if HVDC technology is used. The main features, technological options, and barriers for its deployment are presented in Chap. 6, "HVDC in the Future Power System."
- Transmission and generation have a codependent structure that should be considered for long-term planning. Renewables are mostly installed in the geographical areas where its fundamental resource is abundant, which often happens in remote areas that lack enough transmission resources. Given that the deadlines involved in transmission are much longer than in generation (often around a decade versus a couple of years in some cases), it is exciting to consider the co-optimization of both generation and transmission concurrently. Chapter 7, "Transmission Expansion Planning Outside the Box: A Bilevel Approach," presents this perspective.
- The transmission network ultimately serves demand, carrying power to the distribution grid. Both demand and the distribution network are experiencing an intense transformation, mostly centered around the distributed generation. Chapter 8, "The Impact of Distributed Energy Resources on the Networks," presents the status of the distribution network challenges and an overview of the impact of smart grids at the transmission level.
- Even with more sophisticated power-flow models, important considerations such as system stability are difficult to incorporate in to transmission analyses. However, the extent of the reinforcements currently considered as well as the additional challenges of non-dispatchable renewables and the growing importance of power electronics, make it necessary to improve our understanding of the impact of transmission plans on the stability of the system. Chapter 9, "Stability Considerations in TEP," deals with these issues and presents some guidelines for action.
- The impact of energy storage systems on power system operation and planning are discussed in Chap. 10, "Energy Storage Systems in TEP."
- The key role of the transmission network in power systems as a natural monopoly, as well as its environmental and social impacts, means that it is a heavily regulated activity. Chapter 11, "The Regulation of the Expansion of Electricity Transmission," presents a general overview of the regulatory aspects of transmission and its expansion as well as their consequences for the future transmission grid.

We believe that this book will provide a general perspective that will help both practitioners and academics and become an effective reference in this topic.

Madrid, Spain Sara Lumbreras
Kermanshah, Iran Hamdi Abdi
Madrid, Spain Andrés Ramos

Acknowledgments

The editors of *Transmission Expansion Planning* extend their sincere thanks to the invaluable experience of all researchers who cooperated in the book and, more specifically, to all the authors and reviewers of the book. We extend sincere thanks to the reviewers, including individuals who have consistently and expeditiously delivered comprehensive, discerning, and valuable reviews.

The list of respectful reviewers is sorted in the following:

Álvaro Sánchez-Miralles	Comillas Pontifical University, Spain
Andrea Mazza	Politecnico di Torino, Italy
Aurelio García-Cerrada	Comillas Pontifical University, Spain
Carlos Mateo	Comillas Pontifical University, Spain
D.A. Tejada-Arango	Endesa, Spain
Fernando Postigo	Comillas Pontifical University, Spain
Francisco M. Echavarren	Comillas Pontifical University, Spain
Gerald Sanchís	RTE, Paris, France
Gianfranco Chicco	Politecnico di Torino, Italy
I.C. Gonzalez-Romero	Comillas Pontifical University, Spain
Javier García-González	Comillas Pontifical University, Spain
Javier Renedo	Comillas Pontifical University, Spain
João Gorenstein Dedecca	Trinomics, Netherlands
Luis Olmos	Comillas Pontifical University, Spain
Luis Rouco	Comillas Pontifical University, Spain
Lukas Sigrist	Comillas Pontifical University, Spain
Mansour Moradi	Azad University, Kermanshah Branch, Iran
Michel Rivier	Comillas Pontifical University, Spain

Quanyu Zhao	EDF Energy Services, China
Quentin Ploussard	Argonne National Laboratory, USA
Reza Hemmati	Kermanshah University of Technology, Iran
S. Pineda	University of Málaga, Spain
Sonja Worgin	Comillas Pontifical University, Spain

Madrid, Spain — Sara Lumbreras

Kermanshah, Iran — Hamdi Abdi

Madrid, Spain — Andrés Ramos

Contents

Chapter 1
Introduction: The Key Role of the Transmission Network

Sara Lumbreras, Hamdi Abdi, Andrés Ramos, and Mansour Moradi

Nomenclature

Indices		
i, j	Nodes in the network	
ij	Lines in the network (joining a given pair of nodes)	
EL, CL	Existing lines, candidate lines (sets)	
n	Load level	
y	1 year	
p	Period	
s	Sub-period	
u	Unit (generation)	
t	Thermal generation unit	
h	Hydro (or pumped hydro) generation unit	
Parameters		
	Generation	
$\underline{GP}_u, \overline{GP}_u$	Minimum and maximum generation limits for a unit	MW

S. Lumbreras (✉) · A. Ramos
Institute for Research in Technology (IIT), ICAI School of Engineering, Comillas Pontifical University, Madrid, Spain
e-mail: slumbreras@comillas.edu; Andres.Ramos@upcomillas.es

H. Abdi
Engineering Faculty, Electrical Engineering Department Razi University, Kermanshah, Iran
e-mail: hamdiabdi@razi.ac.ir

M. Moradi
Young Researchers and Elite Club, Islamic Azad University, Kermanshah Branch, Kermanshah, Iran

S. Lumbreras et al. (eds.), *Transmission Expansion Planning: The Network Challenges of the Energy Transition*, https://doi.org/10.1007/978-3-030-49428-5_1

$\overline{GC}_h$	Maximum consumption of a unit for pumped storage	MW
FC_t, VC_u	Fixed cost and variable cost of a unit	EUR/h, EUR/MWh
SU_t	Startup cost for thermal units	EUR
η_h	Efficiency of a pumped-storage hydro unit	p.u.
I_{ph}	Inflows for a hydro reservoir	hm^3
$\underline{R}_h$, $\overline{R}_h$	Minimum and maximum levels allowable for a hydro reservoir	hm^3
	Demand	
D_{ypsni}	Demand	MW
DUR_{psn}	Duration of a demand scenario	h
R_{ps}	Operating reserve requirements	MW
$CENS$	Cost of not served energy	EUR/MWh
$CPNS$	Cost of not served power	EUR/MW
	Transmission	
FCT_{ij}	Investment cost of a transmission line (annualized according to a French amortization schedule)	EUR
$\overline{F}_{ij}$	Power capacity of a transmission line	MW
$\overline{F'}_{ij}$	Upper bound for the disjunctive constraint of a transmission line	MW
R_{ij}, X_{ij}	Parameters of a transmission line in the DCPF model: resistance and reactance	p.u.
S_B	Base power parameter (as the system is expressed in p.u.)	MW
	Costs	
α, β, γ	Weight coefficient of the different components of the objective function: investment and operation costs and reliability costs	
Variables		
	Generation system	
gp_{ypsnu}, gc_{ypsnu}	Generator production for a unit in general or consumption for a pumped hydro unit	MW
u_{ypst}, su_{ypst}, sd_{ypst}	Variables that signal the commitment, startup, or shutdown of a thermal unit [0,1]	p.u.
r_{yph}	Level of hydro reservoir	hm^3
s_{yph}	Spillage	hm^3

	Demand	
ens_{ypsni}	Energy not served MWh	
pns_{yps}	Porwer not served MW	
	Transmission system	
ic_{yij}	Installed capacity of candidate line in each year {0,1}	p.u.
f_{ypsnij}	Power flow	MW
θ_{ypsni}	Voltage angle in a DCPF model	rad

1.1 Motivation

The transmission network, bridging the generation and distribution of energy, plays a pivotal role in the power system. This system must be capable of long-term and continuous energy transfer [1], which, given the long distances covered, leads to networks that are built and operated in AC or DC at high voltage levels [2]. The increasing growth in energy consumption has led to significant changes in the types of consumption as well as power generation sources, most importantly increasing the penetration of renewables in recent decades [2–4]. This trend is set to continue: Europe aims to achieve a 30% improvement in energy efficiency and a 27% increase in the use of renewable energies by making a 40% reduction in greenhouse gas (GHG) emissions by 2030, with the ultimate goal of completely eliminating GHG from power generation by 2050 [4–6]. Renewables can be located anywhere on the grid, causing dramatic and direct changes in the need for transmission. Besides, the need for tighter market integration calls for the expansion of cross-border capacity. These two main drivers have led to an increased need for transmission expansion that should be undertaken in the next couple of decades.

The network is divided into two levels: transmission and distribution. Transmission lines extend across relatively long distances, of the order of hundreds of kilometers, and have large capacities. The final stage of power delivery until the final consumers is undertaken by the distribution network. In Europe, the transmission network relies on high-voltage AC and DC lines of 132, 220, and 400 kV, while the distribution network uses voltages below 36 kV. The previous decades have witnessed an increase in power-transmission needs, of which there are two main drivers:

- The large-scale integration of renewables, where policy targets establish ambitious penetration goals. The European *Winter Package* aims at a 40% emission reduction, a 30% improvement in energy efficiency, and a 27% increase in renewables by 2030 [1]. In the US, a similar role is provided by the renewable portfolio standards, which introduce state-dependent renewable targets [2]. Two extreme alternatives are possible for the integration of these renewables: new capacity will either be placed as large, utility-scale plants (which are becoming increasingly cost efficient) or be deployed in a decentralized manner. In any case, the intensive investment will be necessary either to connect the new large-scale plants and reinforce the new main paths for the flows or to allow the network to provide support for a decentralized system design.

- The development of regional markets, which will demand more cross-border flows that will require new interconnection capacity. Strong interconnection capacity will mitigate market power as it will allow for increasing competition.

These two main drivers are complemented by the following secondary goals:

- Relieving congestion, which leads to price differences among different areas. Congestion can arise in an international context (and therefore appear concerning the development of regional markets) or at a smaller, national scale
- Improving reliability, given that the transmission network can increase the security of supply by providing alternative paths for power flows in the event of failures
- Connecting new generating power plants, either of conventional or renewable technologies
- Solving local constraints
- Reducing losses, an objective which is probably the least important of the list, but in certain cases, it can, however, be economically efficient to install new lines to save loss costs

This new transmission capacity may come in the form of reinforcements but also as a supergrid extending over long distances, particularly offshore. This is presented in detail in Chap. 5.

In addition to having a deep impact on system operation, the fact that the permitting and construction process is so long (often taking more than a decade) means that transmission network expansion planning (TNEP) or, in its more general form, transmission expansion planning (TEP) is a crucial exercise. This book reviews the main facets of this process, with the introduction in this chapter providing an overview of the problem as well as the main formulation choices and solution strategies available. This chapter is not intended to provide an extensive literature review but rather an introduction to the problem complemented with a few illustrative references. The authors refer to their publication [3] for a recent, exhaustive review.

1.2 General Formulation of a TEP Problem

1.2.1 General Formulation for a TEP Problem

This section presents a general formulation for the transmission expansion problem that assumes a DCPF. Although different case studies will need some adaptations, this formulation can easily be adapted to cope with the idiosyncrasies of any relevant project. Besides, it can be readily implemented into optimization solvers, which—as will be discussed below—are one of the most relevant techniques in this context.

1.2.2 *Objective Function*

The model minimizes the sum of three kinds of costs: investment costs (of the new transmission lines), operation costs, and reliability costs. The relative importance assigned to these components is represented as α, β, and γ.

The investment cost is the sum, for all the years in the planning horizon, of the fixed annual cost FCT_{ij} of lines multiplied by the decision to install them or not, ic_{yij}:

$$\alpha \sum_{yij} FCT_{ij} ic_{yij} \tag{1.1}$$

Operation cost is the sum of fixed, variable, and startup costs plus the penalties associated with energy not served and any deficits in operating reserve:

$$\beta \begin{bmatrix} \sum_{ypsnt} DUR_{psn} VC_t gp_{ypsnt} + \sum_{ypsnt} DUR_{psn} FC_t u_{ypst} + \sum_{ypst} SU_t su_{ypst} + \\ \sum_{ypsni} DUR_{psn} CENS\, ens_{ypsni} + \sum_{yps} CPNS\, pns_{yps} \end{bmatrix} \tag{1.2}$$

Reliability costs are also evaluated and included in the objective function, for instance, using an N-1 criterion:

$$\gamma \sum_{ypsni} DUR_{psn} CENS\, ens_{ypsni} \tag{1.3}$$

1.2.3 *Constraints*

Commitment relations between startup and shutdown variables:

$$u_{ypst} - u_{yps-1t} - su_{ypst} + sd_{ypst} = 0 \;\; \forall ypst$$

Power balance:

$$\sum_{g \in i} gp_{ypsnu} - \sum_{h \in i} \frac{gc_{ypsnh}}{\eta_h} + ens_{ypsni} = D_{ypsni} - \sum_j f_{ypsnji} + \sum_j f_{ypsnij} \;\; \forall ypsni \tag{1.4}$$

Voltage balance in DCLF:

$$\begin{array}{ll} f_{ypsnij} = \left(\theta_{ypsni} - \theta_{ypsnj}\right) \frac{S_B}{X_{ij}} & \forall ypsnij, ij \in E \\ \left| f_{ypsnij} - \left(\theta_{ypsni} - \theta_{ypsnj}\right) \frac{S_B}{X_{ij}} \right| \le \overline{F}'_{ij} \left(1 - ic_{yij}\right) & \forall ypsnij, ij \in C \end{array} \quad (1.5)$$

Capacity constraints:

$$\begin{array}{ll} \left| f_{ypsnij} \right| \le \overline{F}_{ij} & \forall ypsnij, ij \in EL \\ \left| f_{ypsnij} \right| \le \overline{F}_{ij} ic_{yij} & \forall ypsnij, ij \in CL \end{array} \quad (1.6)$$

We consider that new transmission lines are not retired, so they will be in place from the moment they are installed to the end of the planning horizon:

$$ic_{yij} \le ic_{y'ij} \; \forall yy'ij, ij \in C, y' > y \quad (1.7)$$

Operation reserve constraint:

$$\sum_t \overline{GP}_t u_{ypst} + \sum_h \overline{GP}_h + pns_{ys} \ge \sum_i \left[D_{yps1i} + R_{ps}\right] \forall yps \quad (1.8)$$

Reservoir constraint:

$$r_{yp-1h} + I_{ph} - \sum_n DUR_{psn} \left[gp_{ypsnh} - gc_{ypsnh}\right] - s_{yph} = r_{yph} \; \forall yph \quad (1.9)$$

Other bounds on system operation variables:

$$\begin{array}{ll} \underline{GP}_g u_{ypsu} \le gp_{ypsnu} \le \overline{GP}_g u_{ypsu} & \forall ypsnu \\ 0 \le gc_{ypsnh} \le \overline{GC}_h & \forall ypsnh \\ \underline{R}_h \le r_{yph} \le \overline{R}_h & \forall yph \\ 0 \le ens_{ypsni} \le D_{ypsni} & \forall ypsni \end{array} \quad (1.10)$$

This formulation can be extended to accommodate HVDC transmission (which is not subject to voltage angle constraints) or phase-shifting transformers (PSTs), which introduce an additional degree of flexibility in the angle constraints.

1.3 Modeling Options

The wide array of works that has been devoted to TEP considers different versions of this problem. This section describes the most important modeling options that can be considered when planning transmission, accompanied by some reference examples for illustration purposes.

As explained above, the criteria that guide the expansion must include network investment cost and operation cost, which includes generation costs, emission costs, or ENS. It is also possible to include other objectives that include the environmental impact of the plan. Besides, in a market context, aggregate social welfare or the reduction of market power can be included as an additional criterion. In most models, all these objectives are added into a single function, but alternative approaches include goal programming (GP) [4], fuzzy decision theory [5], and analytic hierarchy process (AHP) [6].

It is necessary to model system operation with an adequate level of detail to capture the most relevant dynamics between transmission investment and system operation. The model used to consider power flows is key in this context. The main choices, from more courses to more detailed, are transportation models (which include only an energy balance) [7], DCPF, which are the most widely used [8, 9], and ACPFs [10]. The latter, due to the complexity brought by their nonlinearity, are not usually included in classical optimization settings and have appeared mainly in conjunction with metaheuristics as explained below.

One of the most relevant determinants in TEP is whether the transmission is considered in isolation or together with the expansion of generation. Even though in most contexts generation is a liberalized business and it is not planned in a centralized manner, solving the two problems simultaneously can give interesting insights, particularly when large penetrations of renewables are anticipated at large distances from current demand centers [11–13]. This is presented in more detail in Chap. 7. Besides, most models disregard the impact of market power in the generation and instead consider a centralized, cost-based operation. Some exceptions to this simplification do exist [14, 15]. Some approaches consider decentralized transmission decisions, with agents deciding investments in an independent manner [16–18] or forming coalitions with demand or generation companies [19]. The particularities of transmission regulation are presented in Chap. 10.

Uncertainty affects the most relevant factors that determine transmission expansion planning. The consideration of several scenarios for the expansion of generation would be a particularly relevant source of uncertainty. In particular, the impact of distributed generation is especially important and is presented in Chap. 8. As mentioned above, the process for planning, permitting, and building a transmission line can extend for over a decade. Meanwhile, planning and building a new generation plant can take only a couple of years, particularly in the case of renewables. Besides, generation costs depend on global price dynamics, and the long-term evolution of fuel prices is particularly difficult to predict. These examples correspond to *nonrandom uncertainties* given that they cannot be described by a probability distribution and should be defined by scenarios that would usually be developed by experts. On the contrary, *random uncertainties* can indeed be described using a probability distribution. Examples of the latter type are the availability of renewable generation, hydro inflows, demand, or contingencies. Even though uncertainty is increasingly relevant, most transmission planning projects are deterministic. When they do incorporate uncertainty, most approaches resort to stochastic optimization, which focuses on the expected value of the total cost of the expansion plan [5, 20–22]. The main alternative to stochastic optimization is robust optimization, which considers the worst-case scenario [23–25]. Alternatively, fuzzy decision analysis views the scenarios of an uncertain problem in the same manner as a multicriteria problem would deal with its different criteria [26, 27].

Besides, TEP is naturally a multistage problem, as decisions are taken for several time horizons that get updated periodically. However, incorporating several horizons is considerably complex, and only a minority of works consider it [5, 28]. Sequential static approaches, which consider several time horizons but without their dynamic interactions, are far more common [20, 29–31], and even more authors consider only a static version of the TEP problem [14, 32, 33].

The selection of candidate lines to be included is usually undertaken manually by the planner. These candidate lines are later incorporated into the optimization problem. However, the manual selection of candidates can lead to suboptimal solutions. We refer the reader to Reference [34] for a method to deal with this issue. Besides, the inclusion of FACTS as investment possibilities is also a key modeling decision. In particular, only a few works incorporate explicitly PSTs [35] or HVDC [35–38], which is studied in depth in Chap. 6.

1.4 Solution Methods

A wide array of methods have been applied to the TEP problem. We present an overview of the most relevant ones, articulated in two main categories: classical and nonclassical.

1.4.1 Classical Optimization

If the discrete nature of investments is ignored (i.e., we consider it possible to install fractions of transmission lines, continuously) and a transportation model is used for power flows, linear programming can be applied. This leads to an approximate solution of TEP that can deal efficiently with large-scale problem instances [7].

If a discrete investment is considered, mixed-integer programming (MIP) is one of the most widely applied approaches. It allows using a DCLF as a representation of power flows by introducing a disjunctive constraint to enforce the voltage law only if a line has been installed (this is described in the formulation provided above). Simplified versions of losses, such as losses proportional to flow, can also be accommodated in this formulation [39, 40]. As mentioned above, this brings a good compromise between representing the reality of the system and being manageable in terms of computation time, and for this reason, this is the most found approach in the classical optimization context.

Some cases might justify using a more complex, ACLF model for power flows. This makes it necessary to resort to mixed-integer nonlinear programming (MINLP), which leads to much longer computation times and cannot guarantee convergence in general. For these reasons, the number of applications that resorts to this is very limited [41].

In the cases where uncertainty is considered through stochastic programming, if the number of incorporated scenarios is large, solving the problem can become very demanding. It is possible to apply decomposition methods and additional acceleration techniques to enhance the performance of optimization. A comprehensive review of these approaches can be found in [42]. Besides, some reduction techniques can be applied to retain the most important features of a problem instance while disregarding nonessential information. This is covered extensively in Chap. 4.

1.4.2 Nonclassical Techniques

Nonclassical techniques do not offer convergence guarantees. However, they are extremely practical in very large problems where classical techniques cannot be solved in affordable times. Metaheuristics can improve a solution iteratively, often drawing inspiration from natural processes. Genetic algorithms, for instance, mimic natural selection to find increasingly better solutions. They have been applied to a wide range of problems including TEP [43].

Other nonclassical techniques such as expert systems, which work by generalizing a set of sample problem instances, or heuristic searches [9], have also been applied to TEP successfully [44]. Chapter 2 deals explicitly with these methods.

1.4.3 Iterative Methods with Human Interaction

Any model is based on simplifications which might make its solutions unrealistic. Applying optimization to TEP often means that considerations such as reactive power cannot be taken into account. Chapter 9 presents the most relevant of these features. Therefore, before implementing the resulting plan, it is necessary to evaluate it in detail and assess its feasibility according to the most detailed criteria available. The most commonly followed strategy is to iterate between two different functional modules:

- A TEP module, which is used to provide an optimal expansion plan or a discrete set of plan alternatives.
- An evaluation module that includes all the relevant criteria that could not be included in the TEP module because of manageability, such as ACLF or environmental considerations.

The optimal plan resulting from the TEP module is evaluated in all the relevant criteria. If it is deemed satisfactory, then it is selected for implementation. If it is not, then a new constraint can be derived and included for a new iteration of the

TEP module. Chapter 3 deals with undertaking TEP in large-scale, real problems and presents with further details on this.

The remainder of this book covers in deeper detail the most relevant aspects of the TEP problem.

References

1. EC–European Commission, Clean energy for all Europeans. Com **860**, 2016 (2016)
2. Barbose G.: US Renewables Portfolio Standards: 2016 Annual Status Report. (2016)
3. S. Lumbreras, A. Ramos, The new challenges to transmission expansion planning. Survey of recent practice and literature review. Electr. Power Syst. Res. **134**, 19–29 (2016)
4. Alseddiqui J. & Thomas R. J.: Transmission expansion planning using multi-objective optimization," in *Power Engineering Society General Meeting*, 2006. *IEEE*, pp. 8 (2006)
5. M. Moeini-Aghtaie, A. Abbaspour, M. Fotuhi-Firuzabad, Incorporating large-scale distant wind farms in probabilistic transmission expansion planning—Part I: Theory and algorithm. Power Syst. IEEE Trans. **99**, 1–1 (2012)
6. E.A.C.A. Neto et al., An AHP multiple criteria model applied to transmission expansion of a Brazilian southeastern utility, in *Transmission and Distribution Conference and Exposition: Latin America (T&D-LA), 2010 IEEE/PES*, (2010), pp. 575–580
7. A. Marin, J. Salmeron, Electric capacity expansion under uncertain demand: Decomposition approaches. IEEE Trans. Power Syst. **13**(2), 333–339 (1998)
8. M.V.F. Pereira et al., A decomposition approach to automated generation/transmission expansion planning. IEEE Trans. Power Apparatus Syst. **PAS-104**(11), 3074–3083 (1985)
9. S. Binato, G.C. de Oliveira, J.L. de Araujo, A greedy randomized adaptive search procedure for transmission expansion planning. Power Syst. IEEE Trans. **16**(2), 247–253 (2001)
10. A.P. Meliopoulos et al., Optimal long range transmission planning with AC load flow. Power Apparatus Syst. IEEE Trans. **PAS-101**(10), 4156–4163 (1982)
11. B. Alizadeh, S. Jadid, Reliability constrained coordination of generation and transmission expansion planning in power systems using mixed-integer programming. Generation Trans Distribut. IET **5**(9), 948–960 (2011)
12. D. Pozo, E.E. Sauma, J. Contreras, A three-level static MILP model for generation and transmission expansion planning. Power Syst. IEEE Trans. **28**(1), 202–210 (2013)
13. Liu A. et al: Co-optimization of transmission and Other supply resources. Prepared for the Eastern Interconnection States' Planning Council, National Association of Regulatory Utility Commissioners, Washington, DC (2013)
14. L.L. Garver, Transmission network estimation using linear programming. Power Apparatus Syst. IEEE Trans. **PAS-89**(7), 1688–1697 (1970)
15. Gupta N., Shekhar R. & Kalra P. R.: A probabilistic transmission expansion planning methodology based on roulette wheel selection and social welfare. ArXiv Computer Science, (2012)
16. Contreras J.: A cooperative game theory approach to transmission planning in power systems. *Department of Electrical Engineering and Computer Sciences, University of California at Berkeley,*http://www.uclm.es/area/gsee/JavierC/tesis/thesis_JC.pdf (1997)
17. J. Contreras, F.F. Wu, A kernel-oriented algorithm for transmission expansion planning. Power Syst. IEEE Trans. **15**(4), 1434–1440 (2000)
18. Cagigas C. & Madrigal M.: Centralized vs. competitive transmission expansion planning: The need for new tools. In *Power Engineering Society General Meeting,* 2003, *IEEE* Vol. 2 pp. 1017 (2003)
19. Contreras J. & Wu F. F.: Coalition formation in transmission expansion planning. In *Power Engineering Society 1999 Winter Meeting, IEEE,* vol. 2 pp. 887 (1999)

20. C. Serna, J. Duran, A. Camargo, A model for expansion planning of transmission systems a practical application example. Power Apparatus Syst. IEEE Trans. **PAS-97**(2), 610–615 (1978)
21. Latorre G. et al: PERLA: A static model for long-term transmission planning. Modeling options and suitability analysis. in Proc. 2nd Spanish-Portuguese Conf. Elect. Eng. (1991)
22. T. Akbari, S. Zolfaghari, A. Kazemi, Multi-stage stochastic transmission expansion planning under load uncertainty using benders decomposition. International Review of Electrical Engineering **4**(5), 976–984 (2009)
23. V. Miranda, L.M. Proenca, Probabilistic choice vs. risk analysis-conflicts and synthesis in power system planning. Power Syst. IEEE Trans. **13**(3), 1038–1043 (1998)
24. Buygi M. O. et al: Market-based transmission planning under uncertainties. in Probabilistic Methods Applied to Power Systems, 2004 International Conference On, pp. 563–568 (2004)
25. P. Maghouli et al., A scenario-based multi-objective model for multi-stage transmission expansion planning. Power Syst. IEEE Trans. **26**(1), 470–478 (2011)
26. H. Sun, D.C. Yu, A multiple-objective optimization model of transmission enhancement planning for independent transmission company (ITC). Power Eng. Soc. Summer Meeting. IEEE **4**, 2033–2038 (2000)
27. El-Keib A. A., Choi J. & Tran T.: Transmission expansion planning considering ambiguities using fuzzy modeling. in Power Systems Conference and Exposition, 2006. PSCE '06. 2006 IEEE PES, pp. 207–215 (2006)
28. Yoshimoto K., Yasuda K. & Yokoyama R.: Transmission expansion planning using neuro-computing hybridized with genetic algorithm. in *Evolutionary Computation,* 1995., *IEEE International Conference On* pp. 126 (1995)
29. C. Dechamps and E. Jamoulle, "Interactive computer program for planning the expansion of meshed transmission networks," Int. J. Electr. Power Energy Syst., vol. 2, (2), pp. 103–108, 4, 1980
30. Barbulescu C. et al: Congestion management driven transmission expansion planning. *Universities' Power Engineering Conference (UPEC),* Proceedings of 2011 46th International, pp. 1–6 (2011)
31. Costeira M. & Saraiva J. T.: Discrete evolutionary particle swarm optimization for multiyear transmission expansion planning. in 17th Power Systems Computation Conference, Stockholm, Sweden, (2011)
32. R.J. Bennon, J.A. Juves, A.P. Meliopoulos, Use of sensitivity analysis in automated transmission planning. Power Apparatus Syst. IEEE Trans. **PAS-101**(1), 53–59 (1982)
33. P. Buijs, R. Belmans, Transmission Investments in a Multilateral Context. Power Syst. IEEE Trans. **27**(1), 475–483 (2012)
34. S. Lumbreras, A. Ramos, P. Sánchez, Automatic selection of candidate investments for Transmission Expansion Planning. Int. J. Electr. Power & Energy Syst. **59**(0), 130–140 (2014)
35. G. Blanco et al., Real option valuation of FACTS investments based on the Least Square Monte Carlo method. Power Syst. IEEE Trans. **26**(3), 1389–1398 (2011)
36. A.O. Ekwue, B.J. Cory, Transmission system expansion planning by interactive methods. Power Apparatus Syst. IEEE Trans. On **PAS-103**(7), 1583–1591 (1984)
37. Tande J. O. G., Korpås M., Warland L., Uhlen K., Van Hulle F., Impact of TradeWind Offshore Wind Power Capacity Scenarios on Power Flows in the European HV Network (2008)
38. Garzillo A. et al: Offshore grids in Europe: The strategy of Ireland for 2020 and beyond. in AC and DC Power Transmission, 2010. ACDC. 9th IET International Conference On, pp. 1–7 (2010)
39. S.T.Y. Lee, K.L. Hicks, E. Hnyilicza, Transmission expansion by branch-and-bound integer programming with optimal cost – capacity curves. Power Apparatus Syst. IEEE Trans. **PAS-93**(5), 1390–1400 (1974)
40. S. Lumbreras, A. Ramos, Transmission expansion planning using an efficient version of benders decomposition. A case study. IEEE Power Tech, Grenoble (France) (2013)
41. M. Rahmani et al., Efficient method for AC transmission network expansion planning. Electr. Power Syst. Res. **80**(9), 1056–1064 (2010)

42. Lumbreras S. & Ramos A.: How to solve the transmission expansion planning problem faster: Acceleration techniques applied to benders' decomposition. IET Generation, Transmission & Distribution, 2016
43. H.A. Gil, E.L. da Silva, A reliable approach for solving the transmission network expansion planning problem using genetic algorithms. *Electr. Power Syst. Res***58**(*1*), 45–51., 5/21 (2001)
44. R.K. Gajbhiye et al., An expert system approach for multi-year short-term transmission system expansion planning: An Indian experience. Power Syst. IEEE Trans. **23**(1), 226–237 (2008)

Chapter 2
Metaheuristics for Transmission Network Expansion Planning

Gianfranco Chicco and Andrea Mazza

2.1 Introduction

The typical partitioning of the solution methods for the transmission network expansion planning (TNEP) problem considers classical mathematical programming, application of heuristic rules and metaheuristic models [35, 38]. Metaheuristic algorithms are a viable option for the solution of optimisation problems and are of interest because of their capability to solve non-convex, non-linear, integer-mixed optimisation problems, such as the TNEP problem. In particular, these methods are useful to address discrete optimisation, in which the integer part of the problem cannot be eliminated, as it happens in the TNEP in which the transmission lines to add cannot be fractioned [17], and for a large system, this leads to a combinatorial explosion of a number of variables. With respect to the heuristic rules, which are simple and enable the inclusion of detailed modelling, but generally are driven by sensitivity and experience and are not mathematically rigorous [49], the metaheuristic algorithms are guided by solution strategies that should guarantee their convergence to the global optimum of the problem. However, it has to be pointed out that a formal proof of convergence to the global optimum is available only for some metaheuristics. Moreover, the available proofs generally indicate asymptotic convergence (i.e. after an infinite number of iterations). This is mathematically relevant, however, in engineering terms; metaheuristics do not guarantee the convergence to the global optimum and provide no indications on how far the solutions are from the global optimum.

This chapter addresses the application of metaheuristics to the TNEP problem, on the basis of selected literature contributions (mostly journal articles). Advantages

G. Chicco (✉) · A. Mazza
Dipartimento Energia "Galileo Ferraris", Politecnico di Torino, Torino, Italy
e-mail: gianfranco.chicco@polito.it; andrea.mazza@polito.it

S. Lumbreras et al. (eds.), *Transmission Expansion Planning: The Network Challenges of the Energy Transition*, https://doi.org/10.1007/978-3-030-49428-5_2

of using metaheuristics are the simplicity of implementation and the availability of a set of good solutions – not only a single solution [27]. The latter aspect may be of interest in today's systems, with growing uncertainties and higher complexity introduced by distributed energy resources, energy markets and policies [16]. Furthermore, the metaheuristics use only the results of power system solvers, with no need of converting the power system model in the optimisation solver [25]. On the other hand, the metaheuristic solver has to be customised to include the problem constraints.

In the last years, there has been a proliferation of metaheuristic algorithms applied to many engineering problems. In many cases these algorithms have been applied without clearly demonstrating their superiority with respect to other solvers, because of the use of simple and weak metrics to compare the solutions, such as the best solution found, the mean solution and so forth. In this way, part of the research on metaheuristics has been switched to the mere testing of new algorithms, shifting the attention on the real innovations and progresses occurring in the metaheuristics field [68].

According to the survey results indicated in Chicco and Mazza [5], power system planning topics, and the TNEP problem in particular, have been less involved in the massive testing of metaheuristic solvers and variants than other typical problems in the power system area (such as economic dispatch or optimal power flow). However, the number and variety of algorithms implemented to solve the *single-objective* TNEP are still quite high, and weak comparison criteria have been adopted in many cases. Conversely, for the *multi-objective* TNEP, the number of metaheuristics applied has been much more limited. This fact is interesting, considering that many efficient non-metaheuristic methods can solve the single-objective TNEP, while for the multi-objective TNEP, the metaheuristics have proven to be appropriate and fully competitive in many applications.

In the next sections of this chapter, the basic concepts concerning the decision variables and metaheuristic principles are first recalled. Then, the single-objective TNEP and multi-objective TNEP are addressed separately, highlighting specific aspects of the solvers used. Finally, the last section contains the conclusions.

2.2 Decision Variables and Metaheuristic Principles

2.2.1 Decision Variables

The decision variables that appear in all the contributions addressed are the circuits added between two nodes in each corridor (also indicated as right of way). In addition, other decision variables can be considered in single-objective or multi-objective TNEP problems, also depending on the inclusion of generation sources or demand-based strategies in the problem formulation.

In the single-objective TNEP, two main aspects are considered, i.e. the cost of the system and reliability-related aspects. In the first case, the decision variables may be:

1. The outputs of the generators in the different time periods ([20, 27, 34, 63, 71])
2. The proposed installed capacity and the predefined capacity [24]
3. The proposed installed capacities for different generation companies, whose profit is maximised into a low-level problem [26]
4. The installed capacities of the new generation together with the cost of the fuel infrastructure [73]

The reliability of the system is taken into account as:

(a) Loss of load/load shedding under normal condition [10, 17, 18, 21, 27, 36, 42, 51, 52, 72]. The loss of load under normal condition is null for feasible solutions, so it is a constraint integrated into a penalised objective function.
(b) Interruption costs seen from the point of view of generators, customers and transmission owner [24].
(c) Load management imposed by the central dispatching centre [73].

Other aspects addressed are the costs associated to CO_2 emissions [71] and the increase of the wind generation share [26, 46, 53, 54].

In the extended multi-objective approach presented in Jadidoleslam et al. [33], which combines TNEP with the planning of wind capacity, the added capacity of wind plants is taken as a further decision variable, and new circuits between existing nodes not previously connected are also allowed. In the multi-objective approach presented in Hu et al. [31], the set of decision variables includes also the candidate gas pipelines and gas compressors.

2.2.2 Metaheuristic Principles Applied to the TNEP Problem

The term *heuristic* is generally used to identify a tool that helps the user in discovering something. In particular, heuristic methods are useful if they are able to provide solutions when the solution space is unknown, or too large, or highly irregular. The term metaheuristic is also typically used, by adding the prefix *meta* that represents the presence of a higher-level strategy that drives the search. Many metaheuristics are based on translating the representation of natural phenomena or physical processes into computational tools.

In general, the metaheuristics can be partitioned into *single-solution update* methods (in which a succession of solutions is calculated, each time updating the solution only if the new one satisfies a predefined *criterion*) and *population-based* methods (in which many entities are simultaneously sent in parallel to solve the same problem). The metaheuristics may be characterised by considering a set of underlying principles (i.e. acceptance, decay, elitism, immunity, parallelism, selection, self-adaptation and topology) that form a common basis for the various methods or embed the structural differences among the methods [4]. In many cases different metaheuristics have been constructed without verifying whether they have new features or underlying principles to propose [68]. In other cases, metaheuristics are simply tested on TNEP trying to find a solution that reaches or improves previous results, then claiming that the solution method is superior to the other ones. This claim contains a major drawback, that is, the confusion among a good solution found with a method and the overall performance of the method. As discussed in Chicco and Mazza [5], "if the solutions found on a specific problem by using one solver are better than with another solver, this does not mean that the solver is better". In the next part of this chapter, these ideas are applied to address the solutions of the TNEP problem.

2.2.3 Scheme of the Population-Based Metaheuristics

Population-based metaheuristics are used in the large majority of the cases to solve the TNEP problem. The flow chart presented in Fig. 2.1 shows a relatively general scheme for the solution of a population-based problem. The data input is assumed to be appropriate for the problem to be solved. On the basis of this scheme, the details of the TNEP solution approaches are discussed with reference to the various blocks. The variable *iter* is the iteration number. Without loss of generality, in the scheme of Fig. 2.1, the creation of the new population has been inserted after the stop criterion, while in some references the stop criterion is checked after the new population has been created. The decision-making to determine the final solution is needed for the multi-objective approaches that generate multiple compromise solutions, to provide an indication on the most appropriate solution.

In the next sections, the use of metaheuristics to solve single-objective and multi-objective TNEP problems is addressed separately. On the basis of the scheme presented in Fig. 2.1, it is illustrated how to take into account the specific aspects of the TNEP in the various steps of the solution procedure.

2.3 Metaheuristics for the Single-Objective TNEP

2.3.1 Metaheuristics List

Numerous metaheuristics have been applied to the TNEP. A non-exhaustive list of metaheuristics used in the contributions addressed in this chapter (in their standard forms or in customised variants) includes:

AC: Ant colony [36]
BI: Bio-inspired with PSO [42]

Fig. 2.1 Scheme of a population-based metaheuristic

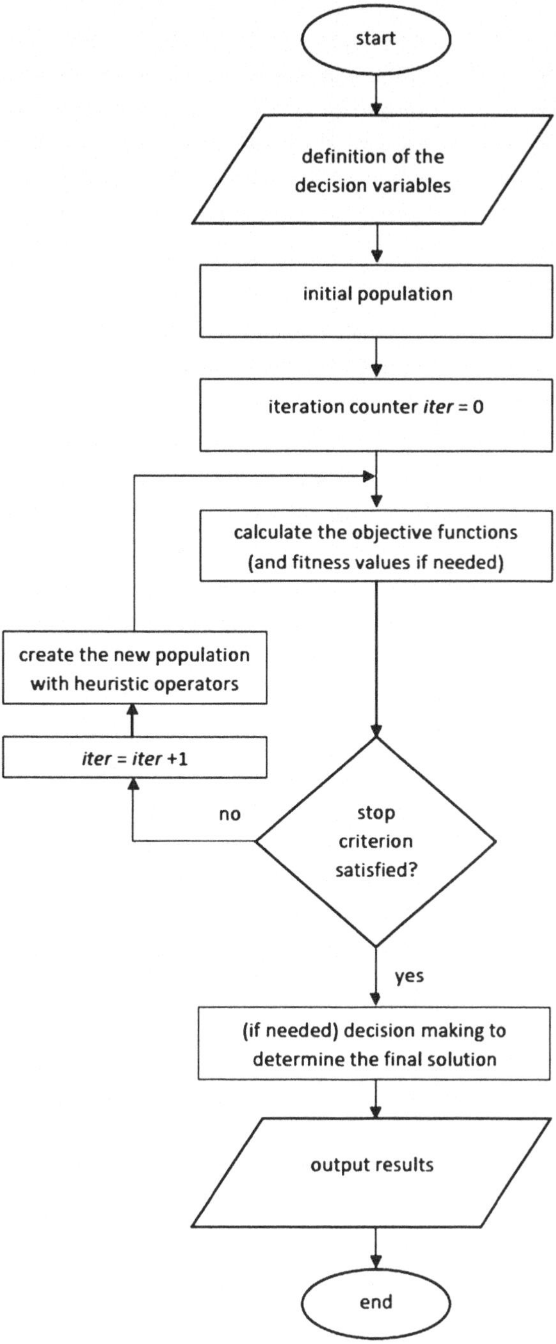

COA:	Chaos optimal algorithm [49]
DE:	Differential evolution [20, 72, 73]
EP:	Evolutionary programming [37]
FF:	Firefly [52]
GA:	Genetic algorithm [10, 18, 21, 24, 40, 48, 50, 63, 71]
GABC:	Gbest-guided artificial bee colony [54]
HAS:	Harmony search algorithm [51]
ICA:	Imperialist competitive algorithm [46]
MGBMO:	Modified gases Brownian motion optimisation [53]
PSO:	Particle swarm optimisation [26, 29, 34, 47, 65, 66]
SA:	Simulated annealing [17]
SR:	Simulated rebounding [27]
TS:	Tabu search [63]

2.3.2 *Initial Population*

In the single-solution update methods (e.g. the SA method in [17]), a single initial solution is provided. In the population-based methods, the most used way is to create the initial population at random. There are various exceptions, motivated in different ways. Techniques to obtain higher solution diversification are used in Hinojosa et al. [27], in Kamyab et al. [34] for the definition of the initial swarm, in Rastgou and Moshtagh [52] in the initial feasibility process and in Gallego et al. [18] with the implementation of an efficient constructive heuristic algorithm (ECHA). In Da Silva et al. [10], the initial population is found by constructing a number of head individuals by using a linear programming (LP) approach with continuous decision variables and selecting the initial population from them [10]. Similar concepts are applied in Gil and da Silva [21], in which the initial population is found from head individuals determined by the solution with continuous variables. Other approaches use the tree search algorithm [48] and clustering [50]. In the AC implementation in Leite da Silva et al. [36], different sequences are formed starting from the last year. In Leite da Silva et al. ([37], an intelligent initialisation procedure (InI) is formulated, and an interesting discussion is provided to indicate that the InI procedure could make the execution faster; however the success rate and the quality of the solutions are higher with the random initialisation. In Qu et al. [49], the initial values for the COA are generated in the range [0,1].

2.3.3 *Objective Functions and Customisation for TNEP*

The general objective of the methods addressed in this section is the cost of the transmission infrastructure. However, in the single-objective formulations, other

Table 2.1 Test and real networks used in the selected literature references

References	Networks	Objectives	Initial population	Comparison metric
[10]	BS, BSE, CO	LL	Random (head)	No improvement
[17]	G6, BS, BNNE	LL	Single initial	Best
[18]	BS, BNNE, CO	LS	ECHA	Best
[20]	IE30	G	Random	Best
[21]	BS, BNNE	LL	Random (head)	Best
[24]	IE5, IE24, IE118	G, LL	Random	Best
[26]	N6, EC, CCI	G	Random	Reserve margin
[27]	G6, EC, CH	G, LL	Different	Rate of success
[29]	G6, IE24	CS	Random	Best
[34]	G6, IE24	G	Random	Swarm diversity
[36]	B6, BST		Different	Various indices
[37]	IE24, BS	G	Random & InI	Various indices
[40]	AZ		Random	(None)
[42]	IE24, BS, CO	LL	Random	Best
[46]	IE24, IE118	G	Random	Best
[47]	G6, IE24	G	Random	Best
[48]	G6, BS		Tree search	Best
[49]	G6, BS		Values in [0,1]	(None)
[50]	ZA		Clustering	(None)
[51]	IE24, IE118	LL	Random	LMP
[52]	IE24, IE118, IR	LL	Feasibility	Best
[53]	G6, IE24, IE25	G	Random	Best
[54]	IE24, CO, BS	G, EV	Random	Best
[63]	C10, C18	G	Random	Best
[65]	G6, AR		Random	Fitness
[66]	G6, AR		Random	Fitness
[71]	G6, AR	G, E	Random	Costs
[72]	G6, IE24, BNE	LL	Random	Best et al.
[73]	G6, IE24	F, DM	Random	Fr-test, box plots

terms appear, typically expressed as costs as well, referring to various aspects (the abbreviations are the ones used in Table 2.1):

CS: Congestion surplus
DM: Demand management
E: CO_2 emissions
EV: Electric vehicles
F: Fuel infrastructure
G: Generation
LL: Loss of load
LS: Load shedding

In addition, there are many specific aspects considered in the construction of the solution methods, some of which are briefly recalled here. Two peculiar aspects of the TNEP problem that should be taken into account are the *incoherence* of the system after the addition of a transmission circuit and the *economies of scale*. These two aspects have an effect on the expected loss of load after the investment. In fact, the incoherence leads to an increase of the power transmitted on at least one circuit after the installation of a new circuit with respect to the power flow pre-investments. On the other hand, the economy of scale can lead to a larger reduction of loss of loads than the one obtainable by applying more small investments having the same capital cost. These two aspects have been detailed in Gil and da Silva [21], where the authors introduced the *loss of load limit curve*. The idea is to avoid the use of high penalty factors at the beginning of the optimisation procedure (which certainly leads to suboptimal solutions), by using *low* penalty factors to produce solutions that are infeasible (they are affected by loss of load). Starting from those solutions, it is possible to build the loss of load limit curve that contains an approximate cost of the optimal solutions. Then, by starting from the infeasible solution presenting the lowest amount of loss of load and modifying the penalty factor, the GA can determine the optimal solution.

An improvement in the analysis of TNEP problems is the incorporation of multiple contingencies in the GA for evaluating the loss of load. The evaluation of the condition with multiple contingencies requires an efficient calculation tool, for avoiding excessive computation time. In Gupta et al. [24], the calculation of *generalised line outage distribution factor* (GLODF) is used. The corresponding GLODF matrix allows reducing the memory used to store information and contributes to drive the optimisation process towards solutions with higher reliability.

The chronological aspects in TNEP are taken into account by considering multiperiod problem formulation. In Leite da Silva et al. [36], this aspect has been considered by using a heuristic method (in this case AC) for choosing the best investment according to the conditions of the last year of the planning horizon. Then, the investments for the previous years are coordinated with the ones related to the last year, by finally evaluating also the interruption costs associated to the different time sequences.

The calculation of the operational cost associated to a single investment proposal can be inaccurate, and thus an hourly demand should be considered. However, the calculation of the hourly OPF can be computationally too expensive. For that reason, the cost value has been evaluated in Mortaz et al. [47] through the use of multivariate interpolation, which allows calculating the variation of the operation costs (incorporated into the objective function) by varying parameters such as the fuel cost and demand. The method is included in the optimisation process (based on PSO) as a tool for evaluating the objective function considered as feasible with a predefined range of values of the parameters taken into account. The values of the parameters are taken uniformly distributed, and in the case of particles characterised by different values, the multivariate interpolation is applied.

A further practical aspect is related to the *budget constraint*. In fact, even though the minimisation of the costs is the most common objective function in TNEP problems, the result of the minimisation can be higher than the available budget. This particular aspect can be addressed by inserting the budget constraint in the optimisation problem, as in Shayeghi et al. [66], where the authors maximise the *adequacy* of the system by using the PSO as the solution algorithm.

2.3.4 Stop Criterion

The iterative process used in the metaheuristics cannot guarantee that the global optimum has been reached. Therefore, the iterations tend to continue by generating new solutions. A suitable *stop criterion* (or *termination* criterion) is then needed. Many papers use as stop criterion the maximum number of iterations. However, this criterion is generally not appropriate, because two situations may happen [4]:

1. *Early stopping*: the process could stop when significant improvements are still occurring.
2. *Unnecessarily late stopping:* the process could stop when the solution had practically no variation in many of the last iterations.

A better stop criterion is to terminate the iterations when no change in the objective function occurs after a given number of successive iterations. This *adaptive* stop criterion (also indicated as stagnation) is the most appropriate for a metaheuristic. The maximum number of iterations may be left together with the adaptive stop criterion as a last-resource option.

Notwithstanding the clear convenience of the adaptive stop criterion, most of the contributions considered adopt only the maximum number of iterations. The adaptive stop criterion is exploited in Georgilakis [20] to switch from one method to another, Gil and da Silva [21], Gupta et al. [24], Hooshmand et al. [29], Leite da Silva et al. [36, 37], Mahdavi et al. [40] and Poubel et al. [48]. In addition, specific stop criteria are used depending on the algorithm executed. For example, in Gallego et al. [17], the SA stops when the minimum "temperature" parameter is reached; in Hinojosa et al. [27], the SR stops when no new solution can be reached; in Moradi et al. [46], the ICA stops when only one empire remains; in Qu et al. [49], the iterations stop when the optimal solution appears for a predefined number of times (this is viable for relatively small systems); and in Sadegheih and Drake [63], the TS stops when no changes occur.

2.3.5 Test Systems for Case Study Applications

The TNEP problem has been solved by using multiple test and real system models. The second column of Table 2.1 shows the various networks used in the literature references considered, synthesised with the following identifiers:

AR: Azerbaijan regional [40]
B6: 6-node test [36]
BS: Brazilian Southern [58, 59]
BST: Brazilian sub-transmission [36]
BNE: Brazilian Northeastern [61]
BNNE: Brazilian North-Northeastern [60]
BSE: Brazilian Southeastern [10]
CCI: Chilean Central Interconnected [27]
C10: Chinese 10-bus [74]
C18: Chinese 18-bus [63]
CO: Colombian [15]
EC: Ecuadorian [27]
G6: Garver 6-bus [19]
IE5: IEEE 5-bus [23]
IE24: IEEE 24-bus [56]
IE25: IEEE 25-bus [14]
IE30: IEEE 30-bus [6]
IE118: IEEE 118-bus [30]
IR: Iran 400 kV [52]
N6: 6-bus [57]
ZR: Zanjan Regional [50]

The G6 network is a classical benchmark, as well as the IEEE networks, in which sometimes a few modifications have been introduced to take into account current developments, e.g. with renewable generation. Historically, some South American networks have been used for comparisons. A few other test and local real networks have been introduced more recently.

2.3.6 Comparisons Among the Solution Algorithms

Most of the contributions considered contain comparisons among the metaheuristic and other methods used for TNEP. The comparisons have been carried out by using different indicators.

For the single-objective TNEP, the simplest metric, e.g. the best value obtained, is used in many cases. The comparisons carried out with this metric are rather weak, because the best solution could be obtained by chance during the solution process, without meaning that the metaheuristic used is better than others [5]. In Torres and Castro [72], the set of indicators is extended from the best to the average and worst solutions, still with a lack of robustness in the comparison. Other criteria are the rate of success [27] and the percentage of best solutions achieved [72]. Only a few papers contain indications on other quality metrics that can represent more detailed and sound comparisons. The set of indicators used in Leite da Silva et al. [36, 37], including quality index, mean quality index, average index for the top ten plans, success rate and mean time of success runs, is a positive attempt to find more elaborated performance indicators. The Friedman test (Fr-test) and the box plots shown in Verma et al. [73] are another useful effort in the direction of providing comparisons with better statistical significance and the possibility of ranking the solution algorithms. To get further insights, the concept of first-order stochastic dominance introduced in Chicco and Mazza [5] can be exploited to construct indicators denoted as optimisation performance indicator based on stochastic dominance (*OPISD*). These indicators make it possible to rank the solutions obtained with heuristic methods, both when the global optimum is known (in this case, the effectiveness of the metaheuristic may be tested) and when the global optimum is unknown (making a relative comparison among the cumulative distributions of the solutions). Further contributions adopt qualitative comparisons based on specific outputs such as the swarm diversity [34], locational marginal price [51], load, generation and reserve margin [26] and various costs [71].

2.3.7 Hybridisation of Metaheuristic Solvers and Other Solutions

Various hybridisations have been presented by coupling metaheuristics between them or by coupling a metaheuristic with another algorithm. One of the most successful hybridisations is the evolutionary PSO (EPSO), which combines the positive characteristics of evolutionary programming and PSO, to obtain an evolutionary model with self-adaptation in which the particle movement operator from PSO is introduced to produce diversity [43]. The EPSO has been applied to TNEP in its discrete form (DEPSO) in Costeira da Rocha [8, 9]. In Chung et al. [7], a multi-objective problem with three objectives (investment cost, reliability and environmental impact) has been transformed into a single-objective problem by summing up the objectives. Then, a GA has been applied to generate the solutions, together with a fuzzy decision analysis to select the best solution. A more recent hybrid approach presented in Shivaie and Ameli [67] combines Melody search and the Powell heuristic, also using information-gap decision theory to handle the planning risks of planning depending on severe uncertainty. The general concepts indicated about the initial population, the stop criterion and the comparisons with other methods are still valid for these hybrid versions.

2.4 Metaheuristics for the Multi-objective TNEP

2.4.1 Relevance of Metaheuristics and Objective Functions for Multi-objective TNEP

For the single-objective TNEP, the metaheuristic approach is not the prevailing one, as many other classical mathematical optimisation algorithms and heuristic rules are used. Conversely, in the multi-objective TNEP, the situation is practically reverted, and the metaheuristics become the leading solution techniques because of their versatility in handling different conflicting objectives.

In general, the solvers used in multi-objective problems are partitioned into those based on weighted sums (which convert the multi-objective problem into a single-objective one, applicable to convex domains), goal programming approaches and Pareto front-based approaches. In the case of TNEP, the weighted sum approach cannot be used because of the non-convexity of the domain. The goal programming approach, applied in Xu et al. [76], needs to know a priori information on the preferences of the decision-maker [39]. The Pareto-based approach has been widely adopted in the current literature.

Selected papers are considered to illustrate the main aspects of the objective functions used to solve the multi-objective TNEP problem and the solution approaches (Table 2.2). The metaheuristics used are the multi-objective versions of population-based methods. The objective functions considered are:

CC: Congestion costs
CEC: Carbon emission costs
CENS: Cost of energy not supplied
CL: Curtailed load
ENS: Energy not supplied
NIC: Network investment costs
PC: Production costs
NSWLS: Number of scenarios without load shedding
WPPC: Wind power plant capacity

A detailed view of the formulations of these objective functions is outside the scope of this chapter. In some cases, conventional optimisation methods are used to compute the values of the objective functions that are used to create the Pareto fronts. Examples are the quadratic optimisation problems solved in Maghouli et al. [39] to compute the congestion cost of each alternative and the amount of load shedding, the two linear programming problems solved in Jadidoleslam et al. [33] for market clearing and calculating the *ENS* in a bi-level programming model and other methods indicated in Sect. 2.4.5.

A further aspect refers to the possible incorporation of uncertainty in the model. The earlier contributions [39, 75] did not address uncertainty. The more recent contributions take into account uncertainty in different ways, with reference to wind

Table 2.2 Characteristics of the multi-objective problems for TNEP

reference	number of objectives	min {NIC}	min {ENS}	min {CENS}	min {PC + CEC}	max {WPPC}	min {CC}	min {CL}	max {NSWLS}	metaheuristic
[75]	3	■		■			■			Improved SPEA
[39]	3	■					■	■		NSGA-II
Moeini et al. 2012a	3	■		■			■			NSGA-II
[28]	2	■	■							CNSGA-II
[69]	2	■							■	SPEA2
[31]	2	■			■					NSGA-II
[33]	3	■	■			■				MOSFLA

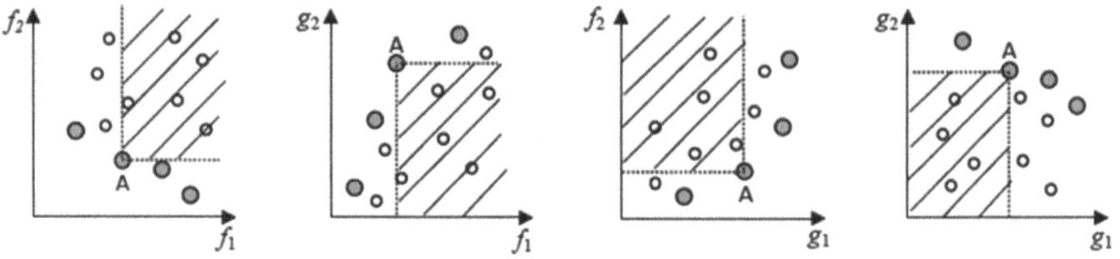

Fig. 2.2 Concept of Pareto dominance. Functions f_1 and f_2 are minimised. Functions g_1 and g_2 are maximised

power [31], wind and load [33, 44], hydro generation [69] and the probabilistic reliability criterion used in Hiroki and Mori [28].

In the next part of this section, the conceptual framework to analyse conflicting objectives through the creation of Pareto fronts is first recalled. On the basis of the general scheme presented in Fig. 2.1, it is illustrated how to take into account the specific aspects of the TNEP in the various steps of the solution procedure.

2.4.2 *Conflicting Objectives and Pareto Front Construction*

Multi-objective problems are defined by considering different objectives simultaneously. In order to set up a multi-objective problem, the objectives considered have to be in *conflict* with each other. Namely, when an objective is improved, at least another objective has to become worse. In this way, the multi-objective optimisation provides as results not only the best values for the individual objectives but also a number of *compromise solutions* seen as feasible decision-making alternatives. In order to represent the compromise solutions, the concept of *dominance* is applied. A solution is *non-dominated* when no other solution does exist with better values for all the individual objective functions. The compromise solutions are then found as the non-dominated solutions of the multi-objective problem. These non-dominated solutions form the *Pareto front*. The resulting concept of Pareto dominance depends on the nature of the objective function, that is, on whether the objective function has to be maximised or minimised. Figure 2.2 shows in the two-objective case how a given solution point dominates other points for different combinations of objective functions to be minimised (f_1 and f_2) or maximised (g_1 and g_2). In each plot, the filled area corresponds to the points dominated by point A indicated in the figure, and the filled circled points are the ones located on the Pareto front.

The construction of the Pareto front is also a way to establish whether two or more objectives are conflicting with each other. In fact, for non-conflicting objectives, the Pareto front degenerates into a single point.

In practical applications, it could be infeasible to calculate the entire Pareto front. In these cases, the *best-known* Pareto front is determined as the computable set of non-dominated solutions.

In the classical construction of the Pareto front, all compromise solutions have the same importance, so that a solution ranking mechanism has to be implemented to identify the most appropriate solution from the Pareto front (Sect. 2.4.7).

An alternative approach is to use fuzzy-based dominance degrees [2], in which the rank of each solution is directly identified in each Pareto front [33]. In this case, for $k = 1, \dots, K$ objectives, the solution point $\mathbf{c}_1 = \{c_{k,1}\}$ is dominated by the solution point $\mathbf{c}_2 = \{c_{k,2}\}$ at the degree of dominance $\kappa(\mathbf{c}_1, \mathbf{c}_2)$ given by

$$\kappa\left(\mathbf{c}_1, \mathbf{c}_2\right) = \frac{\prod_{k=1}^{K} \min\left\{c_{k,1}, c_{k,2}\right\}}{\prod_{k=1}^{K} c_{k,2}} \tag{2.1}$$

The degree of dominance ranges from 0 to 1. If $\kappa(\mathbf{c}_1, \mathbf{c}_2) = 1$, the solution point $\mathbf{c}_1$ is absolutely dominated by $\mathbf{c}_2$. If $\kappa(\mathbf{c}_1, \mathbf{c}_2) < 1$ and $\kappa(\mathbf{c}_2, \mathbf{c}_1) < 1$, the two solution points are non-dominated with each other.

In this framework, the rank index of each solution $\mathbf{c}_i$ belonging to the Pareto front $\mathcal{P}$ (in which the best solutions are the ones with the lowest values) is expressed as:

$$r\left(\mathbf{c}_i\right) = \text{mean}_{\mathbf{c}_j \in \mathcal{P}} \left\{\kappa\left(\mathbf{c}_i, \mathbf{c}_j\right)\right\} \tag{2.2}$$

2.4.3 Concepts Referring to the Pareto Front for Multi-objective Metaheuristics

The main concepts used in the solution methods are the ones borrowed from the general multi-objective solvers, that is:

- *Pareto front ranking* (or *non-dominated sorting*): the non-dominated solution points form the top-ranked Pareto front (rank $r = 1$); then, by removing these points, the resulting set of non-dominated solution points forms the front with rank $r = 2$. This procedure continues for the successive ranking (Fig. 2.3).
- *Crowding distance*: the crowding distance represents the *average distance* between the ith solution and the closest solutions belonging to the same Pareto front and estimates the *density* of the solutions located around that solution. The calculation of the crowding distance is performed by considering all the normalised objectives, and the crowding distance may also represent an indication on the perimeter of the cuboid (represented in a two-dimensional solution space with the rectangle shown in Fig. 2.3). A larger crowding distance means that the ith solution is representative of that part of the solution space, and no other can substitute it. So, it is a good candidate to maintain the *diversity* of the solution set. On the other hand, a smaller crowding distance means that several solutions can be representative of that part of the solution space; hence, maintaining all of them during the optimisation process could reduce the diversity of the solution set.

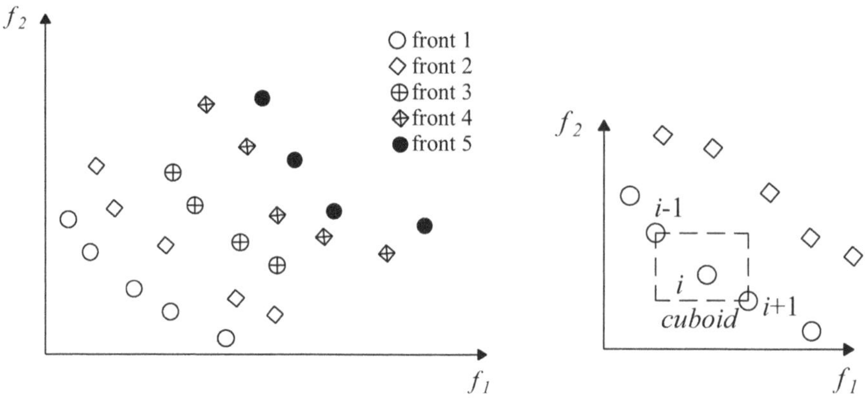

Fig. 2.3 Pareto front ranking and notion of crowding distance

Another indicator used in Wang et al. [75] to express the location uniformity is the spacing index S taken from the unpublished M.S. thesis of Schott [64], formulated for $k = 1, \ldots, K$ objectives as:

$$S = \sqrt{\frac{1}{K-1}\sum_{k=1}^{K}\left(\overline{d} - d_k\right)} \tag{2.3}$$

in which a column vector containing J values equal to unity is denoted as $\mathbf{1}_J$, J being the number of points in the Pareto front $\mathcal{P}$:

$$\overline{d} = \frac{1}{K-1}\sum_{k=1}^{K} d_k \tag{2.4}$$

$$d_k = \min_{\mathbf{c}_i \in \mathcal{P}}\left\{\mathbf{1}_J^{\mathrm{T}}\left|\mathbf{c}_k - \mathbf{c}_i\right|\right\} \tag{2.5}$$

2.4.4 Initial Population

Given the network constraints, a random choice of the initial population would be rather ineffective. For this reason, specific knowledge on the TNEP problem is exploited in the selection of the initial population, in different ways.

In Wang et al. [75], knowledge from TNEP is added to the SPEA algorithm. A random initialisation is used to generate a number of candidate networks, many of which could have isolated nodes. Then, an "isolated node absorbing initialisation" is carried out. In this procedure, a main network is created, and the possible isolated nodes (or groups of nodes) are then connected to the main network through the random addition of new lines. The procedure continues until there is no isolated node. In addition, the so-called "borderline search" strategy is added to avoid branch overload. For any candidate network in which there is an overload, the reduction of overloading can be obtained by adding one line to a corridor, and the line that leads to the highest overloading reduction is added to the candidate network. In this way, the initial population satisfies the constraints. In Sousa and Asada [69], the initial population is composed of feasible network topologies. A modified constructive heuristic algorithm is applied to each unfeasible topology by solving a linear programming problem with the use of a DC power flow model. If the resulting configuration is feasible, it is accepted and applied to each scenario; otherwise, a tournament-based method based on a sensitivity index is used to select a line to add, and the linear programming is solved again, continuing the process until a feasible configuration is found.

In Maghouli et al. [39], the initial solutions are selected at random among the feasible solutions. The initial population is indicated in Moeini et al. [44] and Jadidoleslam et al. [33] to be composed of alternative solutions, without further details. Also in Hiroki and Mori [28] and Hu et al. [31], the initial population is randomly initialised. It can be considered that these solutions have to be feasible, i.e. the problem constraints are satisfied.

2.4.5 Solution Methods and Customisation on the TNEP

The specific knowledge on the TNEP is applied during the generation of the solutions. In particular, the TNEP constraints are applied to the solutions to avoid the creation of unfeasible networks. For all the contributions addressed, the equality constraints are given by the DC power flow equations and the power balance at each node. In Maghouli et al. [39], the DC power flow is calculated in normal and contingency cases. In Hu et al. [31], the operation and security constraints are calculated for the electricity and natural gas networks and for the candidate transmission lines to be added. Furthermore, there are operation constraints of existing and candidate gas pipelines and gas compressors and an energy conversion equality constraint that links the electricity and natural gas systems.

The inequality constraints included generally refer to the maximum number of new lines added to a corridor (or the maximum number of transmission lines in each corridor), the power flow limits at each line and the limits on active power generation outputs. Further specific limits depending on the problem addressed are set on load values [75]; load curtailment [39]; positive profit, maximum risk for wind power installation and maximum wind capacity per site [33]; and rescheduling of generators [28]. An explicit network connectivity constraint is indicated in Hiroki and Mori [28], while in the other contributions, network connectivity is assured with the customised procedures aimed at obtaining feasible solutions.

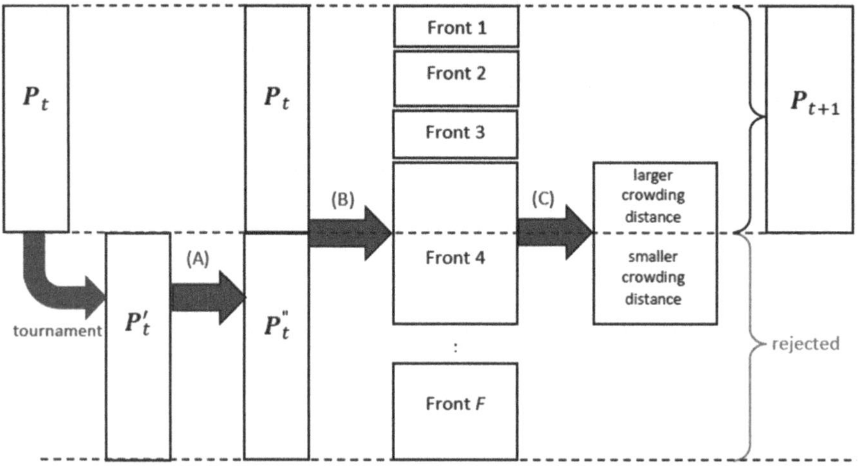

Fig. 2.4 Example of formation of the population at iteration t in NSGA-II
(A) Crossover and mutation on the population $\boldsymbol{P}'_t$
(B) Non-dominated sorting of the combined front $\boldsymbol{P}_t \cup \boldsymbol{P}''_t$
(C) Crowding distance sorting on the ordered Pareto front that leads to exceed the population size (e.g. Front 4 in this case)

The metaheuristic algorithm problems (indicated in Table 2.1) adopted to solve the TNEP are taken from the literature:

- Non-dominated sorting genetic algorithm (NSGA-II, [13]) most used as a metaheuristic solver or as a comparison benchmark for other metaheuristics. An example of formation of the population in NSGA-II is shown in Fig. 2.4. Different fronts are constructed by using non-dominated sorting, of which the first one is the Pareto front. The solutions to form the new population are then picked up from the successive fronts. If the number of solutions in the last front exceeds the size of the population used, the solutions with the smaller values of crowding distance are eliminated.
- Controlled non-dominated sorting genetic algorithm (CNSGA-II, [12]), in which the diversity of the set of solutions is claimed to be improved with respect to NSGA-II by using the reproduction of the solution candidates in the successive iterations.
- Improved version of the Strength Pareto Evolutionary Algorithm (SPEA, [77]), where the improvement indicated in Wang et al. [75] consists of the population initialisation and borderline search mentioned before.
- Strength Pareto Evolutionary Algorithm 2 (SPEA2, [78]).
- Multi-objective shuffled frog leaping algorithm (MOSFLA, [2]).

Specific knowledge on the TNEP is applied in the generation of the solutions with calculation of the objective functions. If the problem formulation does not incorporate N-1 security aspects, as in Wang et al. [75], the N-1 security of the

Pareto front points is checked after the formation of the Pareto front and is used as a mechanism to reduce the number of points. For each alternative solution, in Moeini et al. (2012a), the congestion cost is determined through a probabilistic optimal power flow based on the point estimation method (PEM), and the annualised value of the expected energy not supplied considering the uncertainties in wind power generation and load is obtained from probabilistic linear programming again using PEM. An improved PEM that takes into account the correlations among different wind farms is used in Hu et al. [31] to solve an optimal power flow and determine the expected production costs and the load curtailments.

2.4.6 Stop Criterion

The references addressed adopt the maximum number of iterations as the stop (or termination) criterion. The only (positive) exception is indicated explicitly in Maghouli et al. [39], where the iterative process is terminated when no other non-dominated solution is found in a predefined number of successive iterations. This corresponds to the adaptive stop criterion indicated in Sect. 2.3.4 as the most appropriate one. In Moeini et al. [44], there is a generic indication about the possible consideration of the number of individuals in the first Pareto front, in addition to the maximum number of iterations.

2.4.7 Final Decision from Solution Ranking

Finally, the compromise solutions can be *ranked* to assist the decision-maker in determining the most likely solution from the Pareto front. Some ranking methods such as the analytic hierarchy process (AHP) require translating the personal judgement of the decision-maker into numerical values (from 1 to 9, according to the Saaty scale, [62]) to be used in the computation (the comparisons among the different Pareto front outcomes can be carried out in an automatic way as in [41]). Other tools generally adopted to rank the Pareto front points are the Technique of Order Preference by Similarity to Ideal Solution (TOPSIS, [32]) and fuzzy-based tools [2, 11]. In the selected contributions addressed, fuzzy-based decision-making is the most applied. In Moeini et al. [44] and Jadidoleslam et al. [33], for each objective function $k = 1, \ldots, K$, a fuzzy membership is defined (if all the objective functions have to be minimised) as

$$\mu_{c_{k,i}} = \begin{cases} 0 & c_{k,i} > c_k^{\max} \\ \frac{c_k^{\max} - c_{k,i}}{c_k^{\max} - c_k^{\min}} & c_k^{\min} \le c_{k,i} \le c_k^{\max} \\ 1 & c_{k,i} < c_k^{\min} \end{cases} \tag{2.6}$$

Then, the decision-maker establishes the target levels $\hat{\mu}_k$ for each objective, and the best solution i^* is found by solving the optimisation problem:

$$i^* = \mathrm{argmin}_{\mathbf{c}_j \in \mathcal{P}} \sum_{k=1}^{K} \left| \hat{\mu}_k - \mu_{c_{k,i}} \right|^{\nu} \tag{2.7}$$

where the exponent is an integer number $\nu > 0$. From Deb [11], the use of larger values of ν reduces the sensitivity of the final solution to the target values.

Another TNEP-related criterion is the comparison among non-dominated solutions by using the incremental cost-benefit (*ICB*) ratio [39], given by the ratio between the reduction in the congestion cost with respect to the base case and the investment corresponding to the solution under analysis (as the base case has no investment).

Furthermore, the ranking index *RI* is used in Wang et al. [75] for the Pareto front solutions (for all variables to be minimised). All the variables are normalised by considering the corresponding maximum and minimum values of each objective $k = 1, \ldots K$, to make them comparable. For a given point $\mathbf{c}_i = \{c_{k,i}\}$ located on the Pareto front, its normalised version is:

$$c'_{k,i} = \frac{c_{k,i} - c_k^{\min}}{c_k^{\max} - c_k^{\min}} \tag{2.8}$$

The ranking index *RI* is calculated as the Euclidean distance between the normalised variables:

$$RI_i = \sqrt{\sum_{k=1}^{K} c'_{k,i}} \tag{2.9}$$

The solution with the lowest ranking index *RI* is taken as the best one.

2.4.8 Test Systems for Case Study Applications

The various contributions have used different test and real networks for their case study applications. An 18-bus test system and a 77-bus system have been tested in Wang et al. [75]. The Garver test system [19] has been used in Sousa and Asada [69]. The IEEE 24-bus reliability test system [55] is the most used one, either in its classical version [28, 39, 69], or in a modified version with wind power [33, 45], or integrated with a 15-bus natural gas system [31]. Further local systems have been used from Southern Brazil [69], Iran [39, 45] and China [31].

2.4.9 Comparisons Among the Solution Algorithms

The comparisons among the proposed approaches and other algorithms used in the literature are generally limited. In Moeini et al. [45] and Hu et al. [31], there is no comparison with other algorithms, and all the comparisons refer to "internal" cases developed inside the paper. The approach presented in Maghouli et al. [39], based on NSGA-II, is compared with the expansion plan proposed by the Iranian Grid Management Company. In Wang et al. [75], the comparison is carried out between the original SPEA and the improved SPEA proposed in the paper, concluding that the proposed algorithm finds more solutions than the original SPEA at the same iteration times and also provides better location uniformity. In the other three contributions [28, 33, 69], the proposed method is compared with NSGA-II. In summary, the CNSGA-II method used in Hiroki and Mori [28] maintains better solution diversity in part of the Pareto front with respect to NSGA-II. In Sousa and Asada [69], SPEA2 and NSGA-II exhibit similar solution quality. In Jadidoleslam et al. [33], the results obtained with MOSFLA are indicated to be better than the ones obtained with NSGA-II in Moeini et al. [45].

For multi-objective solvers, classical methods to rank the Pareto front solutions such as AHP and TOPSIS have not been used in the journal contributions addressed. The most exploited approach is based on fuzzy memberships. However, the comparisons among the Pareto fronts obtained have not been a subject of particular attention yet. These comparisons may be done by using quality indicators. An overview of the indicators proposed in the literature is presented in Zitzler et al. [79]. The distance between points located in the Pareto front under analysis and the closest points of the optimal or pseudo-optimal Pareto front can be considered, calculating, for instance, the quality indicator as the average of these distances. In other cases, the quality indicator is assessed with a chi-square-like deviation measure, in order to exploit the Pareto front diversity [70, 79]. An appropriate quality indicator is the *hyper-volume* determined from the Pareto front [3, 80], used both for assessing performance and for guiding the search in various hyper-volume-based metaheuristics [1]. General formulations for an efficient calculation of the hyper-volume from Pareto fronts in multiple dimensions are still unavailable. However, the Pareto fronts indicated in the above sections for the TNEP problem are defined in two and three dimensions. Hypervolume calculations for these cases are available, as illustrated in Guerreiro and Fonseca [22].

2.5 Conclusions

Some main conclusions may be drawn from the contents of this chapter:

- The most successful applications of metaheuristics to the TNEP problem are the ones that solve multi-objective problems. The classical metaheuristic algorithms for multi-objective programming have to be revisited to incorporate the specific knowledge on the technical aspects that concern the network topology.

- The current literature considers single aspects of the TNEP; however an overall approach that incorporates several aspects is still lacking. Modern formulations of the TNEP problems have to be developed to take into account the evolving aspects of the energy systems, markets and sustainable development. The recent contributions have started mixing up various TNEP objectives, and this line is expected to continue and to be reinforced, even though the complexity of the problem formulation and solution could increase.
- There is a need for establishing benchmark functions and benchmark networks. The network structures have to incorporate the main elements that appear in today's systems and take place in the TNEP formulations.
- From the current literature results, there is an apparent need to exploit more robust performance indicators for comparing the solutions obtained from meta-heuristics with a single objective or multiple objectives, to avoid an uncontrolled proliferation of solution algorithms that do not carry methodological insights.

References

1. A. Augera, J. Bader, D. Brockhoff, E. Zitzler, Hypervolume-based multiobjective optimization: Theoretical foundations and practical implications. Theor. Comput. Sci. **425**, 75–103 (2012)
2. S. Benedict, V. Vasudevan, Fuzzy-Pareto-dominance and its application in evolutionary multi-objective optimization. Proc. 3rd Int. Conf. Evol. Multi-Criterion Optim. (EMO), Berlin, Germany, 399–412 (2005)
3. D. Brockhoff, J. Bader, L. Thiele, E. Zitzler, Directed multiobjective optimization based on the weighted Hypervolume Indicator. J. Multicrit. Decis. Anal. **20**, 291–317 (2013)
4. Chicco G, Mazza A (2013) An Overview of the Probability-based Methods for Optimal Electrical Distribution System Reconfiguration. *Proc. 4th International Symposium on Electrical and Electronics Engineering (ISEEE 2013)*, Galati, Romania, 10–12 October 2013
5. G. Chicco, A. Mazza, Heuristic optimization of electrical energy systems: Refined metrics to compare the solutions. Sustain. Energy GridsNetworks **17**, 100197 (2019)
6. Christie R: Power Systems Test Case Archive. [Accessed June 20, 2019]. Available: http://www.ee.washington.edu/research/pstca/pf30/pg_tca30bus.htm (1993)
7. S. Chung, K.K. Lee, G.J. Chen, J.D. Xie, G.Q. Tang, Multi-objective transmission network planning by a hybrid GA approach with fuzzy decision analysis. Elect. Power Energy Syst. **25**, 187–192 (2003)
8. M. Costeira da Rocha, J. Tomé Saraiva, A multiyear dynamic transmission expansion planning model using a discrete based EPSO approach. Electr. Power Syst. Res. **93**, 83–92 (2012)
9. M. Costeira da Rocha, J. Tomé Saraiva, A discrete evolutionary PSO based approach to the multiyear transmission expansion planning problem considering demand uncertainties. Int. J. Electr. Power Energy Syst. **45**(1), 427–442 (2013)
10. E.L. Da Silva, H.A. Gil, J.M. Areiza, Transmission network expansion planning under an improved genetic algorithm. IEEE Trans. Power Syst. **15**(3), 1168–1174 (2000)
11. K. Deb, *Multi-Objective Optimization Using Evolutionary Algorithms* (Wiley, New York, 2001)
12. K. Deb, T. Goel, Controlled elitist non-dominated sorting genetic algorithm for better convergence, in *Proc. of ACM First International Conference on Evolutionary Multi-Criterion Optimization*, (2001), pp. 67–81

13. K. Deb, A. Pratap, A. Agarwal, Meyarivan, A fast and elitist multiobjective genetic algorithm: NSGA-II. IEEE Trans. Evol. Comput. **6**, 182–197 (2002)
14. A.O. Ekwue, B.J. Cory, Transmission system expansion planning by interactive methods. IEEE Trans. Power Syst. **103**(7), 1583–1591 (1984)
15. A.H. Escobar, R.A. Gallego, R. Romero, Multistage and coordinated planning of the expansion of transmission systems. IEEE Trans. Power Syst. **19**(2), 735–744 (2004)
16. L. Gacitua, P. Gallegos, R. Henriquez-Auba, Á. Lorca, M. Negrete-Pincetic, D. Olivares, A. Valenzuela, G. Wenzel, A comprehensive review on expansion planning: Models and tools for energy policy analysis. Renew. Sust. Energ. Rev. **98**, 346–360 (2018)
17. R.A. Gallego, A.B. Alves, A. Monticelli, R. Romero, Parallel simulated annealing applied to long term transmission network expansion planning. IEEE Trans. Power Syst. **12**(1), 181–188 (1997)
18. L.A. Gallego, L.P. Garcés, M. Rahmani, R.A. Romero, High-performance hybrid genetic algorithm to solve transmission network expansion planning. IET Generation Trans. Distrib. **11**(5), 1111–1118 (2017)
19. L.L. Garver, Transmission network estimation using linear programming. IEEE Trans. Power Apparatus Syst. **PAS-89**(7), 1688–1697 (1970)
20. P.S. Georgilakis, Market-based transmission expansion planning by improved differential evolution. Int. J. Electr. Power Energy Syst. **32**(5), 450–456 (2010)
21. H.A. Gil, E.L. da Silva, A reliable approach for solving the transmission network expansion planning problem using genetic algorithms. Electr. Power Syst. Res. **58**(1), 45–51 (2001)
22. A.P. Guerreiro, C.M. Fonseca, Computing and updating Hypervolume contributions in up to four dimensions. IEEE Trans. Evol. Comput. **22**(3), 449–463 (2018)
23. N. Gupta, R. Shekhar, P.K. Kalra, Congestion management based roulette wheel simulation for optimal capacity selection: Probabilistic transmission expansion planning. Int. J. Electr. Power Energy Syst. **43**, 1259–1266 (2012)
24. N. Gupta, R. Shekhar, P.K. Kalra, Computationally efficient composite transmission expansion planning: A Pareto optimal approach for techno-economic solution. Int. J. Electr. Power Energy Syst. **63**, 917–926 (2014)
25. R. Hemmati, R.A. Hooshmand, A. Khodabakhshian, State-of-the-art of transmission expansion planning: Comprehensive review. Renew. Sust. Energ. Rev. **23**, 312–319 (2013)
26. R. Hemmati, R.A. Hooshmand, A. Khodabakhshian, Coordinated generation and transmission expansion planning in deregulated electricity market considering wind farms. Renew. Energy **85**, 620–630 (2016)
27. V.H. Hinojosa, N. Galleguillos, B. Nuques, A simulated rebounding algorithm applied to the multi-stage security-constrained transmission expansion planning in power systems. Int. J. Electr. Power Energy Syst. **47**, 168–180 (2013)
28. K. Hiroki, H. Mori, An efficient multi-objective Meta-heuristic method for probabilistic transmission network planning. Proc. Comput. Sci. **36**, 446–453 (2014)
29. R.A. Hooshmand, R. Hemmati, M. Parastegari, Combination of AC transmission expansion planning and reactive power planning in the restructured power system. Energy Convers. Manag. **55**, 26–35 (2012)
30. Illinois Institute of Technology, IEEE 118 Bus Test System. Available: http://motor.ece.iit.edu/Data
31. Y. Hu, Z. Bie, T. Ding, Y. Lin, An NSGA-II based multi-objective optimization for combined gas and electricity network expansion planning. Appl. Energy **167**, 280–293 (2016)
32. C.L. Hwang, K. Yoon, *Multiple Attribute Decision Making: Methods and Applications* (Springer-Verlag, New York, 1981)
33. M. Jadidoleslam, A. Ebrahimi, M.A. Latify, Probabilistic transmission expansion planning to maximize the integration of wind power. Renew. Energy **114**(B), 866–878 (2017)
34. G.R. Kamyab, M. Fotuhi-Firuzabad, M. Rashidinejad, A PSO based approach for multi-stage transmission expansion planning in electricity markets. Int. J. Electr. Power Energy Syst. **54**, 91–100 (2014)

35. T.S. Kishore, S.K. Singal, Optimal economic planning of power transmission lines: A review. Renew. Sust. Energ. Rev. **39**, 949–974 (2014)
36. A.M. Leite da Silva, L.S. Rezende, L.A. da Fonseca Manso, L.C. de Resende, Reliability worth applied to transmission expansion planning based on ant colony system. Int. J. Electr. Power Energy Syst. **32**(10), 1077–1084 (2010)
37. A.M. Leite da Silva, M.R. Freire, L.H. Honório, Transmission expansion planning optimization by adaptive multi-operator evolutionary algorithms. Electr. Power Syst. Res. **133**, 173–181 (2016)
38. S. Lumbreras, A. Ramos, The new challenges to transmission expansion planning. Survey of recent practice and literature review. Electr. Power Syst. Res. **134**, 19–29 (2016)
39. P. Maghouli, S.H. Hosseini, M.O. Buygi, M. Shahidehpour, A multi-objective framework for transmission expansion planning in deregulated environments. IEEE Trans. Power Syst. **24**(2), 1051–1061 (2009)
40. M. Mahdavi, H. Shayeghi, A. Kazemi, DCGA based evaluating role of bundle lines in TNEP considering expansion of substations from voltage level point of view. Energy Convers. Manag. **50**(8), 2067–2073 (2009)
41. A. Mazza, G. Chicco, A. Russo, Optimal multi-objective distribution system reconfiguration with multi criteria decision making-based solution ranking and enhanced genetic operators. Int. J. Electr. Power Energy Syst. **54**, 255–267 (2014)
42. I. Miranda de Mendonça, I. Chaves Silva Junior, B. Henriques Dias, A.L.M. Marcato, Identification of relevant routes for static expansion planning of electric power transmission systems. Electr. Power Syst. Res. **140**, 769–775 (2016)
43. V. Miranda, N. Fonseca, EPSO - best-of-two-worlds meta-heuristic applied to power system problems, in *Proc. of the 2002 Congress on Evolutionary Computation (CEC'02)*, vol. 2, (2002), pp. 1081–1085
44. M. Moeini-Aghtaie, A. Abbaspour, M. Fotuhi-Firuzabad, Incorporating large-scale distant wind farms in probabilistic transmission expansion planning; part I: Theory and algorithm. IEEE Trans. Power Syst. **27**, 1585–1593 (2012a)
45. M. Moeini-Aghtaie, A. Abbaspour, M. Fotuhi-Firuzabad, Incorporating large-scale distant wind farms in probabilistic transmission expansion planning. Part II: Case studies. IEEE Trans. Power Syst. **27**, 1594–1601 (2012b)
46. M. Moradi, H. Abdi, S. Lumbreras, A. Ramos, S. Karimi, Transmission expansion planning in the presence of wind farms with a mixed AC and DC power flow model using an imperialist competitive algorithm. Electr. Power Syst. Res. **140**, 493–506 (2016)
47. E. Mortaz, L.F. Fuerte-Ledezma, G. Gutiérrez-Alcaraz, J. Valenzuela, Transmission expansion planning using multivariate interpolation. Electr. Power Syst. Res. **126**, 87–99 (2015)
48. R.P.B. Poubel, E.J. De Oliveira, L.A.F. Manso, L.M. Honório, L.W. Oliveira, Tree searching heuristic algorithm for multi-stage transmission planning considering security constraints via genetic algorithm. Electr. Power Syst. Res. **142**, 290–297 (2017)
49. G. Qu, H. Cheng, L. Yao, Z. Ma, Z. Zhu, Transmission surplus capacity based power transmission expansion planning. Electr. Power Syst. Res. **80**(1), 19–27 (2010)
50. H.K. Rad, Z. Moravej, An approach for simultaneous distribution, sub-transmission, and transmission networks expansion planning. Int. J. Electr. Power Energy Syst. **91**, 166–182 (2017)
51. A. Rastgou, J. Moshtagh, Improved harmony search algorithm for transmission expansion planning with adequacy–security considerations in the deregulated power system. Int. J. Electr. Power Energy Syst. **60**, 153–164 (2014)
52. A. Rastgou, J. Moshtagh, Application of firefly algorithm for multi-stage transmission expansion planning with adequacy-security considerations in deregulated environments. Appl. Soft Comput. **41**, 373–389 (2016)
53. C. Rathore, R. Roy, A novel modified GBMO algorithm based static transmission network expansion planning. Int. J. Electr. Power Energy Syst. **62**, 519–531 (2014)
54. C. Rathore, R. Roy, Impact of wind uncertainty, plug-in-electric vehicles and demand response program on transmission network expansion planning. Int. J. Electr. Power Energy Syst. **75**, 59–73 (2016)

55. Reliability Test System Task Force, The IEEE reliability test system-1996. IEEE Trans. Power Syst. **14**, 1010–1020 (1999)
56. M.J. Rider, A.V. Garcia, R. Romero, Power system transmission network expansion planning using AC model. IET Gener, Trans Distrib. **1**, 731–742 (2007)
57. J.H. Roh, M. Shahidehpour, L. Wu, Market-based generation and transmission planning with uncertainties. IEEE Trans. Power Syst. **24**(3), 1587–1598 (2009)
58. R. Romero, A. Monticelli, A hierarchical decomposition approach for transmission network expansion planning. IEEE Trans. Power Syst. **9**(1), 373–380 (1994)
59. R. Romero, A. Monticelli, A zero-one implicit enumeration method for optimizing investments in transmission expansion planning. IEEE Trans. Power Syst. **9**(3), 1385–1391 (1994b)
60. Romero R., Gallego R.A., Monticelli A (1995) Transmission System Expansion Planning by Simulated Annealing, Power Industry Computer Applications - PICA 95, Salt Lake City, May 1995
61. R. Romero, A. Monticelli, A. García, S. Haffner, Test systems and mathematical models for transmission network expansion planning. IEE Proc-Gener Transm Distrib. **149**(1), 27–36 (2002)
62. T.L. Saaty, A scaling method for priorities in hierarchical structures. J. Math Psychol. **15**(3), 234–281 (1977)
63. A. Sadegheih, P.R. Drake, System network planning expansion using mathematical programming, genetic algorithms and tabu search. Energy Convers. Manag. **49**(6), 1557–1566 (2008)
64. Schott JR: Fault Tolerant Design Using Single and Multi-Criteria Genetic Algorithm Optimization. Master's Thesis, Department of Aeronautics and Astronautics, Massachusetts Institute of Technology, Cambridge, MA (1995)
65. H. Shayeghi, M. Mahdavi, A. Bagheri, An improved DPSO with mutation based on similarity algorithm for optimization of transmission lines loading. Energy Convers. Manag. **51**(12), 2715–2723 (2010)
66. H. Shayeghi, M. Mahdavi, A. Bagheri, Discrete PSO algorithm based optimization of transmission lines loading in TNEP problem. Energy Convers. Manag. **51**(1), 112–121 (2010b)
67. M. Shivaie, M.T. Ameli, Strategic multiyear transmission expansion planning under severe uncertainties by a combination of melody search algorithm and Powell heuristic method. Energy **115**(1), 338–352 (2016)
68. K. Sörensen, Metaheuristics—The metaphor exposed. Int. Trans. Oper. Res. **22**(1), 3–18 (2015)
69. A.S. Sousa, E.N. Asada, Long-term transmission system expansion planning with multi-objective evolutionary algorithm. Electr. Power Syst. Res. **119**, 149–156 (2015)
70. N. Srinivas, K. Deb, Multiobjective optimization using nondominated sorting in genetic algorithms. Evol. Comput. **2**(3), 221–248 (1994)
71. Y. Sun, C. Kang, Q. Xia, Q. Chen, N. Zhang, Y. Cheng, Analysis of transmission expansion planning considering consumption-based carbon emission accounting. Appl. Energy **193**, 232–242 (2017)
72. S.P. Torres, C.A. Castro, Specialized differential evolution technique to solve the alternating current model based transmission expansion planning problem. Int. J. Electr. Power Energy Syst. **68**, 243–251 (2015)
73. P. Verma, K. Sanyal, D. Srinivsan, K.S. Swarup, Information exchange based clustered differential evolution for constrained generation-transmission expansion planning. Swarm Evol. Comput. **44**, 863–875 (2019)
74. X. Wang, J.R. McDonald, *Modern Power System Planning* (McGraw-Hill International, London, 1994)
75. Y. Wang, H. Cheng, C. Wang, Z. Hu, L. Yao, Z. Ma, Z. Zhu, Pareto optimality-based multi-objective transmission planning considering transmission congestion. Electric Power Syst. Res. **78**(9), 1619–1626 (2008)
76. Z. Xu, Z.Y. Dong, K.P. Wong, A hybrid planning method for transmission networks in a deregulated environment. IEEE Trans. Power Syst. **21**(2), 925–929 (2006)

77. E. Zitzler, L. Thiele, Multiobjective evolutionary algorithms: A comparative case study and the strength Pareto approach. IEEE Trans. Evol. Comput. **3**(4), 257–271 (1999)
78. E. Zitzler, M. Laumanns, L. Thiele, in *SPEA2: Improving the Strength Pareto Evolutionary Algorithm*, ed. by Swiss Federal Institute of Technology, (2001) Technical Report
79. E. Zitzler, L. Thiele, M. Laumanns, C.M. Fonseca, V. Grunert da Fonseca, Performance assessment of multiobjective optimizers: An analysis and review. IEEE Trans. Evol. Comput. **7**(2), 117–132 (2003)
80. E. Zitzler, J. Knowles, L. Thiele, Quality assessment of Pareto set approximations, in *Multiobjective Optimization: Interactive and Evolutionary Approaches*, ed. by J. Branke, K. Deb, K. Miettinen, R. Slowinski, (Springer, Berlin Heidelberg, 2008)

Chapter 3
Transmission Network Expansion Planning of a Large Power System

Gerald Sanchís

3.1 Transmission Network Planning at the National Level

3.1.1 Introduction – Historical Approach

Since the beginning of the use of the electricity as energy supply to the population, the transmission network has aimed at transporting bulk power to load centers across large geographic areas.

Thus, very early and very quickly, more and more important electrical networks and systems were developed.

Compared to capital-intensive production assets, relatively cheap transport assets for building power grids offer significant savings in terms of production costs.

Consequently, the direct connection between power plants and their immediate outlet was quickly completed for their integration into electrical systems, more economical and more resistant to each component failure. These electrical systems themselves then merged into larger ones.

At this period, generation and transmission planning was generally performed by integrated utilities. They applied long-term, national energy policies and pushed for large centralized power plants.

During the period of expansion, the transmission network was easily developed, following the implementation of generation and the development of the load centers.

With the development of the unbundling between generation and transmission which occurred in the 1990s, generation planning was separated from transmission planning. Transmission planning is nowadays mainly performed by a transmission

G. Sanchís (✉)
RTE, Paris, France
e-mail: gerald.sanchis@rte-france.com

S. Lumbreras et al. (eds.), *Transmission Expansion Planning: The Network Challenges of the Energy Transition*, https://doi.org/10.1007/978-3-030-49428-5_3

system operator (TSO). It uses supposed production planning because competing producers are unwilling to disclose their long-term strategic plans.

Thus, the unbundling has induced a more complex process for the development of the transmission network. The system planners must make their own assumptions about the development, technology, and location of the different production plants. At the same time, the electricity demand has evolved with the application of efficiency measures for example.

In order to better cope with these uncertainties on generation and on demand, system planners now use a scenario methodology. This methodology is particularly relevant for addressing long-term perspectives.

It is necessary to explain why it is better to create multiple scenarios in the context of system planning, rather than just one scenario. First of all, the uncertainty linked to the future location of electricity production and consumption depends on prices, technological advances, geopolitical developments, climate change, etc. Second, data on production projects is less and less shared as producers compete. And when this information is available, long construction delays are an additional source of uncertainty. Finally, it is important to take into account the effect of different policy choices, as well as the different options for implementing policies, such as the reduction and efficiency of energy demand, the integration of renewable energy sources (RESs), reduction of CO_2 emissions, and dismantling of polluting units.

3.1.2 Key Drivers for Network Development

Despite this constantly changing context, the drivers of network development are essentially the same as before in recent decades, allowing efficient and reliable development of load and generation. Yet, they become more complex with the consequential changes in the legal and regulatory framework. TSOs must nowadays address simultaneously a series of major challenges, such as:

- To carry out their main missions to respond to long-term energy policy; integrate large quantities of renewable energy sources, among others, often concentrated and in isolated locations; make easier the market integration; and, above all, ensure the security of supply while of course securing the financial means to achieve these objectives.
- To face more and more uncertainties and a more complex regulatory and legal background. In particular, the authorization procedures, emanating from different authorities, lead to slower planning and development processes. However, the main concern is the lack of social acceptance which seriously delays or compromises the implementation of transmission projects.
- To take charge of production disturbances and load balance and reduce the costs of mitigation measures.
- To allow access to the network for all users, generators, operators, and customers, making easier pooling and scaling effect.

3.1.3 Scenario as a Basis for the System Planner

The system planning process starts with the construction of meaningful scenarios, the laying down of technical and economic assumptions, and the identification of possible solutions.

A scenario can be defined as a description of possible future including alternative environments, in order to provide an overview of future perspectives to decision-makers in complex situations.

Scenarios combine uncontrollable uncertainties and controllable options. The uncontrollable uncertainties cannot be controlled by the decision-makers. They concern the evolution of different domains, such as technology, financial issues, public perception, etc. The controllable options can be chosen by the decision-makers. They address different issues such as political decisions, regulation rules, subsidies, etc.

For the TSOs, the scenarios at the national and regional levels help to decide on extensions of the network development network in coherence with the political and regulatory objectives. Thus, the TSOs must find the optimal balance between the needs of the network and the acceptable costs in order to ensure the political objectives at reasonable costs and with the expected quality of service.

In this regard, the scenarios developed by the TSOs should reflect the possible evolution of parameters such as fuel prices, geopolitical evolutions, climatic conditions, the market, regulations, technological breakthroughs, and other uncertainties.

The scenario methodology helps to assess the impact of network choices, facilitating the comparison between different solutions in the long-term perspective.

The performance of a generation and load scenario can be assessed with reference to the following main issues:

- Adequacy forecast: the ability of the generation system to meet the load with a suitable security of supply.
- Economic consistency: the cost assessment including generation and network costs.
- Political targets: the power system must meet the energy policy targets.

A probabilistic assessment can be carried out whenever necessary, but it should be mentioned that certain deterministic indicators, which are easy to calculate, are also available.

Figure 3.1 shows the main different elements of the generation adequacy analysis. The remaining capacity (RC) represents the difference between the reliably available capacity (i.e., the average available power from the installed generation capacity) and the load. RC is the part of net generating capacity (NGC) left on the system to cover any unexpected load variation and unplanned outages at a reference point.

When the RC is positive, it means that some spare generating capacity is likely to be available on the power system under normal conditions.

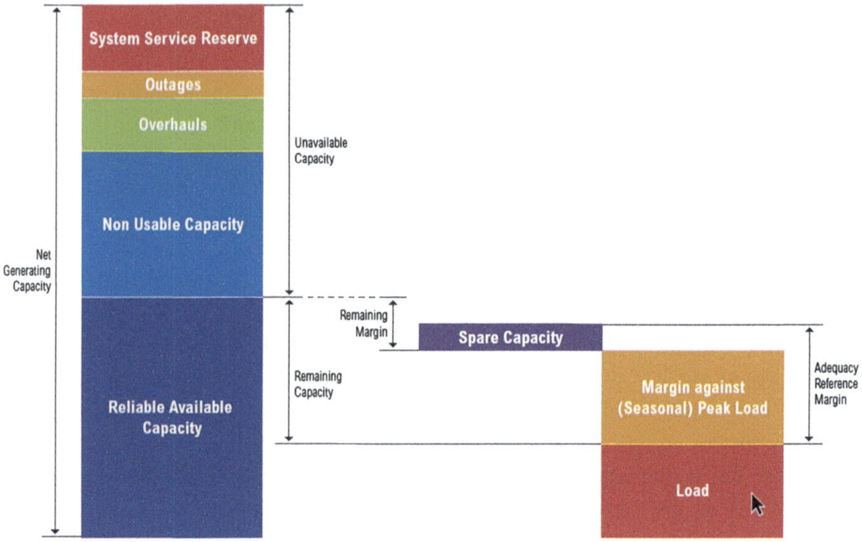

Fig. 3.1 Generation adequacy analysis. (ENTSO-E source [1])

When the RC is negative, it means that the power system is likely to be short of generating capacity under normal conditions.

3.1.4 *Methodology for Network Expanding*

After forecasting demand and generation evolution with reasonable accuracy through scenarios, the system planner performs market simulations and then grid development studies.

Market simulations aim at defining the type and the location of generation that should address the demand at any time.

They assess the impact of the new transmission project. Thus they compare the situations with and without the transmission network project. The comparison is about the economic efficiency, the ability of the system to schedule plants to their intrinsic merit order, the overall resulting variable generation costs, as well as the overall amount of CO_2 emissions and volumes of spilled energy. An economic optimization is conducted for a full year taking into account several constraints, such as the flexibility and availability of thermal units, wind and solar profiles, load profile and uncertainties, and transmission capacities (TCs) between countries.

Thus, the hourly in-feed of all generation resources aggregated by marketplace is calculated. This is classically done by minimizing the overall generation cost while respecting all relevant constraints of the generation technologies and maximum exchanges between different market areas. This simulation serves as a data basis for more detailed and in-depth analysis of potential grid constraints.

The network development studies determine the possible bottlenecks in the power system and must assess the appropriate network capacity reinforcements. The capacity reinforcements must ensure the supply as well as the system stability and security at any time while respecting network constraints. The network development studies involve some specific requirements. In order to avoid any cascading effect, the system is designed and operated with the principle that the failure of a single network element, such as transmission line, transformer, generating unit, etc., will not result in any disturbance of the grid user or violate any system and equipment limitation in frequency, voltage, and current. On an AC system, power flows must be adjusted immediately. This is the so-called N-1 security criterion [1].

3.2 Transmission Expansion Planning for Large Systems

3.2.1 Introduction

As explained above, the development expansion planning is nowadays mainly managed by TSOs. However, the scope is generally limited to the area covered by the TSOs. Nevertheless, the justification of interconnections between countries requires a more global vision.

Across the world, cross-border electricity connections are yet not very numerous. Their development slowly started in Europe just after the second World War. Figure 3.2 shows the electricity trade, imports/exports, between neighboring OECD Europe countries since 1974.

The total amount of imports are similar to the exports, meaning that the load flow through the interconnections is well balanced for a full year.

Thus, the imports of electricity grew from almost 90 TWh in 1974 to 475 TWh in 2017, representing an average annual growth rate of 4%, compared to 2% growth

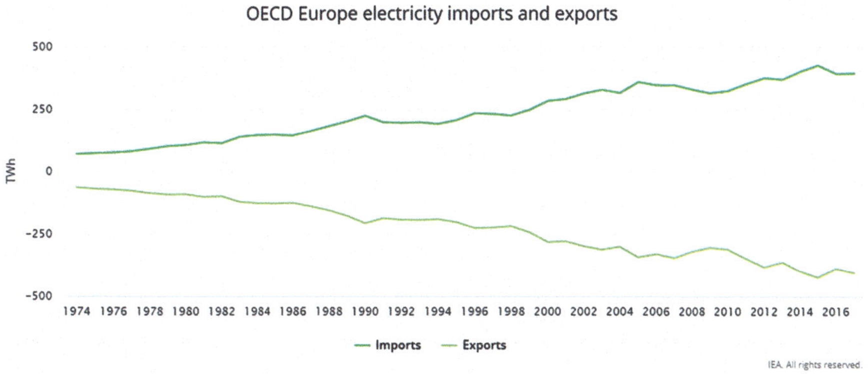

Fig. 3.2 OECD Europe electricity imports and exports (IEA source [2])

in overall electricity supply. Compared to the total demand (3600 TWh in 2017), the total imports represent 13%. This ratio should increase with the Clean Energy Package decided in 2018 by the European Union and the push for the development of the interconnections.

Figure 3.3 shows the top 20 countries worldwide for the amount of electricity imports and exports in 2018.

These figures highlight the location of the main active interconnections worldwide: in North America (between Canada and the USA), in South America (between Brazil and Paraguay), in Africa (between South Africa and Mozambique), and in Europe (between all European countries).

Historically, the supply of electric power was a national issue. The cross-border exchange was not developed for business purposes, but mainly for the security of supply. Figure 3.4 shows the exchange of electricity in 1955 between European countries.

As mentioned, although all countries were interconnected, only few countries had non-balanced exports/imports. Moreover, the total quantity of electricity annually exchanged was very low compared to the present situation (for instance, 1000 lower for France [4]). The end of the 1990s was the beginning of a move toward a single European electricity market. In this respect, the development of the interconnections was promoted as enabler. The development of Renewable Energy Sources is a new and additional driver for the promotion of the interconnections.

3.2.2 European Experience of Transmission Expansion Planning for a Large System

With the implementation of the European Union Regulation EC 714/2009, the European Network Transmission System Operators for Electricity (ENTSO-E) has been set up in 2009. The regulatory framework set out the tasks of ENTSO-E and the obligation for the TSOs to cooperate at the EU and regional level and drove the development of many deliverables [5].

Among other things, there is the obligation to coordinate regional network development planning and the biennial publication of a 10-year network development plan (TYNDP [1]). In 2014, ENTSO-E published the first TYNDP, giving visibility on transmission needs until 2030, including four possible scenarios (visions) that differ greatly in the areas of renewables, energy efficiency, and demand-side measures.

New methods for planning the system were investigated. In this regard, a "top-down" approach has been introduced in order to include EU 20.20.20 targets.

For the even longer term looking toward 2050, the e-Highway2050 project was launched in 2012 and achieved 3 years later [6]. The following part of this chapter presents the main key findings of this EU Project which address the topic of transmission expansion planning.

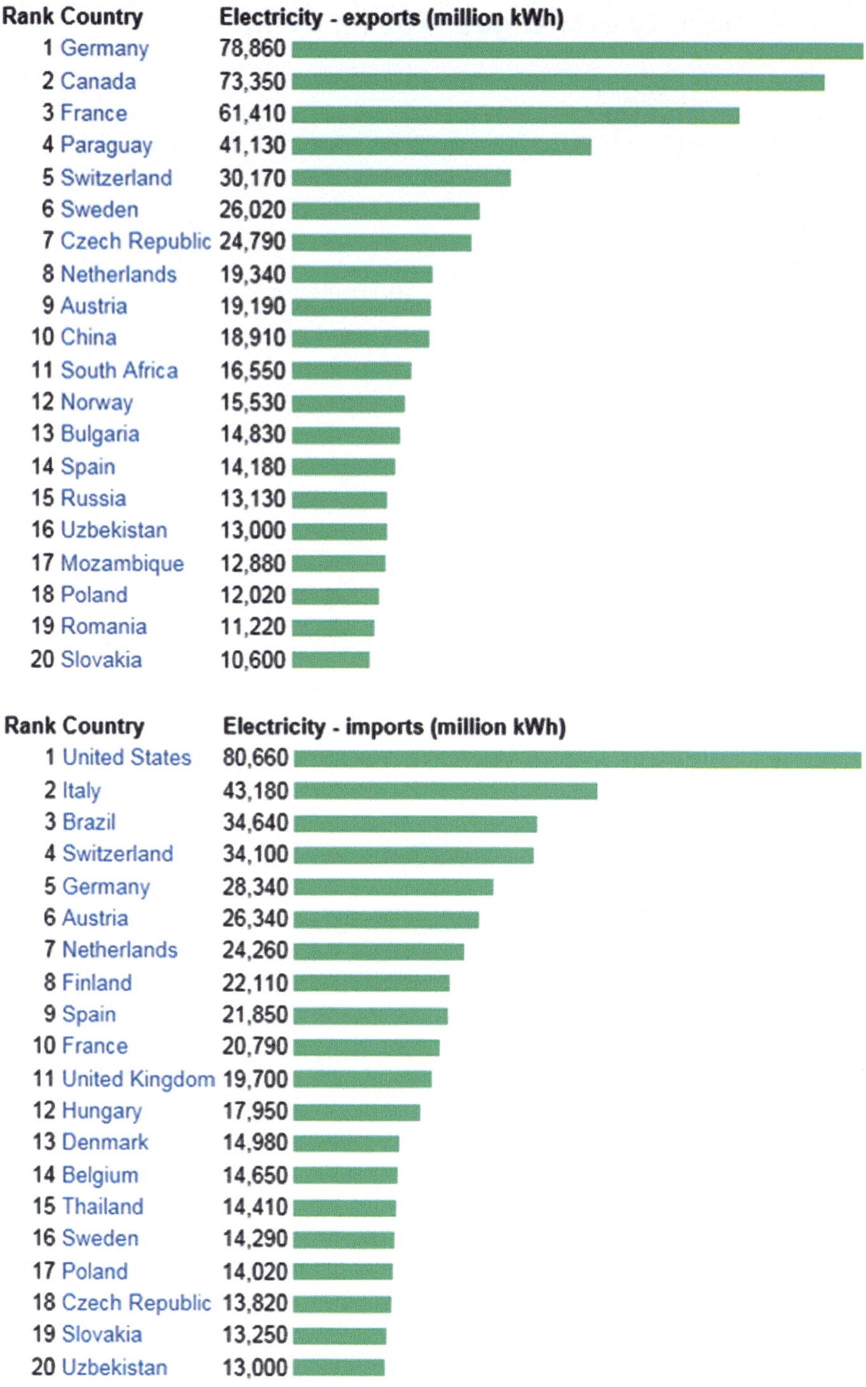

Fig. 3.3 Top 20 countries for imports/exports of electricity (IEA source [2])

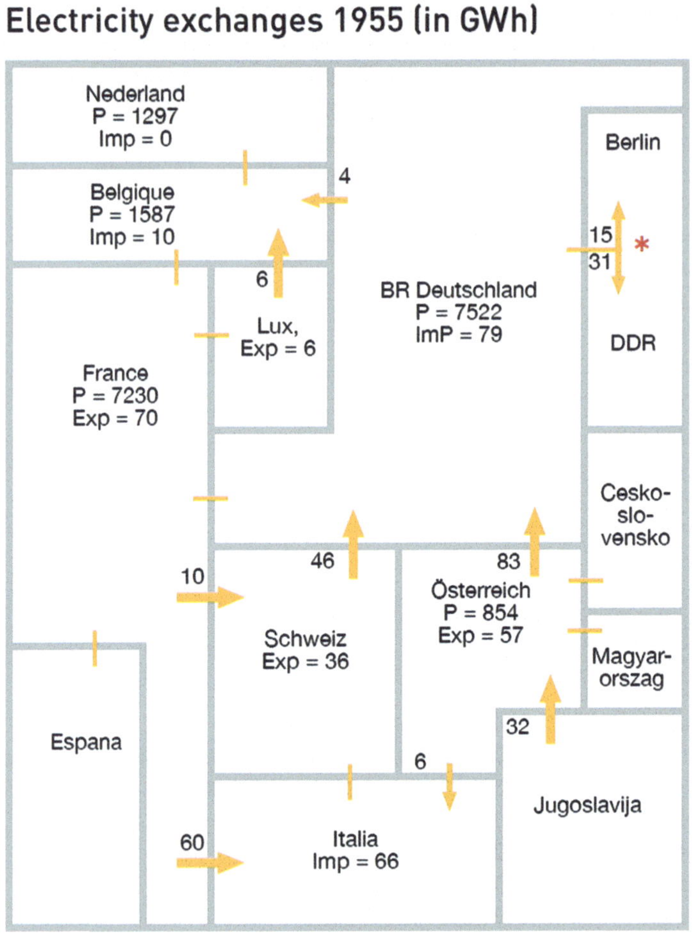

Fig. 3.4 Exchange of electricity in 1955 between European countries (UCPTE source [3]).

3.2.3 Motivation for e-Highway2050 Project

With the objective of reducing greenhouse gas (GHG) emissions to 80%–95% below 1990 levels by 2050, the European Union has analyzed the implications for the energy sector in the Energy Roadmap 2050 taking into account different scenarios.

In this perspective of these very ambitious targets for GHG emission reduction, the Renewable Energy Sources should represent the main part of the energy mix in Europe and be developed in different locations, often far away from major consumption sites. Electricity should be transported over longer distances, across national borders, to be delivered where the consumption needs arise. Such a long-distance and meshed transmission network at the pan-European level introduces the opportunity of an innovative concept of "Electricity Highway System" (EHS).

In order to define which transmission grid should be useful to face this evolution, the European Commission launched a call for a research project linked to the Seventh Framework Programme (FP7). The project aimed at delivering a top-down methodology to support the planning of a pan-European EHS capable of meeting European needs for electricity transmission up to 2050 [6].

The project developed methods and tools to support the planning of Electricity Highway System, based on various future power system scenarios, including options for a pan-European grid architecture under different scenarios, taking into account benefits, costs, and risks for each of them. The newly developed top-down methodology, which addressed the transition planning between now and 2050, was built around five main steps:

- The development and application of an approach to design different long-term scenarios in terms of energy generation mix, exchanges, and consumption scenarios
- The power location, using the assumptions about generation mix, energy exchanges, and consumption for each scenario, at country and local levels
- The system simulations, taking into account the different generation and demand profiles, in order to identify the possible weak points of the transmission grid in case of no reinforcement
- The identification of optimized grid architecture in 2050, while taking into account storage, demand-side management, and transmission technologies available by 2050
- The development of implementation routes from now to 2050 of the pan-European transmission system, covering each of the studied scenario and optimized by taking into account social welfare, environmental constraints, as well as grid operations issues

3.2.4 Scenario Building Process

The scenario building process in e-Highway2050 was performed in seven main steps (see Fig. 3.5).

The approach started by defining a number of uncertainties that should influence the future but could not be controlled by the decision-makers and technical and non-technical options that could be chosen by the decision-makers. Boundary conditions were specified (with upper and lower limit values) for the different uncertainties and options.

Examples: in the technology domain, the maturity of carbon capture and storage (CCS) was an uncertainty; on the contrary, the development of distributed generation was an option. In the economic domain, the growth domestic product (GDP) growth was an uncertainty, as well as the demography.

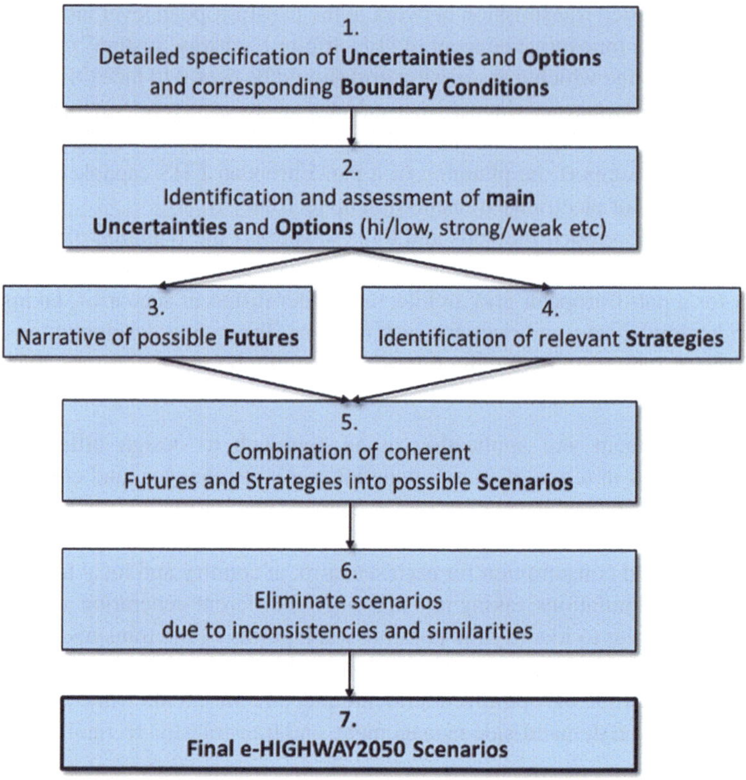

Fig. 3.5 Scenario building process

To limit the possible combinations to a tractable number, the main uncertainties and options were selected, for which the boundaries were specified in the form of numerical values (min, max, average). The main uncertainties were combined into possible futures that were narrated in a verbal way. Finally, five futures were set up: "Green Globe," "Green EU," "EU-Market," "Big Is Beautiful," and "Small Things Matter." They took into account energy and climate possible policies and assumptions on technological development and on economic and sociopolitical perceptions.

In parallel, the main options were combined into relevant strategies for the implementation of EHS in different possible futures. In the project, six strategies were developed ("Market led," "Large-scale RES," "Local solutions," "100% RES," "Carbon-free CCS and nuclear," and "No nuclear"), the main differences being the level of deployment for centralized/decentralized generation, the development of RES electricity, nuclear, fossil fuel, as well as the level of imports/exports with countries outside Europe. A combination of a future and a strategy made a possible scenario. Multiple scenarios were then generated by trying several different strategies within the same future or by testing one strategy in many possible futures.

After a detailed process of evaluation, selection, and elimination, a set of final e-Highway2050 scenarios was proposed.

The main criteria used for the selection process read as follows:

- A scenario was relevant when it challenged the entire existing European electricity system, not just the grid.
- The selected scenarios should substantially differ from each other, in coherence with the identified boundary conditions.
- Some of the scenarios should challenge the electricity system in a way which differed from the current state of affairs.

Finally, the seven-step filtering process led to the following five scenarios:

- S1 – Large-scale RES: focus on the deployment of large-scale RES technologies. A high priority is given to centralized storage solutions accompanying large-scale RES deployment.
- S2 – High GDP growth and market-based energy policies: internal EU market, EU wide security of supply, and coordinated use of interconnectors for cross-border flows exchanges in EU. CCS technology is assumed mature.
- S3 – Large fossil-fuel deployment with CCS and nuclear electricity: electrification of transport, heating, and industry is considered to occur mainly at the centralized (large scale) level. No flexibility is needed since variable generation from photovoltaic (PV) and wind is low.
- S4 – 100% RES electricity: 100% based on renewable energy, with both large-scale and small-scale links with North Africa. Thus, both large-scale storage technologies and small-scale storage technologies are needed to balance the variability in renewable generation.
- S5 – Small and local: the focus is on local solutions dealing with decentralized generation and storage and smart grid solutions mainly at the distribution level.

3.2.5 Approach of e-Highway2050

In consistency with the methodology introduced in §1.4, the e-Highway2050 project has included a simplified European grid model that allows forecasting generation and demand on a regional ("cluster") level.

Indeed, the whole European transmission network includes almost 10.000 electrical nodes (Fig. 3.6), which were not tractable for the project. Therefore, a clustering approach has been introduced, reducing the level of description of the grid.

The geographical clustering process was performed by splitting Europe and its countries into smaller parts, relevant for the system modeling. The basis for this analysis has been the Nomenclature of Territorial Units for Statistics (NUTS3 regions) which is set by Eurostat (Fig. 3.7).

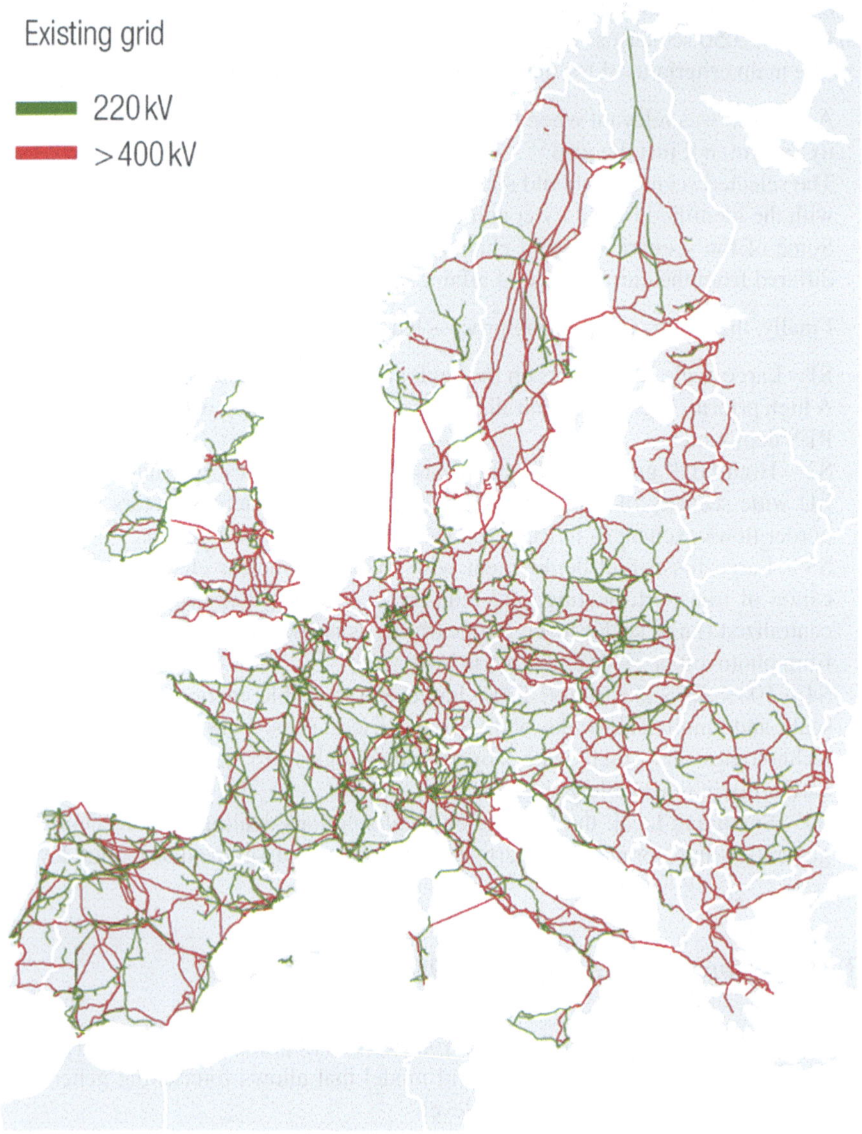

Fig. 3.6 Present European transmission grid (e-Highway2050 source [6])

The clustering is based on real system characteristics in order to represent the underlying transmission system appropriately. The clusters were still valid for all scenarios of the project.

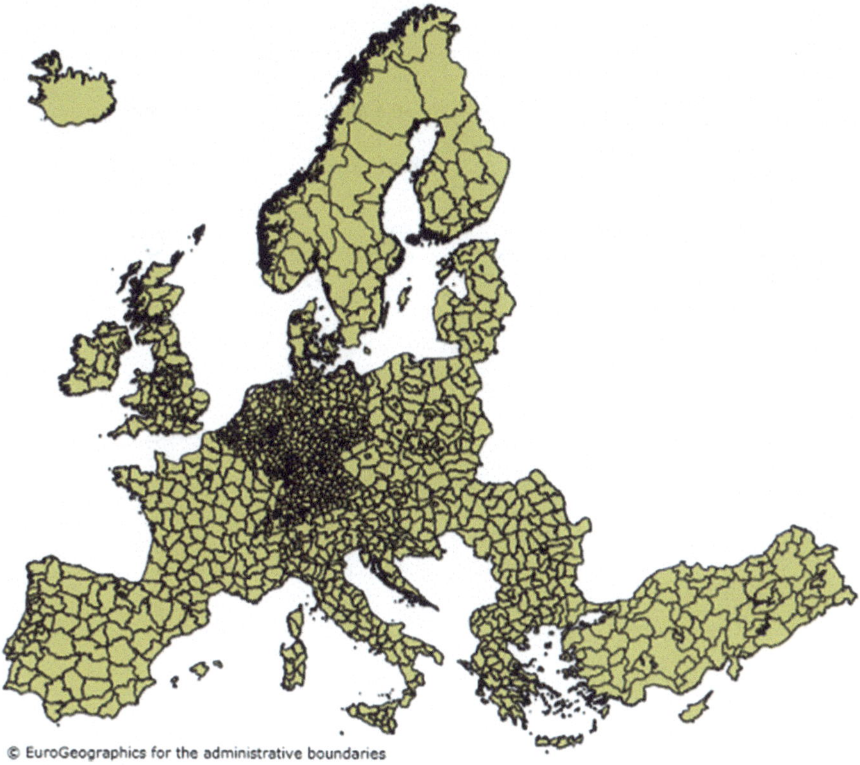

Fig. 3.7 NUTS regions [7]

The clustering model includes different criteria such as population, potential of RES generation, land use – availability of the areas – and installed generation capacity (thermal and hydro).

In the first round, an algorithm (K-means) was applied considering the first four measurable criteria. The optimization aims at joining together the incremental NUTS regions in a way that homogenous clusters are reached. In the second round, the TSO expertise was taken into account for the optimization of the clusters.

3.2.6 Grid Reduction

Based on the completed cluster definition, which has been described above, a grid simplification process has been conducted. The purpose was to reduce the European grid from several thousands of nodes to just one per cluster, to represent the main flows between clusters. Basically, the grid reduction process consists of the calculation of the grid transfer structure and the calculation of transfer capacities.

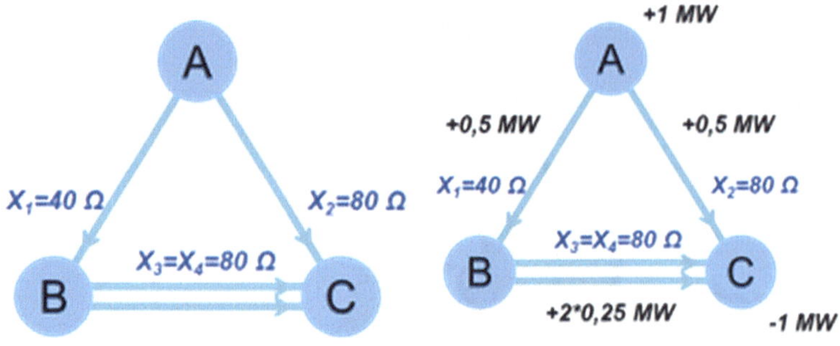

Fig. 3.8 Example of distribution of power flows

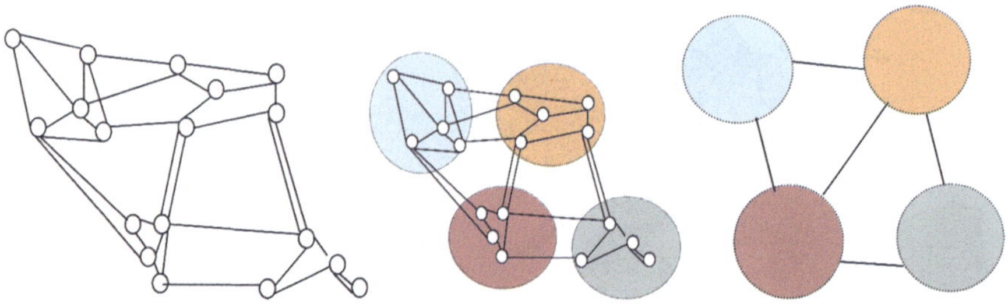

Fig. 3.9 Illustration of steps in grid reduction

In a meshed grid, the distribution of power flows depends on the impedances of the different lines. Assuming the transportation of 1 megawatt (MW) from fictional point A to fictional point C, not only the direct line between A and C but also the lines between A and fictional point B and between B and C is loaded (compare Fig. 3.8). Therefore, a grid simplification has to find a way to estimate the flows on the different lines.

In the present case, A, B, and C are not substations, but a set of substations resulting from the clustering. To reduce the problem size and respect the uncertainties involved with the long time perspective, several grid nodes have been summarized in one virtual ling, meaning a "copper-plate" inside the cluster.

This process led to a simplified network illustrated in Fig. 3.9 below, where all substations of a given cluster are unified in an equivalent node, and all links between two clusters were unified in an equivalent link. Therefore, one of the main challenges was to find a method to estimate the flows in this simplified grid, that is, the sum of flows that in detailed grid consideration would be exchanged between two clusters.

In general, there are several approaches to estimate the power flow distribution between the clusters in electricity grids. For the project, two methods, namely, the equivalent impedance approach and the Power Transfer Distribution Factor (PTDF) approach, were examined in depth to find a method for calculating the distribution. The equivalent impedance method, which has been chosen, presents accurate methods for describing the load flow across borders, but to model completely the network, maximum flow limitations need to be added. This implies to determine the N-1 safe physical transmission capacity between clusters in addition to the impedance calculations.

Finally, the project chose the equivalent impedance method as the method for grid simplification in this project. It was found that the reliability of results was similar to the PTDF method and sufficient for the purposes of this project. Besides that, as already described, the equivalent impedance model was advantageous as it allowed for rough estimations of the impacts of updates in the zonal network. This reduced the time and the efforts necessary to provide sound research results, as in some cases full recalculations of the model could be avoided.

The equivalent impedance method assumes that the equivalent grid behaves like a detailed network, and, thus, the equivalent line has equivalent impedance. However, on top of that equivalent impedance, at the same time, the method proposes to define an initial loop flow to account for the asymmetries generated by the clustering between load and generation within the different substations composing the cluster.

In order to evaluate both equivalent impedances and these loop flows, the method uses several thousands of snapshots with a wide range of loads and generation. Furthermore, the resulting flow needs to be calculated. In the next step, a reduced network of clusters with only one link between the clusters is modeled. The goal of the equivalent impedance method is to find the admittance matrix for the system and the referring flow, so that estimated flows best match the sample flows using a least squares approach.

The underlying optimization problem of the impedance method can be expressed as follows:

Given a sample of s snapshots of flows $F_{ij,s}$ on each equivalent link i, j, the optimal admittance matrix Y and the optimal set of initial loop flows T_0 can be determined such that:

$$\mathrm{Min}\,\{Y|\,T_0\}\sum\nolimits_{s,i,j}\left(F_{ij,s}-T_{ij,s}\right)^2$$

under constraints :

$$\forall s : Y.\Theta_s = I_s$$

$$\forall s, \forall I, \forall j \quad T_{ij,s} = Y_{ij}\left(\Theta_{is} - \Theta_{js}\right) + T_{oij}$$

Where:

- T_{ijs} is the estimated Flow i, j for snapshot s when applying Y matrix and T_0 vector.
- F_{ijs} is the real flow (full-grid direct current [DC] load flow calculation) between i, j for snapshot s.
- I_s the cluster injection vector for snapshot s.

- θ_{is} is the angle for cluster i for snapshot s.
- θ_s is the angle vector for snapshot s.
- T_{0ij} is the initial loop flow between clusters i, j.

In a nutshell, the method determines the Y and T_0 minimizing the error on the sample between estimated flows and target flows, coming from a full-grid direct current (DC) load flow calculation. In order to solve this problem, a specific algorithm has been developed. The quality of the results of the equivalent impedance method is implied by its design. The method includes an error estimator, which enabled users to have a critical view on the equivalent network and thereby also on the clustering previously conducted. It was concluded that the method provided the estimated flow for each interconnector and each snapshot which enabled all kinds of error analyses on a local and a global scale.

As described above, the method reduced the full network model of initially thousands of nodes to less than a hundred nodes. The basis for the analysis was provided by the geographic clustering conducted before the grid simplification, which has been described above. That is, the method was applied to the synchronous continental AC clustered grid as depicted below, based on a CIM network and modeling a 2030 network.

Figure 3.10 shows the final results of the grid reduction. The gray lines represent the existing corridors in 2015, and the red ones represent the new corridors to be realized by 2030. The width of the lines shows the Transmission Capacities (TCs) of the corridors.

3.2.7 *Quantification of Scenarios*

Starting from the description of the five scenarios, the quantification methodology defined the installed capacities of each generation technology at the cluster level while meeting the targeted energy mix of the scenario and the system adequacy in a copperplate model, meaning that generation should meet the demand without any grid constraint at any time. The quantification was progressively assessed at the macro level, then at the country level, and finally at the cluster level.

Based on the assumptions about the evolution of the population, the GDP, the transfer of uses toward electricity, and efforts on energy efficiency, a yearly volume of demand has been calculated for each country and for each scenario.

The installed capacities were defined in each macro-area based on weighting distribution keys which combined information about the potential of generation capacities and demand in a given macro-area. The weights of the distribution keys were scenario dependent.

Then a system simulation over the whole Europe without considering internal grid constraints allowed tuning the installed capacities and storage, in order to reach a sufficient level of adequacy and to improve the imbalances of the macro-areas

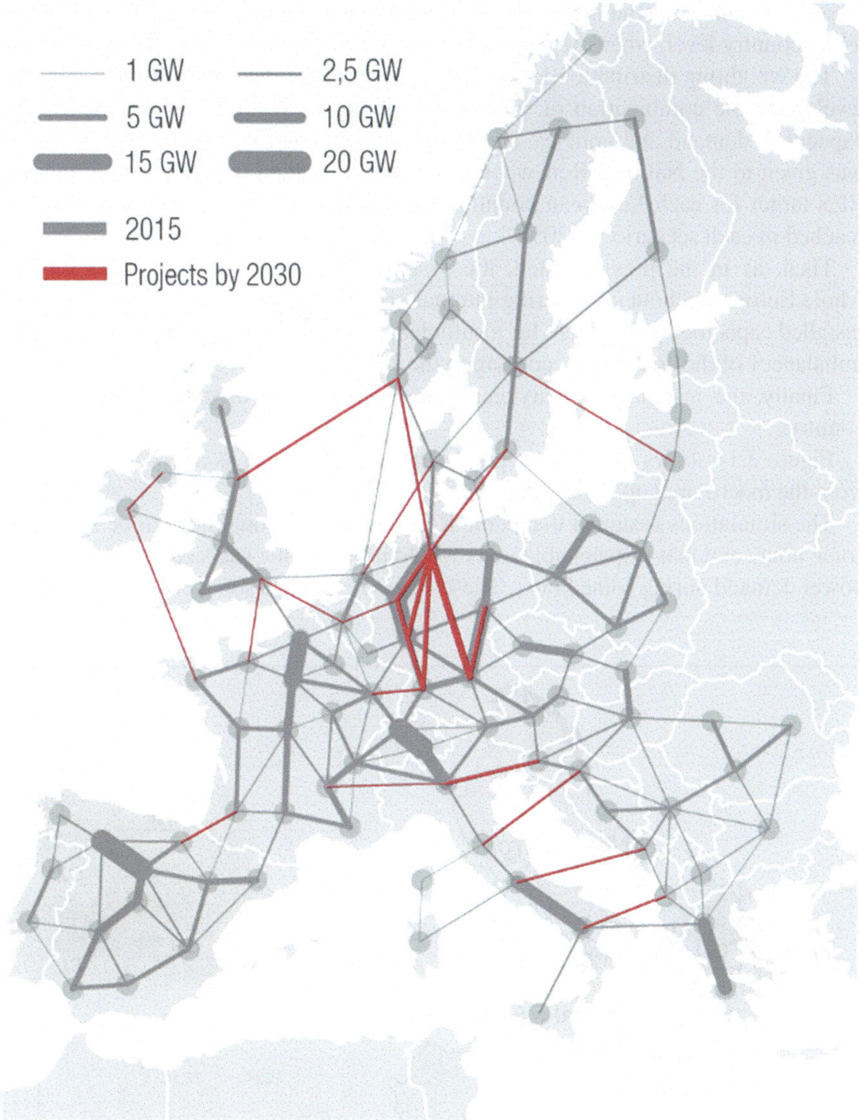

Fig. 3.10 Map of European cluster with TCs [6]

according to the scenarios. For the simulations, the availability of the dispatchable plants, including biomass, nuclear, and fossil generation, was assumed around 80%.

At this stage of the process, no country or national considerations were taken into account. Only the perspectives of the whole Europe were considered.

Afterward, the priority was given to European perspectives but national policies and trends were considered while splitting the installed capacities of a macro-area among its countries. The installed capacities of each macro-area were broken down to the country level, where 36 European countries are considered (ENTSO-E area).

The weighting distribution keys used to go from the macro level to the country level were the combination of information about the potential of the generation capacities, demand and policies, and trends of each country. A particular attention was given to the National Renewable Energy Action Plan for 2020, providing the RES target for each European country. These plans set the minimal values to be reached in each scenario for 2050.

Then, as in the previous step, the system simulation was performed for the whole Europe, without considering internal grid constraints, in order to provide the installed capacities and storage for a sufficient level of adequacy and to improve the imbalances of the countries according to the scenarios.

Finally, the installed capacity generation values were distributed across each country.

Figure 3.11 shows the three-step scenario quantification, as a top-down process from the macro-area up to the cluster level.

The simulations assumed that some RES generation would have to be spilled at some hours and that dispatchable units would probably not succeed to face the high power demand during some hours. These issues resulted in some energy not served.

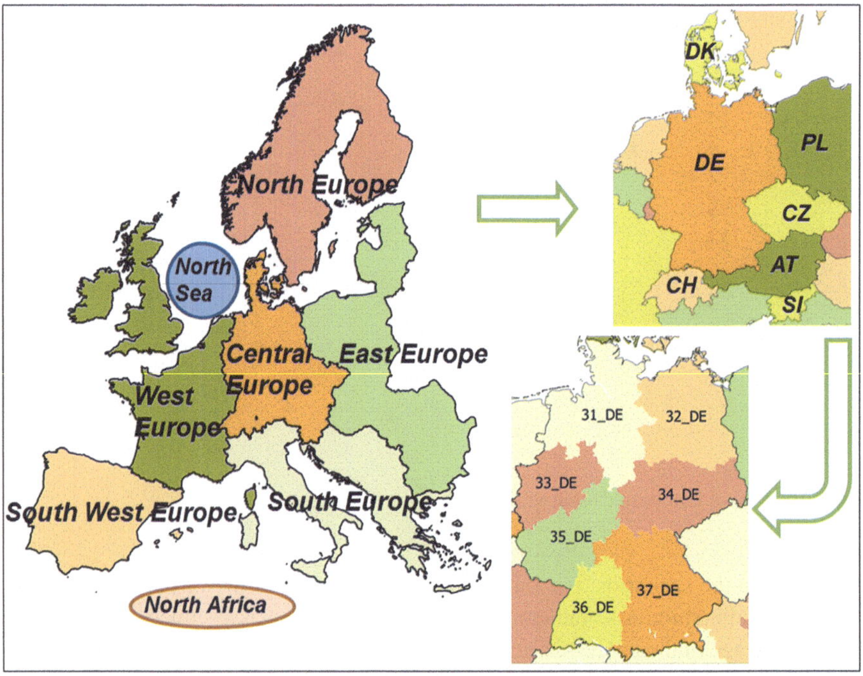

Fig. 3.11 e-Highway2050 top-down approach for scenario quantification [6]

Table 3.1 Allowed imbalances per country in each scenario

Scenario	S1	S2	S3	S4	S5
Range of imbalances per macro-area/country	20–180%	20–180%	30–130%	70–130%	20–180%

To reach an acceptable level of adequacy and to stay close to the targeted energy mix, some increase of the generation capacities should be required.

For this task, an algorithm using the ANTARES [8] tool was used. The first set of 99 Monte Carlo (MC) years was simulated returning adequacy outputs, such as the loss of load duration (LOLD). The LOLD is a quite convenient way to assess how far the system is from a reasonable adequacy. If the LOLD is higher than 3 hours, meaning that extra capacity is needed, incremental changes in installed capacities are then applied in the algorithm.

Simulations with ANTARES were performed again. This process was repeated until the adequacy criteria (LOLD < 3 h) were valid and unsupplied energy was low. Once all technologies have been increased, some storage was added to the system. Finally, if every bound has been reached and adequacy was still not met, peaking units were added at the end. These units should be used a few hours per year but were necessary to ensure adequacy. They represented either peaking thermal units or DSM.

For clarification, the imbalance is defined as the ratio between the generation and the demand of an area. For each scenario, a maximal acceptable imbalance per macro-area and country was defined according to the description of the scenarios (see Table 3.1 below). After simulations, if the maximal imbalance was not respected in some areas, some generation units were redistributed.

Finally, for each scenario, the installed capacities have been defined at the different geographical levels and for different perspectives:

- Installed capacities
- Energy mix
- Imbalances (the ratio of yearly generation over yearly demand)

As example, Figs. 3.12 and 3.13 show the results for one scenario (S3: Large fossil fuel, with CCS and nuclear).

3.2.8 System Simulations and Grid Development

The system simulations aim at detecting grid constraints and then proposing reinforcement to solve them. System simulations were carried out with Antares, which includes a market simulator linked to the grid.

Antares performs probabilistic simulations, including the stochastic method for the assessment of the load, the Renewable Energy Sources, or the availability of

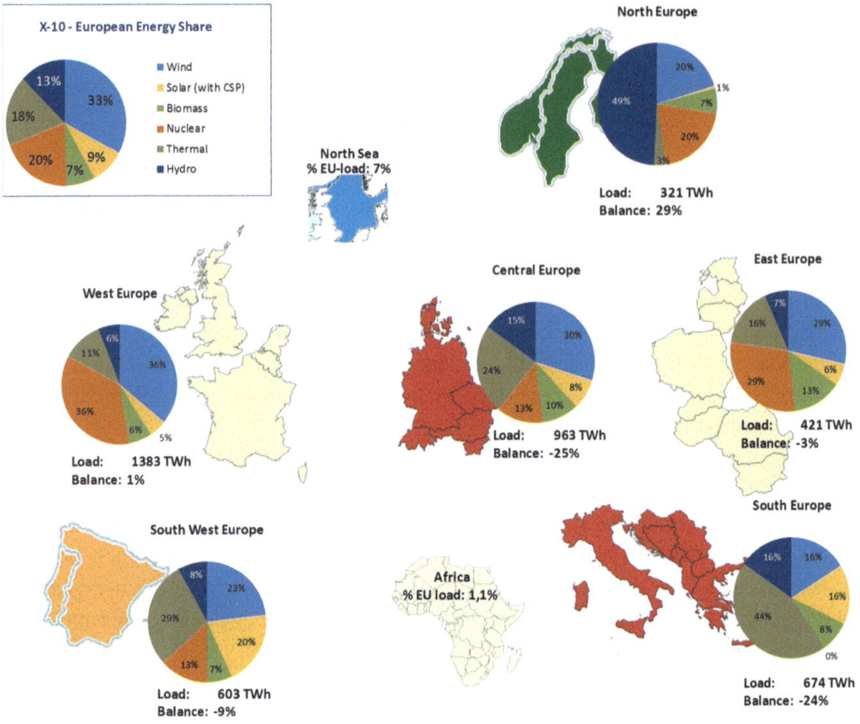

Fig. 3.12 Energy mixes and the imbalances for scenario S3 (Large fossil fuel, with CCS and nuclear), at the macro level [6]

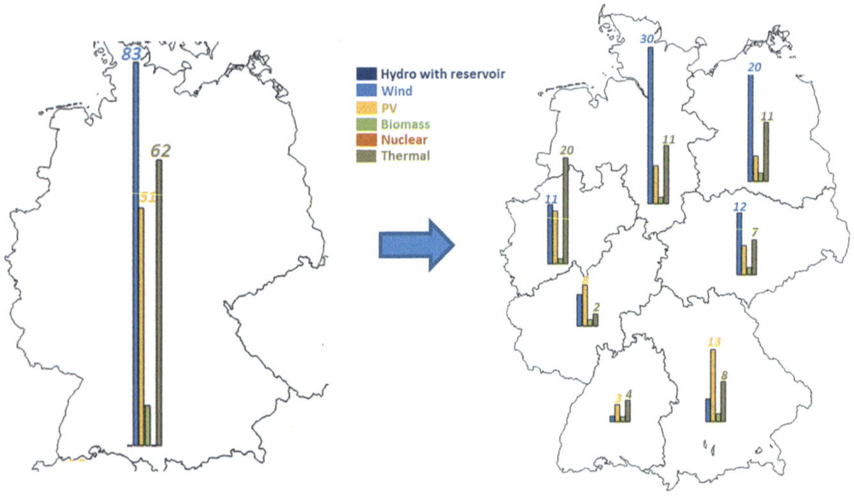

Fig. 3.13 Installed capacities for scenario S3, at the country and local level, and for Germany [6]

thermal units. It analyzes the adequacy and costs of the system. With this tool, the grid is described in terms of a limited capacity of transmission between clusters, taking into account the repartition of the flows with the impedances of the real transmission lines.

The simulations combine a detailed modeling of generation, demand, and grid. The generation and the use of storage are optimized in order to minimize the overall cost of the system. The probabilistic simulations are carried out for 99 Monte Carlo years, including the combination of 11 different solar and wind "years" with 3 different demand "years" (temperatures) and 3 different hydrology (e.g., dry, average, wet) "years." Thus this set of 99 MC years can capture the variability of the solar, wind, and hydro generation, availability of power plants, as well as the level of demand. The optimization takes into account the grid characteristics, the equivalent impedances, and the equivalent physical capacities. The time step resolution is 1 h and simulation covering a period of 1 year.

The network constraint effects were measured by the difference of the results of the two simulations. The first one is a copperplate simulation, in which transmission grid is assumed to be without constraints, i.e., where network capacities are set to infinite. On the other hand, simulation with grid constraints, in which capacities are limited to the "starting grid" at first and then the "starting grid" with transmission requirements (TR) in the reinforcement process.

The copperplate simulation provides the maximum capacity that could be achieved by grid reinforcement to ensure system security and optimize operating costs.

The "starting grid" is the grid already in place or already decided. For the e-Highway2050 project, it came from the TYNDP (version 2014) and included the projects foreseen until 2030. Thus, the "starting grid" simulation provides the location of the grid constraints in the case of the implementation of 2050 demand and generation development, in the 2030 transmission network.

The "transmission requirements" (TR) represents the needed increase of capacity between two clusters. At this stage, they do not presume the technological option and detailed route. TR represented the "reinforcements" expected in the "starting grid."

The analysis pointed out the grid constraints, their significance, their localization, and the most critical periods of the year. With the grid development analysis, Transmission Requirements were suggested and tested in an iterative process. Once all the reinforcements were identified, they were transposed into possible technologies to assess their cost and verify the profitability of the complete set of reinforcements called an "architecture."

The set of results in such a probabilistic approach was immense (8760 h $\times$ 99 MC years) for each cluster (~100) and each power line (>200). To perform the analysis of this very large database, the process used relied on a progressive approach in both time and space dimensions. This approach aimed at understanding where bottlenecks occur in the system, when, and how.

Thus, the e-Highway2050 project analyzed in depth different indicators for each scenario, at different time and geographical scales. The indicator ENS (energy not served or unsupplied energy) represents the volumes of energy not served due to network limitations. The indicator extra spillage or delta spillage depicted the must-run energy (e.g., renewables) that cannot be consumed due to the congested grid. The indicator thermal redispatch included, with positive redispatch in a given cluster, means that, due to grid constraints, local thermal generation was increased to secure the load. The thermal redispatch included as well negative redispatch. It means that thermal generation optimized in the copper plate simulation was reduced. Negative redispatch occurred in countries/clusters with capacities in competitive technologies, e.g., nuclear and biomass, and mainly countries/clusters with excess of energy. The indicator marginal cost variation (MCV) of the links displayed the potential benefits for the system for an extra MW available on a given inter-cluster link. It pointed out the first bottlenecks in the system.

To analyze the key hourly indicators over the year, "heat maps" were used (see Fig. 3.14). They showed the hourly distribution of the values per weeks (week 1 to week 52) and hours (from 0) during 7 days in each week. The values displayed are averages over the 99 MC years. By this display, seasonal and daily patterns of the indicators can be detected. All indicators are presented in MW (or MWh/h) on all diagrams.

In this respect, the "heat maps" pointed out the critical weeks of the year regarding ENS, spillage, and redispatch. They were analyzed in average over the 99 MC years but also for specific Monte Carlo years to assess the synchronicity and amplitudes of the phenomenon. For those weeks, synchronous surplus and deficit areas were identified. "Surplus areas" corresponded to the area in which there were unused renewable and/or cheap thermal generations that could be released by grid reinforcement. On the contrary, "deficit areas" face ENS and/or positive redispatch due to grid limitations. It was necessary to ensure the synchronicity of those phenomena as they could occur in different Monte Carlo years or different periods of the day, and thus connecting these areas could not solve the problems.

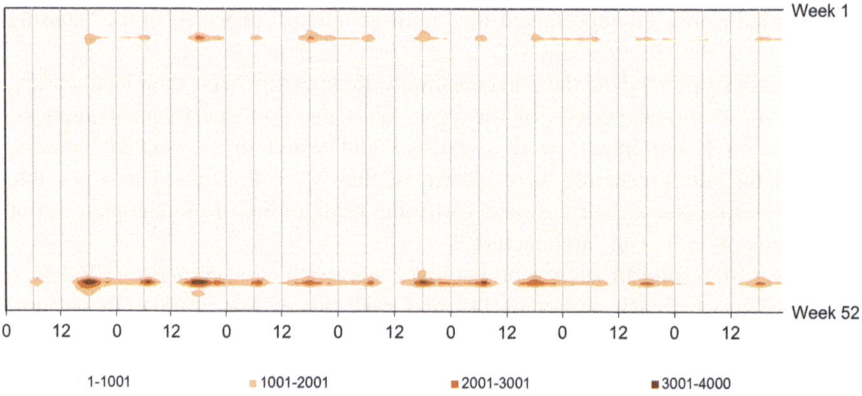

Fig. 3.14 Example of "heat map"

Based on the constraint analysis of the characteristic weeks, reinforcements were suggested to connect areas in surplus to areas in deficit. Their sizes were set based on the identification of synchronous volumes at stake. The possibility to stop in the different clusters to collect or distribute energy along the path was considered.

Figure 3.15 illustrates the results of this grid analysis.

At each iteration of the process, a set of reinforcements was studied.

During the first steps of the process, the focus was mainly on solving ENS for the most critical weeks. Indeed, ENS has a very high cost and thus reinforcements are likely to be very profitable. Due to this fact, a sensitivity analysis on the profitability of the proposed grid architecture has been executed for two different levels of ENS costs (10,000 €/MWh and 1000 €/MWh). Moreover, areas with ENS in the most critical weeks usually faced high positive redispatch in other weeks and were thus very good candidates for profitable reinforcements over the whole year.

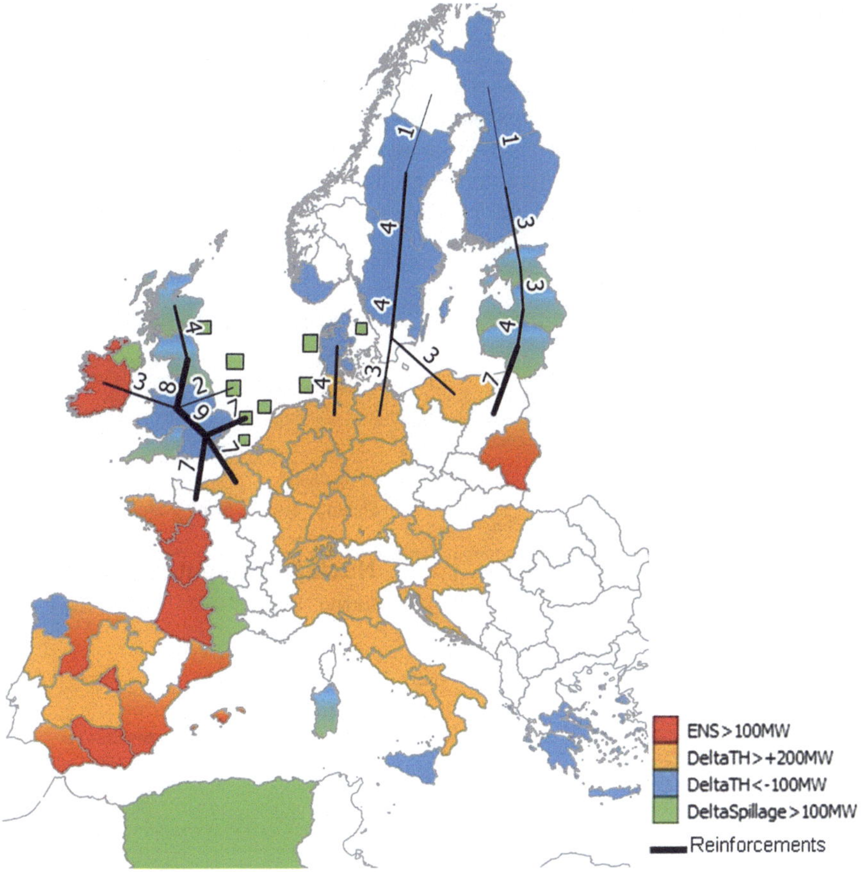

Fig. 3.15 Example of map with average hourly values on the 99 Monte Carlo years in a characteristic week and reinforcements [6]

Grid reinforcements were modeled in the simulations as DC links. Once a set of reinforcements has been tested in the characteristic weeks, an annual simulation was performed on the 99 Monte Carlo years to assess the global impact.

Annual gains on ENS, spillage, and redispatch were calculated. The annual benefit of each "step" of the process was calculated as the sum of generation cost savings and ENS cost reduction.

Figure 3.16 illustrates the benefit of each step, according to the different criteria.

The cost of the tested set of reinforcements was afterward roughly assessed considering only DC cables as the most expensive case. It was compared with the benefits, in order to verify that the investments were profitable whatever technology was used – these investments were referred to as "no regret investments." Some reinforcements might be modified if they were inefficient or oversized (oversizing characterized by very small remaining MCV or flows well below capacity).

Based on the remaining constraints in the characteristic weeks, a new iteration of reinforcement was then defined. If there were no more significant issues, which means only small and spread volumes of ENS, spillage, and redispatch remain, the iterative process was stopped.

The final grid proposal was made of all the transmission requirements (reinforcements) defined in the different steps and defines the final architecture.

As explained previously, Transmission Requirements (TRs) were not related to any specific technology, as they represent the major grid reinforcement needs. However, they were transposed into possible technologies in order to ensure that

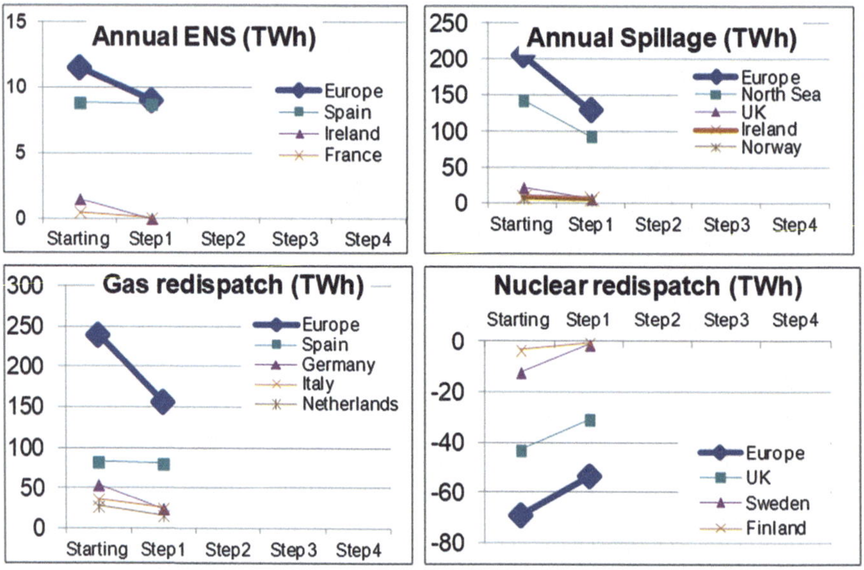

Fig. 3.16 Example of impact of a set of reinforcements on the annual results

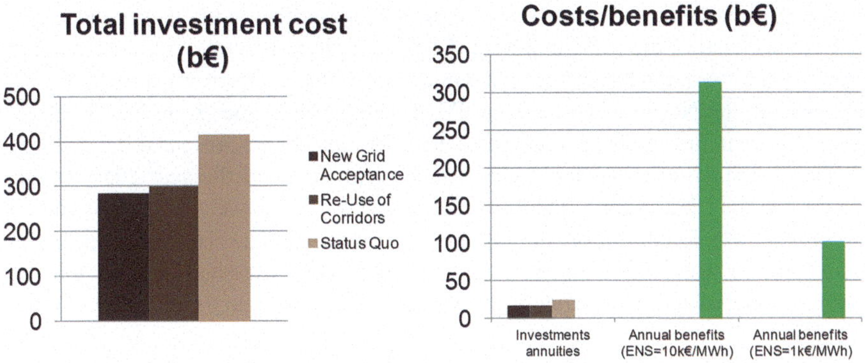

Fig. 3.17 Comparison of investment annuities and annual benefits for final architecture and the 3 strategies

there were technical solutions available and to better assess the cost of the final architecture.

The selection of technologies in 2050 will be highly impacted by the level of public acceptance toward new lines; this will impact the costs of the architectures. Thus, three strategies were considered to encompass a large range of possible costs:

- Status quo: the public opposition against new infrastructure prevents any new OHLs. Only refurbishment of existing lines or new DC cables can be implemented.
- Re-use of corridors: the public accepts new OHL as long as they are close to existing lines. Therefore new AC or DC overhead lines can be implemented when they are in the existing corridors.
- New grid acceptance: the public accepts new OHL and also the development of new corridors. DC cables are also possible but OHL is preferred when possible due to their lower costs.

For the three strategies, the investment annuity of the final architecture was compared to the annual benefit of it, to measure the level of profitability (example in Fig. 3.17).

This assessment of the profitability encompassed the whole architecture and did not evaluate the profitability of partial packages of reinforcements.

3.2.9 Application of the Methodology to One Scenario of the e-Highway2050 Project

The present paragraph presents the application of the methodology to one scenario selected by the e-Highway2050 project, scenario S1: Large-scale RES.

As described previously, the scenario must be quantified in the first round.

In this scenario, the most important RES technology is wind with 20% coming from a single area: the North Sea. The same goes for hydroelectricity for which 35% comes from the Nordic countries and PV of which 50% comes from North Africa. Installed capacities, demand, and imbalances per country are presented in Fig. 3.18.

The system simulations highlighted the critical weeks of the year 2050. In this regard, for the ENS criteria, the most critical weeks are between weeks 46 and 13 (see Fig. 3.19). For the level of spillage, maximum spillage variation happens in weeks 20–26, with maximum in week 25. Maximum "expected" hourly spillage variation (calculated as maximum of average values) is higher than 100 GW (see Fig. 3.19).

Figure 3.20 shows the first reinforcements needed for solving ENS in the critical week of winter (week 48) and their effect of the key criteria.

- ENS in Germany is solved by connection with Norway and Sweden, where high spillage and cheap thermal generation exist.
- ENS in Poland is solved by connection with Finland and the Baltic states, where high spillage and cheap thermal generation exist.

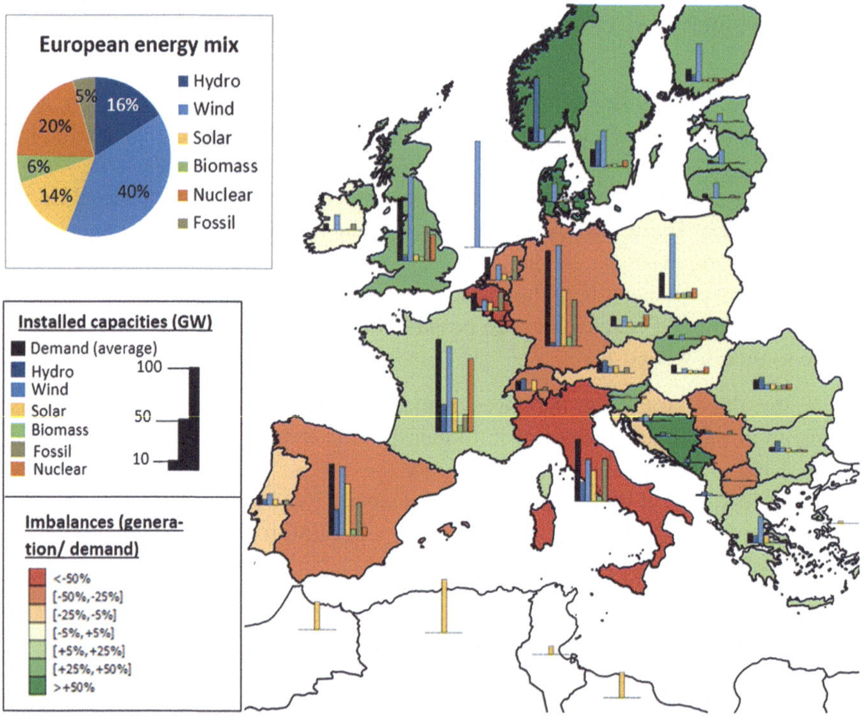

Fig. 3.18 European energy mix, installed capacities, and imbalances per country – Large-scale RES [6]

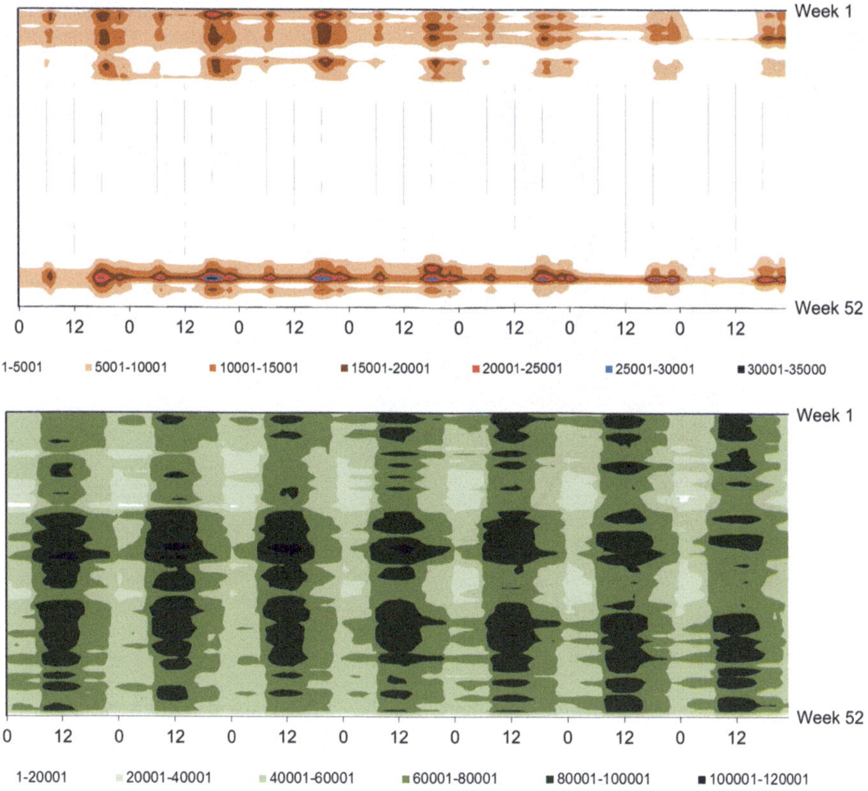

Fig. 3.19 ENS hourly values (MW) for the whole system – S1 and Spillage variation, hourly values (MW) for the whole system

- ENS in France is solved by connection with the UK and Ireland.
- Etc.

Figure 3.21 shows the average values for winter week 48, after applying reinforcements proposed in step 1 above, which connect distant areas with energy in excess to continental Europe, and the effect of step 2 on the annual volumes of ENS, spillage, gas, and nuclear redispatch.

Based on MCVs, the following reinforcements are implemented in step 2 (new reinforcements are drawn in bold on the map above):

- Reinforcements at NS-UK, north UK-Ireland, and UK-Belgium and the Netherlands, to optimize redispatching.
- Reinforcement of the UK-France-Spain corridor, to transport spillage from NS and cheap thermal from the UK toward Spain.
- Reinforcements in the South of France and between France and Spain are added to allow transport of energy from North Africa toward Spain and to solve ENS in the area around Barcelona.

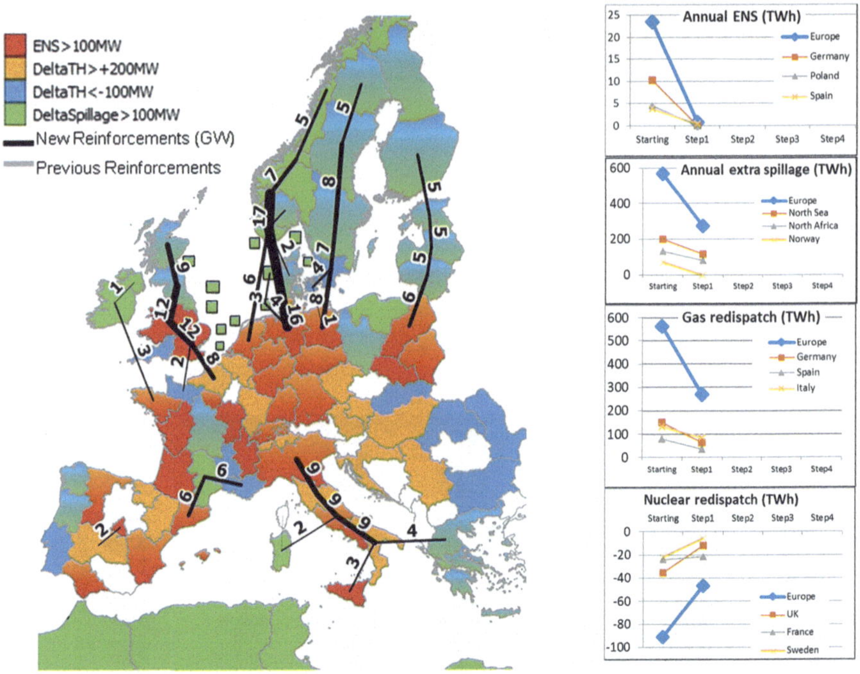

Fig. 3.20 Average hourly values in winter week 48 in the starting grid and reinforcements in step 1, Scenario S1 – Annual volumes of ENS, spillage, gas, and nuclear redispatch

- Reinforcement NS-DE is added to use wind from NS to reduce gas generation in Germany.
- Reinforcement of the link toward Italy to transport still existing excess of energy from North Africa to North Italy.

Figure 3.22 shows the average values for winter week 2 after applying reinforcement implemented in step 1 and step 2. This time it is week 2, since the focus is on the reduction of spillage and expensive gas generation that are more indicative in week 2.

In this step, the impacts of the reinforcements proposed are:

- Reduction of spillage (reduction by 62 TWh)
- Reduction of gas generation (61 TWh)
- Increase of nuclear generation (10 TWh)

Figure 3.23 shows the average values for winter week 2 after applying reinforcement implemented in steps1, 2, and 3.

Based on MCVs, the following reinforcements are implemented in step 4:

- Additional reinforcements from Denmark to Germany.
- Increased reinforcement of the United Kingdom-France-Spain backbone link.

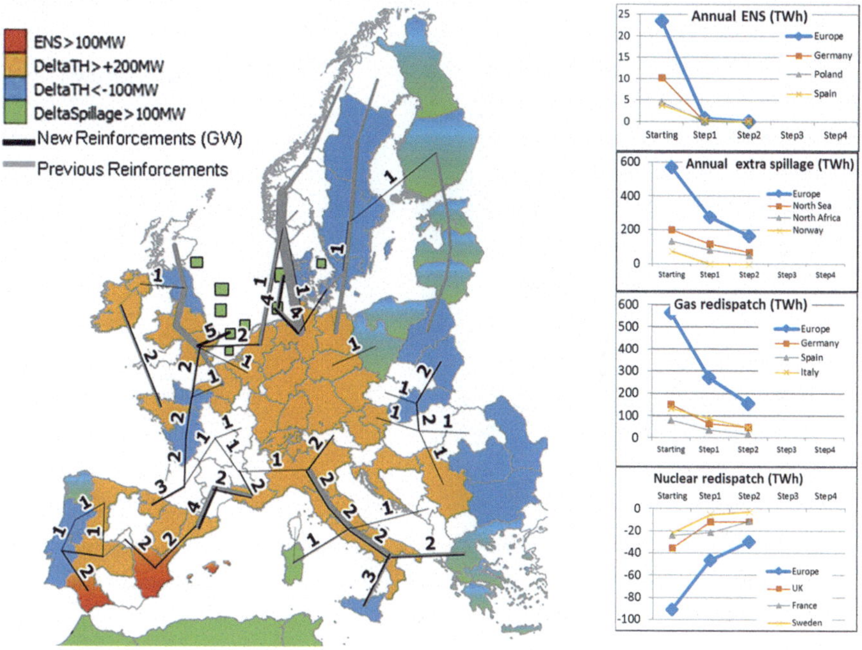

Fig. 3.21 Average hourly values on the 99 Monte Carlo years in winter week 48 after step 1 and reinforcements in step 2 and annual volumes of ENS, spillage, gas redispatch, and nuclear redispatch

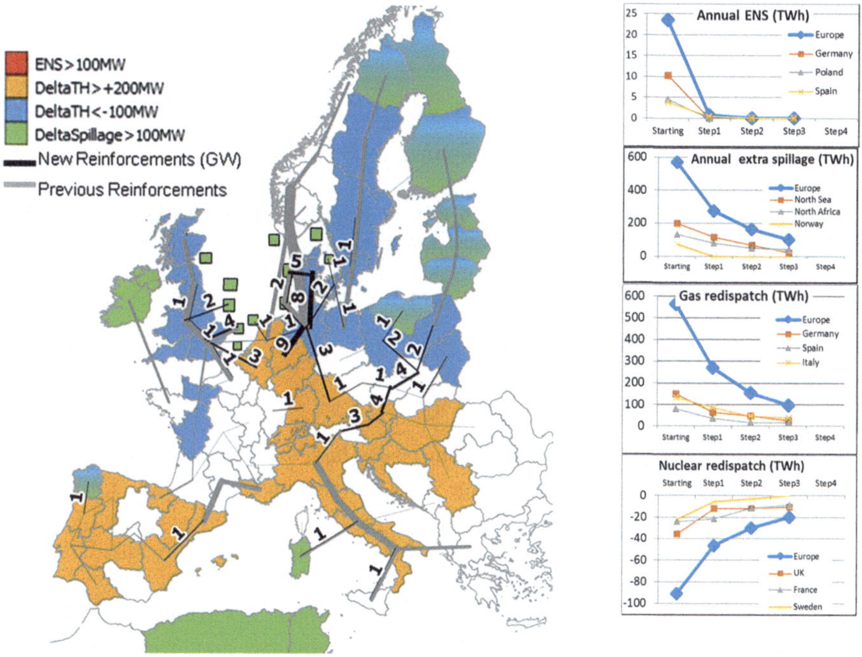

Fig. 3.22 Average hourly values on the 99 Monte Carlo years in winter week 2 after step 2 and reinforcements in step 3 and annual volumes of ENS, spillage, gas redispatch, and nuclear redispatch

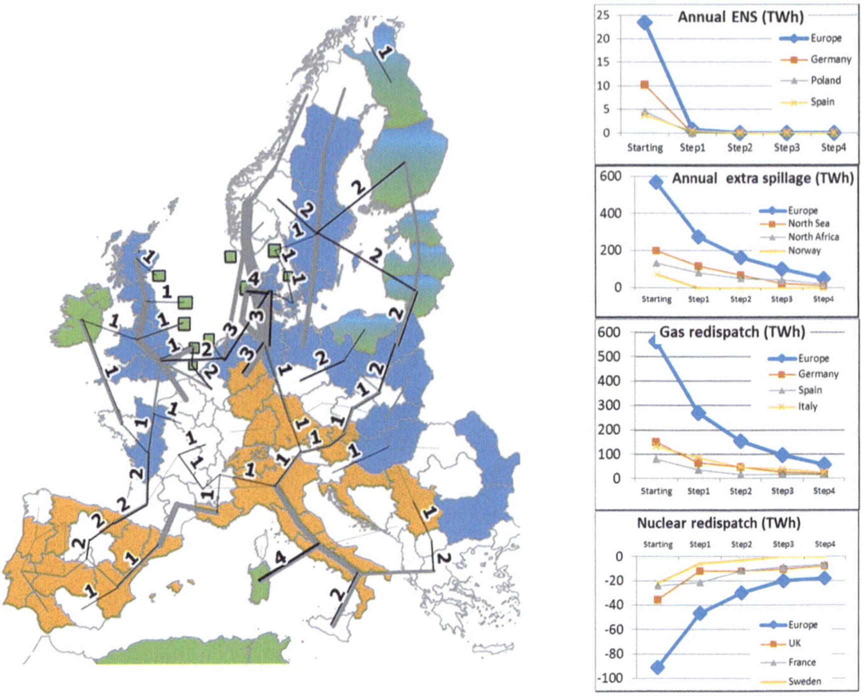

Fig. 3.23 Average hourly values on the 99 Monte Carlo years in winter week 2 after step 3 and reinforcements in step 4, 3 and annual volumes of ENS, spillage, gas redispatch, and nuclear redispatch

- Reinforcement of southern Finland towards Sweden then towards Poland, to transport the spilled RES.
- Reinforcement of Sardinia-Sicily towards mainland Italy, to transport solar energy from North Africa and reduce thermal redispatching in Italy.

With step 4, the impact of the reinforcements proposed are:

- Reduction of spillage (reduction by 50 TWh)
- Reduction of gas generation (35 TWh)
- Increase of nuclear generation (2 TWh)

Finally, after the four steps, significant issues have been solved with almost all ENS suppressed and a level of spillage drastically reduced from 609 TWh to 88 TWh.

The final grid for scenario S1 is in the Fig. 3.24, where reinforcements higher than 1 GW are presented.

Figure 3.24 shows also the cost of the full additional architecture according to the strategies "new grid acceptance," "re-use of corridors," and "status quo." The annual cost of each strategy is compared to the annual benefit.

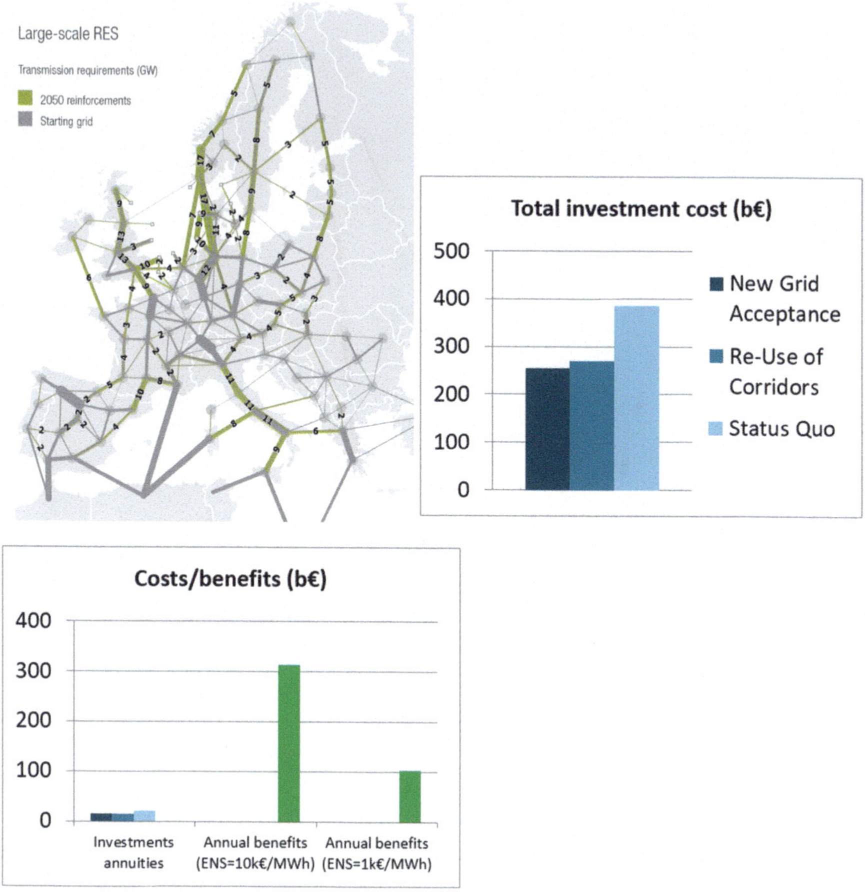

Fig. 3.24 Final grid proposed for scenario S1 – Large-scale RES; total investment cost and cost/benefit comparison

The same methodology has been applied to the five scenarios, providing five different grid architectures. However, even if the scenarios are extremely diverse, some major corridors were common to all of them. Figure 3.25 shows the common reinforcements. In this regard, they appear robust to face the large uncertainties in 2050. Therefore, they are good candidates for mid-term grid investments.

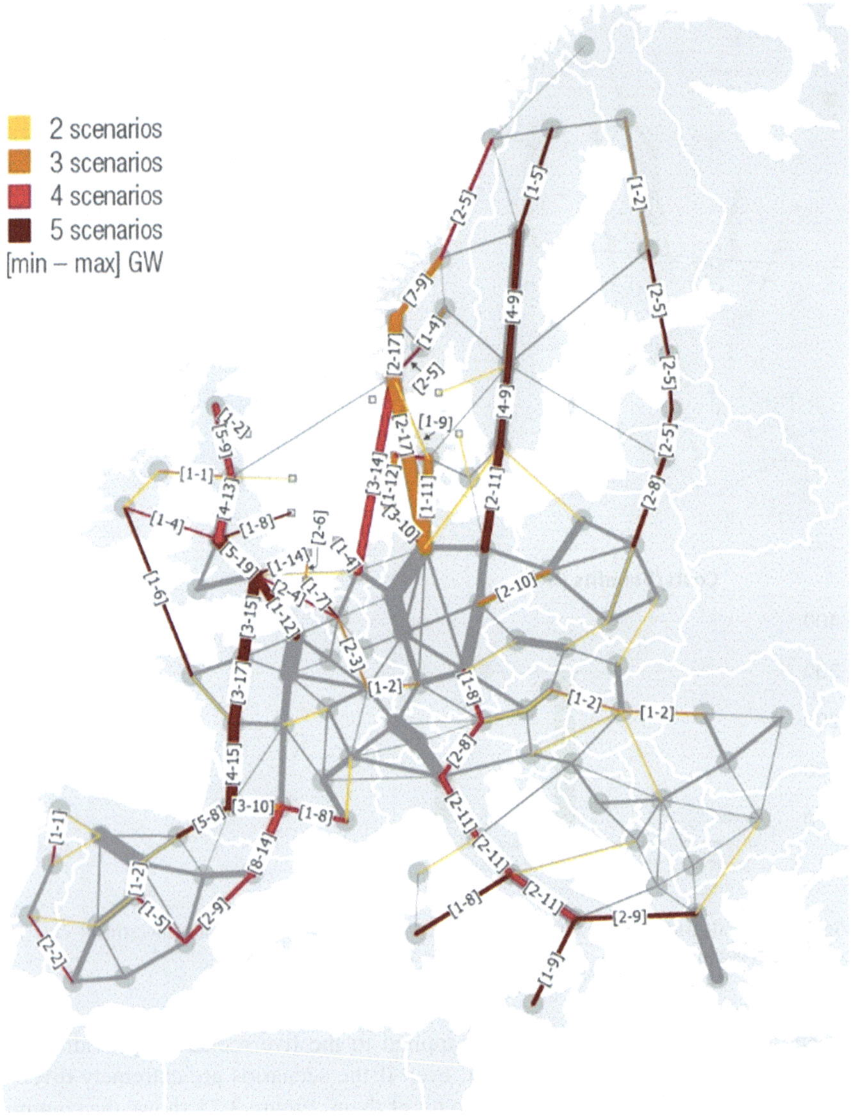

Fig. 3.25 Common reinforcements (widths are according to average reinforcement capacity and the color represents the number of scenarios where the reinforcement is needed)

3.3 Conclusion

The transmission expansion planning method of the European network, developed within the framework of the e-Highway2050 research project, is innovative and the first for the planning of a network with the dimension of a continent.

A new method was needed to identify the future transmission network needed to connect new sources of renewable sources to replace thermal power plants. It was a question of having a top-down approach, starting from generation and demand objectives at the European level to deduce the network necessary for development and more particularly the interconnections to be developed between countries.

This global top-down approach differs from a bottom-up approach which would have consisted in compiling country projections, but without reaching an optimal solution.

This pan-European method developed by e-Highway2050 is now used by the European Network of Transmission System Operators for Electricity (ENTSO-E) for providing transmission expansion planning in the long term in Europe.

With the development of renewable energy sources, the transmission expansion planning of large power systems should become more relevant. Thus CIGRE [9] has also used the methodology of transmission expansion planning developed with e-Highway2050 in a feasibility study on a global grid. It shows the added value of interconnecting the continents in comparison with keeping them as separate electricity systems. The global electric network, compared to non-interconnected grids, enables the use of more wind and solar resources, instead of gas-fired generation, reducing the total cost for the worldwide community and contributing to a strong further reduction of CO_2 emissions.

References

1. TYNDP - Ten-Year Development Plan - ENTSO-E, https://tyndp.entsoe.eu/
2. International Energy Agency IEA databank : https://www.iea.org/statistics/electricity/
3. UCPTE data https://www.entsoe.eu/fileadmin/user_upload/_library/publications/ce/110422_UCPTE-UCTE_The50yearSuccessStory.pdf
4. RTE statistics http://clients.rte-france.com/lang/an/visiteurs/vie/flux_physiques_2.jsp
5. J. Verseille, K. Staschus, The mesh-up – ENTSO-E and European TSO cooperation in operations, planning and R&D. IEEE Power&Energy, January (2015)
6. Gerald Sanchis, Thomas Anderski, Nathalie Grisey, Dragana Orlic, Gianluigi Migliavacca, 'e-Highway2050: A Research Project Analyzing Very Long Term Investment Needs for the Pan-European Transmission System', CIGRE 2016
7. EC Europe, Eurostat data https://ec.europa.eu/eurostat/documents/345175/7451602/NUTS3-2013-EN.pdf
8. Antares simulator: https://antares.rte-france.com/wp-content/uploads/2016/09/160913-Antares_public_long.pdf
9. CIGRE C1.35 Global Electricity Network – feasibility study Technical Brochure n°775, 2019

Chapter 4
Reduction Techniques for TEP Problems

Quentin Ploussard

4.1 Temporal Representation

In modern power systems, there are numerous sources of temporal variability, and the number of operation situations, or "snapshots," to consider in solving the Transmission Expansion Planning (TEP) problem can be huge, which makes this problem difficult to solve [1]. However, several snapshots may have similar impacts on the optimal expansion plan. This raises the need for snapshot selection methods that can detect such similarities. A snapshot selection method aims to identify a smaller set of snapshots that is supposed to represent all the relevant operation situations from the TEP perspective.

The benefits that can be realized by network investments are some of the most relevant drivers for the TEP decisions. This is because the decision to initiate a transmission expansion project directly depends on the reduction in operation cost that the corresponding new assets would incur if they were installed in the system. It is an estimate of these economic benefits, along with reinforcement costs, that drives the decision to construct some specific reinforcements rather than others. By finding the most representative snapshots, which are based on the potential operation cost reductions achieved by network expansion, this selection approach is effectively based on the outcome of network expansion.

The methodology to select the most representative snapshots and their respective weights is divided into the following steps:

1. Computation of a surrogate for the optimal set of network reinforcements by running a relaxed version of the TEP problem
2. Calculation of the hourly benefits achieved by each reinforcement
3. Simplification of the spatial representation of the hourly benefits

Q. Ploussard (✉)
Argonne National Laboratory, Lemont, IL, USA

S. Lumbreras et al. (eds.), *Transmission Expansion Planning: The Network Challenges of the Energy Transition*, https://doi.org/10.1007/978-3-030-49428-5_4

4. Aggregation of the snapshots according to the simplified representation of their associated benefits

4.1.1 Relaxed TEP Problem and Related Optimal Investments

First, a relaxed version of the TEP problem is solved in order to determine a set of promising reinforcements. Later, the potential benefits achieved by each of these reinforcement during each hour will be calculated.

In the relaxed version of the TEP problem, binary investment decision variables are converted into linear ones constrained between 0 and 1. Herein, this problem is referred to as the "relaxed TEP model." This model corresponds to a LP problem for which the computational burden is much less in comparison to the TEP problem that considers binary investment decision variables. The latter is a MILP problem and is referred to as the "non-relaxed TEP model."

Each operating hour, or snapshot, is characterized by benefits, i.e., reductions in operation costs, achieved by the reinforcements computed in the relaxed TEP model. As mentioned previously, these newly built grid elements generally correspond to virtual reinforcements rather than actual discrete ones. The investment cost, capacity, and admittance of these virtual reinforcements are largely proportional to the value of their associated investment decision.

Candidate lines that are not implemented, i.e., the candidate lines for which the optimal investment decision value is 0 in the relaxed TEP model, are disregarded in the calculation of line benefits.

4.1.2 Computation of Line Benefits for Each Snapshot

The approximate reinforcements are previously computed by solving the relaxed TEP model. The potential benefits achieved by each of these reinforcements are calculated based on the operation cost savings realized by the overall set of reinforcements in each operating hour.

The total operation cost savings are calculated as the operation cost difference between the case in which all investment decision values are assumed to be equal to zero (initial optimal power flow [OPF[1]]) and the case in which all investment decision values are assumed to be equal to the optimal investment decision values of the relaxed TEP model.

The benefit generated by a new line during a given hour should be calculated as the operation cost savings it achieved during that hour. In principle, these savings

[1]The optimal power flow is the set of power flows in an electrical network which minimizes the system operation costs.

should be calculated as the operation cost difference between the case in which this line is not built and the case in which the line is built, all other things being equal. However, when a given line is part of a group of lines assumed to be installed simultaneously, i.e., an expansion plan, defining the benefit this specific line can achieve is not an easy task. This is because the operation cost savings achieved by a line depend on what other lines are being installed together with it and what is their order of deployment.

In the literature, there are several approaches that have been proposed to calculate the benefits achieved by a specific line within an expansion plan [2]. One of these approaches involves calculating the benefit $B_{s,ijc}^{\mathrm{PINT}}$ of installing a line (i, j, c) as the operation cost reduction due to the installation of this line when no other reinforcement from the expansion plan has yet been undertaken. This approach is called the PINT approach ("Put IN one at a Time") (4.1) [3]. A more recent approach from ENTSOE[2] involves calculating the benefit $B_{s,ijc}^{\mathrm{TOOT}}$ of installing a line (i, j, c) as the operation cost reduction when all the other reinforcements from the expansion plan have already been undertaken. This approach is called the TOOT approach ("Take Out One at a Time") (4.2) [3].

$$\begin{aligned} B_{s,ijc}^{\mathrm{PINT}} &= \text{Operation costs in s when no line is installed} \\ &- \text{Operation costs in s when only line (i, j, c) is installed} \end{aligned} \tag{4.1}$$

$$\begin{aligned} B_{s,ijc}^{\mathrm{TOOT}} &= \text{Operation costs in s when only line (i, j, c) is installed} \\ &- \text{Operation costs in s when all lines are installed} \end{aligned} \tag{4.2}$$

In practice, however, individual reinforcements from the network expansion plan may be undertaken in any possible order. In addition, the benefits realized by a given reinforcement when it is the first element to be installed can be deemed to be largely complementary of the benefits realized by this reinforcement when it is the last element to be installed within the expansion plan. Indeed, benefits achieved by a reinforcement when it is the only one to be installed (as in PINT) may overestimate the value of this individual line, since those benefits may also be produced by other reinforcements in the expansion plan. Conversely, benefits produced by a reinforcement when it is the last one to be installed within the plan (as in TOOT) may underestimate the value of this line, since those benefits are usually marginal compared to the overall benefits already achieved by all the other installed lines.

Because of this, it seems appropriate to combine the TOOT and PINT approaches when calculating the benefits achieved by each individual reinforcement. The outcome of combining these two approaches is a more accurate estimate of the benefits achieved by each reinforcement than the estimate produced by using PINT

---

[2]The European Network of Transmission System Operators (ENTSOE), representing 43 electricity transmission system operators (TSOs) from 36 countries across Europe.

or TOOT alone. Results obtained using the combination of the two approaches have proven to be reasonably accurate. Thus, the benefits that are being allocated to each individual reinforcements from the network expansion plan (obtained by solving the relaxed TEP model) are calculated as a weighted average of the benefits calculated from the PINT approach and those calculated from the TOOT approach (4.3). In Eq. (4.3), using a value of 0.5 for the coefficient α produced good results for the selection of representative snapshots. As a matter of fact, it is not more probable or less probable for a given reinforcement to be the first to be undertaken within an expansion plan, or to be the last. In practice, most reinforcements within an expansion plan are undertaken in between other reinforcements. However, it is often not computationally feasible to consider all the possible orders of deployment of individual reinforcements within a large expansion plan.

$$B_{s,ijc} = \alpha B_{s,ijc}^{TOOT} + (1-\alpha)\, B_{s,ijc}^{PINT} \tag{4.3}$$

Once benefits achieved by each individual reinforcement during each hour are calculated, we can represent these hours, or snapshots, within the space of reinforcement benefits. In this space, each axis measures the benefits (calculated with Eq. (4.3)) achieved by each individual reinforcement. The dimension of this space is the number of reinforcements that are being considered. Each snapshot is represented by a point in this space and its position is determined by the benefits achieved by individual reinforcement during this snapshot.

In practice, a significant portion of the snapshots are associated to very low benefits in all of the reinforcements. These snapshots are located near the origin in the space of line benefits. Because of this, if the clustering algorithm were run in the space of reinforcement benefits considering all of the snapshots, a large portion of them, the ones close to the origin, would be inaccurately grouped together. Then, in order to account for this, snapshots associated to large benefits in at least one reinforcement, referred to as "relevant snapshots," can be grouped separately from the rest of the snapshots, referred to as "nonrelevant snapshots." Nonrelevant snapshots are clustered according to the associated overall operation cost savings (i.e., not allocated by individual reinforcements) realized in them by deploying the whole network expansion plan. On the other hand, relevant snapshots are clustered according to their position in the space of line benefits. In practice, relevant snapshots can be defined as the smallest possible set of snapshots where a significant part, e.g., 90%, of the total operation cost savings is achieved.

4.1.3 *Dimension Reduction of the Line Benefit Space*

Because of the sparsity in reinforcement benefits across snapshots and the numerous correlation factors among the benefits achieved by several reinforcements, it is advisable to identify the main underlying sources of variability of these benefits. In addition, identifying these main sources of variability can help represent these benefits across snapshots in a more compact way [4]. In spaces with too many dimensions, distances among data points become relatively uniform and meaningless, which makes the clustering of data points more difficult. This raises the need to reduce the dimensionality of the space of reinforcement benefits while preserving most of the information about the data variability.

Principal component analysis (PCA) is applied to the original space of reinforcement benefits to reduce its dimensionality. The PCA algorithm transforms a dataset in a new, more compact, one while minimizing the loss of information. This is achieved by using a new frame of reference in which the axes are referred to as "principal component." The first principal component of the transformed dataset accounts for the largest variability of the original dataset. Each successive principal component has the largest remaining variability and is orthogonal to the previous components. The major part of the variability from the initial dataset is captured using the first principal components of the transformed dataset.

Using this process, the dimensionality of the line benefit space can be substantially reduced without significant loss of information.

4.1.4 Clustering Algorithm

Once the clustering space has been reduced, the final step is to identify k main clusters, i.e., groups, of snapshots within this reduced space and their associated representative snapshot. This can be done using well-known clustering algorithms.

One may suggest the centroid of a cluster as its representative. The centroid of a cluster is defined as a virtual data point whose coordinates are equal to the average coordinates of all the data points making the cluster. However, the centroid of a cluster of snapshots does not correspond, in general, to an existing snapshot. Because of this, the centroid cannot be taken as a representative snapshot. Only one of the snapshots being part of the cluster can be selected as the representative of this cluster. This representative data point is called a medoid.

For this reason, medoid-based clustering algorithms such as the K-medoid algorithm should be applied to identify clusters of snapshots and their representatives, instead of centroid-based algorithms such as the K-means algorithm [5].

In the initial, i.e., non-reduced, TEP problem, each snapshot represents one specific hour of the year. All of the snapshots have the same weight equal to 1. In the reduced TEP problem, the representative snapshot of each of the k clusters is assigned a weight equal to the number of snapshots in this cluster. In other words, in the initial TEP problem, each snapshot s within their original set has a weight $\rho_s = 1$. However, in the reduced TEP problem, in case the snapshot s is representative of a cluster of n snapshots, it is assigned a weight equal to $\rho_s = \mathrm{n}$, whereas in case the snapshot s is not a cluster representative, its associated weight is equal to $\rho_s = 0$ (the snapshot is eliminated from the reduced set). For instance, if the time scope of the TEP problem is a single year, by applying this snapshot selection technique, the number of snapshots is reduced from 8760 snapshots to k

representative snapshots. This way, the size of the TEP problem can be significantly reduced in its temporal dimension while preserving the information relevant for the resolution of the problem.

4.2 Spatial Representation

The spatial dimension, i.e., the network, is certainly the most delicate dimension to reduce in the TEP problem. As a matter of fact, the network is the framework in which candidate lines may or may not be installed. In addition, congestion occurring within the network drives the transmission expansion. Therefore, if not applied carefully, a network reduction procedure may yield a new network representation where the ends of critical candidate lines are no longer represented.

In the context of the TEP problem, it is advisable to keep the most important candidate lines preserved within the reduced model. To achieve this, critical pairs of buses are identified by solving a linear relaxation of the TEP problem. Next, a network partition[3], in which these critical pairs of buses are border buses[4], i.e., connected to inter-area lines, is performed. Last, a bus elimination process is performed to remove the majority of the non-border buses from the original network.

The proposed network reduction methodology is divided into the following steps:

1. Identification of the critical lines and their end buses
2. Calculation of the network partition
3. Elimination of the non-border buses
4. Calculation of the equivalent features of the reduced network

4.2.1 Identification of the Critical Pairs of Buses

Reducing the size of the electrical network can decrease the computational time of solving a TEP problem. However, in a TEP context, it is critical to preserve the lines that are most impacted by the network expansion. The flows computed in the preserved lines of the reduced network should be as close as possible to the ones computed for the original network. Lines that are congested in the network before its expansion may be perceived as critical. However, the congestion of a few lines can be relieved by installing a single line in a new corridor. As a result, it is necessary to capture the potential effects of network expansion on the overall operation of the

---

[3]Given a network, a network partition is a classification of its nodes into groups, or areas.

[4]Given a network and a network partition, border buses are those buses that belong to a certain area of the partition and that are connected through a line to at least one bus from another area.

network rather than the initial congestions that drive this expansion [1]. To capture the effects of the network expansion on system operation, a relaxed version of the TEP problem is solved. A relaxed version of this problem lessens the computational burden in comparison to the original TEP problem. In the relaxed version of the TEP problem, investment decisions are represented by continuous variables instead of binary ones. The solution of this relaxed problem reveals useful information about which new lines, in existing or new corridors, are likely to be built in addition to which lines are congested in the expanded network.

Two independent problems are solved: the OPF is computed before any network expansion, and the relaxed TEP problem is also computed. Pairs of buses are identified as critical if they comply with any of the following conditions:

- The two buses at each end of any line for which the flow reaches 100% capacity in at least one operating situation based on the OPF solution in the network before its expansion
- The two buses at each end of any line for which the flow is above 80% capacity in at least one operating situation based on the solution of the relaxed TEP problem
- The two buses at each end of candidate lines that are partially or completely built based on the solution of the relaxed TEP problem, i.e., candidate lines for which the value of the optimal investment decision variable is strictly greater than 0
- The two buses at each end of power flow controlling devices, such as HVDC, that are already installed or candidate lines

The 80% congestion threshold invokes a tradeoff between the magnitude of the network reduction and the accuracy in representing the network congestion and how it is impacted by the network investments in TEP analyses. In other words, there is a tradeoff between reduction and accuracy. This threshold should be low in order to retain as much information as possible about the network congestions. Thus, the TEP solution is most efficient when considering the reduced network. However, the number of critical buses and, thus, the size of the reduced network might be too large if the congestion threshold is too low. The choice of the 80% congestion threshold was demonstrated to offer a good compromise for our case studies.

4.2.2 Network Partition

4.2.2.1 Minimum Multicut Problem and Appropriate Weight

Our goal is to reduce as far as possible the size of the network while preserving the critical pairs of buses identified above so we can accurately model the flow in those lines. To preserve these pairs of buses, a Gaussian elimination is necessary [6]. In addition, it is required that the two ends of each critical pair belong to different clusters, or areas. This way, we can eliminate non-border buses in each area with little or no impact on the potential power flows going in these critical lines.

In the proposed method, the network is being reduced by eliminating as many non-border buses as possible. At the end of the reduction process, all the border buses should be represented in the reduced network. Therefore, minimizing the size of the network requires finding a partition of the network that minimizes the number of border buses. In addition, all the buses that compose the critical pairs of buses, which are now referred to as "critical buses," should be border buses according to the network partition. This way, we are ensuring that they will be represented in the reduced network. However, a network partition that ensures that critical buses are border buses may also have additional border buses that are not critical buses. The additional border buses, besides the critical ones, correspond to the ends of each additional intercluster link, besides the critical lines, produced by the network partition. Therefore, minimizing the number of border buses involves finding a network partition minimizing the number of additional intercluster links. More specifically, what should be minimized is the number of additional border buses the additional intercluster links may produce. Indeed, if an additional intercluster link is already connected to one critical bus, only one more border bus will be produced instead of two (the critical bus is already a border bus). Next, we describe the formulation of the problem to be solved to compute the target network partition.

Let $(W_{i,j})_{i,j \in \{1,\ldots,N\}}$ be a matrix representing a weighted network of N buses, and let $(s_k, t_k)_{k \in \{1,\ldots,K\}}$ be a list of K "sink-source" pairs of buses in this network. The weighted multicut problem [7] aims to find a network partition that minimizes the sum of the weights of intercluster links while ensuring that, for each pair k, the buses s_k and t_k belong to two different clusters. This problem can be formulated as in (4.4).

$$\min_{\substack{all\ possible\\ clustering}} \left(\sum_{\substack{i\ and\ j\ belong\ to\\ different\ clusters}} W_{i,j} \right) \tag{4.4}$$

s.t. s_k and t_k belong to different clusters for each k.

The multicut problem is particularly suitable for the network partition problem we aim to solve. In our case, the "sink-source" pairs correspond to the critical pairs of buses defined above. As explained previously, it is not the number of intercluster links that we want to minimize, but the number of border buses. This can be achieved by assigning to each link (i,j) in the initial network (i.e., the network before expansion) a weight $W_{i,j}$ that corresponds to the number of noncritical buses at its ends. An example is depicted in Fig. 4.1.

We can assign a weight $W_{i,j} = 0$ to all the pairs of buses that are not directly connected by a line in the initial network. Indeed, assigning these two buses to different clusters does not generate additional border buses. However, if there is a

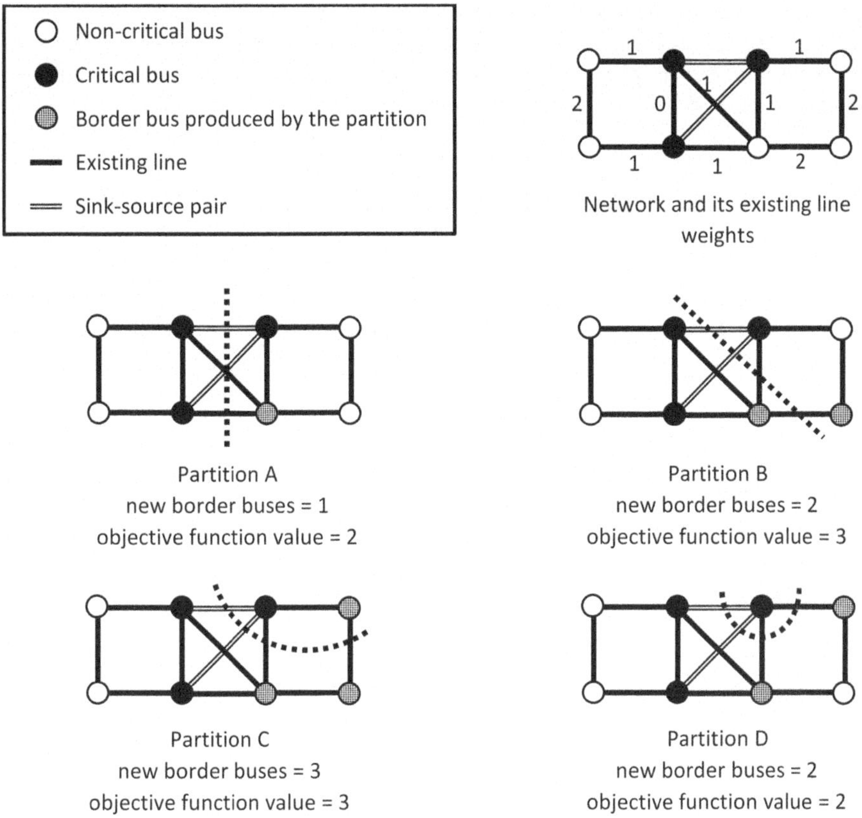

Fig. 4.1 Possible partitions of a network when solving the multicut problem. The weight of existing lines, based on the number of critical buses at their ends, is shown in the first network at the top. The four partitions at the bottom all satisfy the sink-source pair constraints. The objective function value, calculated with Eq. (4.4), is always greater or equal to the number of border buses produced by the partition. In partitions C and D, these two values are equal. The partition leading to the least amount of border buses is clearly partition A. However, the optimal solutions according to Eq. (4.4) are both partitions A and D

line directly connecting buses i and j and $p \in \{0, 1, 2\}$ of these buses are critical, then assigning these buses to two different clusters will generate at most $2 - p$ new border buses.

Ideally, the value of the objective function of the problem described by Eq. (4.4) should be exactly equal to the additional number of border buses of the associated network partition. However, these two values may not necessarily be equal. The number of additional border buses due to the inclusion of a new intercluster link may not be the same as the number of noncritical buses in this link as it happens in partition A in Fig. 4.1. This is because a noncritical bus can be connected to more than one intercluster link and, thus, would be accounted for more than once in the objective function. However, the objective function in (4.4) is an upper estimate of the number of border buses associated with the partition and is reasonably close to this number. Thus, minimizing this objective function also minimizes the number of border buses.

4.2.2.2 Relaxation of the Multicut Problem and a Rounding Algorithm

An equivalent but more practical formulation of the weighted multicut problem defined in (4.4) is that in (4.5), (4.6), (4.7), and (4.8). In this new formulation, distance variables $d(i,j)$ are introduced [8]. If buses i and j belong to the same area, the value of the variable $d(i,j)$ is equal to 0. Otherwise, its value is equal to 1.

$$\min \left(\sum\nolimits_{(i,j)\in E} d(i, j)\ W_{i,j} \right) \tag{4.5}$$

s.t.

$$d(s_k, t_k) \geq 1, \quad k \in \{1, \ldots, K\} \tag{4.6}$$

$$d(u, v) + d(v, w) \geq d(u, w), \quad u, v, w \in \{1, \ldots, N\} \tag{4.7}$$

$$d(u, v) \in \{0, 1\}, \quad u, v \in \{1, \ldots, N\} \tag{4.8}$$

Equation (4.5) is an alternative way to formulate the problem described in (4.4) using distance variables. E defines the set of lines in the initial network, i.e., the set of existing corridors. Equation (4.6) ensures that buses s_k and t_k belong to different areas of the partition. Equations (4.7) and (4.8) guarantee that $d(i,j)$ is satisfying the requirement for distance function by being positive and by verifying the triangle inequality. The problem formulated in (4.5), (4.6), (4.7), and (4.8) is strictly equivalent to the one formulated in (4.4), and their optimal solutions are identical.

Formally, $d(i,j)$ is a binary variable, and the problem described by (4.5), (4.6), (4.7), and (4.8) is an integer programming (IP) problem. This problem is NP-hard and, therefore, difficult to solve for large networks. In the literature, the problem is generally relaxed by linearizing the variable $d(i,j)$. The relaxed problem can be solved in much less computational time than the original IP problem. Then, a rounding algorithm is applied [9]. A rounding algorithm converts the optimal solution of the relaxed problem into an integer solution with a relatively low integrality gap. Lastly, the network partition is defined by the connected components of the initial network, after eliminating all the links (i,j) for which $d(i,j) = 1$. Each connected component of this modified network corresponds to an area of the partition. Efficient algorithms can compute the connected components of an undirected graph in $O(N + |E|)$, i.e., in a computational time proportional to the number of nodes and lines [10].

4.2.3 Bus Elimination

Once a partition and the border buses have been identified, non-border buses are removed through Gaussian elimination [6]. The Gaussian elimination of the non-border buses from a given cluster can be performed via Kron reduction in two different ways: in a single step (multiple buses at a time) or in multiple steps (one bus at a time). The latter method updates iteratively the admittance matrix and the operating condition of the network according to Eqs. (4.9) and (4.10).

$$Y_{i,j}{}' = Y_{i,j} - \frac{Y_{i,k}Y_{k,j}}{Y_{k,k}}, \quad i, j \in \{1, \ldots, k-1, k+1, \ldots, N\} \tag{4.9}$$

$$P_m{}' = P_m - \frac{Y_{m,k}P_k}{Y_{k,k}}, \quad m \in \{1, \ldots, k-1, k+1, \ldots, N\} \tag{4.10}$$

where $(Y_{i,j})_{i,j \in \{1,\ldots,N\}}$ and $(Y_{i,j}{}')_{i,j \neq k}$ are the admittance matrix of the electrical network before and after removing bus k, and $(P_m)_{m \in \{1,\ldots,N\}}$ and $(P_m{}')_{m \neq k}$ are the power injection vector before and after removing bus k.

Computing the Kron reduction one bus at a time presents two advantages. First, compared to the alternative way, it is not necessary to compute an inverse matrix, which is computationally intensive. Second, it is not always desirable to eliminate all the non-border buses from a given area. In some cases, doing so may lead to a dense subnetwork comprising too many lines, which would defeat the purpose of reducing the network. Therefore, if the elimination of the last non-border buses dramatically increases the number of lines in the reduced network, it is advisable to avoid eliminating them. An illustration of this phenomenon is shown in Fig. 4.2.

When eliminating the non-border buses one at a time, the order of elimination is important. One method commonly used [11] is to eliminate the non-border buses with the lowest degree[5] first.

In each area, non-border buses are eliminated iteratively until none of them remains. In each iteration, the number of lines and buses remaining in that area is recorded. Finally, the reduced subnetwork used to represent that area is the one that contains the minimum total number of lines and buses. This corresponds to step 3 in Fig. 4.2.

[5]The degree of a bus is the number of lines connected to it.

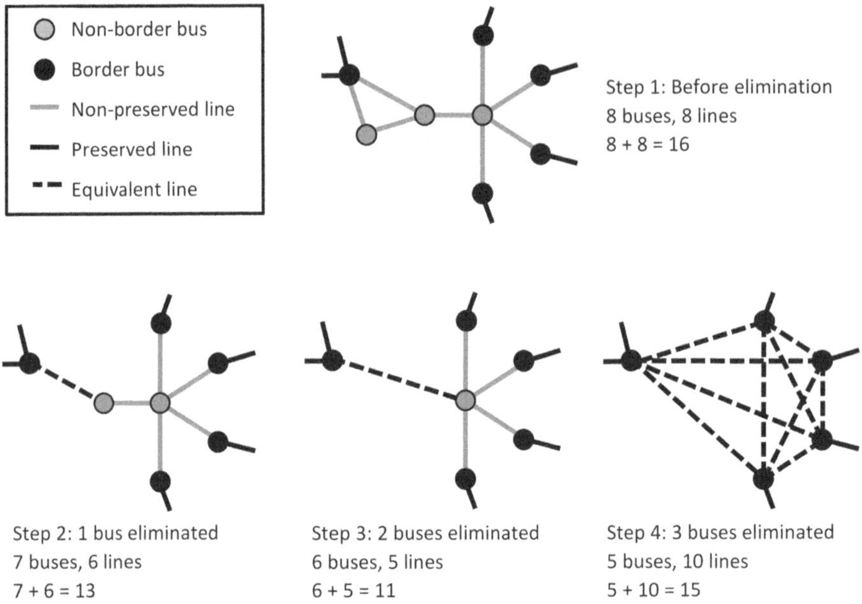

Fig. 4.2 Steps of the Kron reduction process in a given area. At step 4, eliminating all the non-border buses results in a dense subnetwork (any pair of buses is connected by a line). To avoid this, the reduction process should stop at step 3, where the total number of buses and lines is minimum

4.2.4 Equivalencing and Candidate Lines in the Reduced Network

The admittance of the equivalent lines is computed in each iteration according to Equation (4.9). The fraction of power injection (generation minus load) allocated from an eliminated bus to the remaining buses is computed in each iteration according to Equation (4.10). During the elimination of non-border buses, the admittance and capacity of the lines crossing two areas are not affected. However, a new capacity should be computed for the intra-area lines. This calculated capacity should enable very similar transfers of power between buses as in the original network. Methods to calculate this equivalent line capacity are provided in the literature [12].

The advantage of applying the Kron reduction process to the non-border buses within each area is that it preserves the power flows in the lines connecting two different areas. Therefore, if a new line is installed between two areas, its computed power flow should be exactly the same in the reduced and non-reduced network. The candidate lines that can be modeled in the reduced network are the ones whose ends are border buses located in two different areas. The other candidate lines cannot be represented in the reduced network because their installation would not guarantee to preserve power flow in inter-area lines. However, eliminating these candidate lines from the reduced TEP problem should not have a dramatic impact on the accuracy of the TEP solution. Indeed, according to Sect. 4.2.1, these candidate lines are neither partially built nor parallel to any congested line.

4.3 Candidate Grid Elements to Consider

In TEP studies, the set of candidate grid elements, also called search space, is the dimension that has the largest impact on the computational time when solving the problem. The number of combinations of expansion projects increases exponentially with the number of candidate grid elements. This motivates the use of search space reduction methods to identify the most economically favorable expansion projects.

A possible way to guide the search space reduction is to consider the solution of a "hybrid" relaxed TEP model [13]. In this model, the capacity of candidate lines that can be installed in each corridor is continuous and unbounded. This TEP model is called "hybrid" because it assumes the flows in existing and candidate AC lines satisfy the DC Load Flow (DCLF) model and the Transportation Load Flow (TLF) model, respectively. This model is a LP problem, which can be solved relatively quickly. However, the TLF model does not provide sufficient accuracy to represent the flows in candidate AC lines. As a result, the DCLF model must be applied to the candidate AC lines to the fullest possible extent.

We can achieve this using an iterative approach. First, a "hybrid" relaxed TEP problem, in which flows in candidate AC lines satisfy the TLF model, is solved. Then, we use the solution computed in the previous iteration to refine the TEP problem formulation. The problem is reformulated by including additional constraints to guide the flows in candidate AC lines towards satisfying the DCLF model.

The proposed methodology to reduce the TEP problem's search space is described by the following steps:

1. Computation of a relaxed TEP problem that considers an unbounded number of candidate lines in each corridor and assumes the TLF model in candidate AC lines
2. Iterative computation of a relaxed TEP problem that considers a bounded number of candidate lines in each corridor and assumes a relaxed DCLF model in candidate AC lines
3. Identification of the reduced search space according to the optimal values of the investment decision variables in the final iteration

4.3.1 Relaxed TEP Problem with an Unbounded Number of Candidate Lines per Corridor

For simplicity, it is assumed that all candidate lines are AC lines and that candidate AC lines in the same corridors have identical technical features. In other words, all candidate AC lines between the same pairs of buses have the same power capacity and the same admittance. As a result, each individual candidate line can be identified by the corridor *(i,j)* where it can be installed. Analogously, we can address the general case considering parallel AC candidate lines with different features and DC candidate lines.

The overall equivalent admittance $Y_{i,j}^{cand}$ of a set of parallel candidate lines installed in a specific corridor is equal to the admittance $Y_{i,j,1}^{cand}$ of a single candidate line from this corridor multiplied by the number of candidate lines installed x_{ij} (4.11). Similarly, the equivalent capacity $\overline{f_{i,j}^{cand}}$ is equal to the capacity $\overline{f_{i,j,1}^{cand}}$ multiplied by the number of candidate lines installed x_{ij} (4.12).

$$Y_{i,j}^{cand} = x_{ij} Y_{i,j,1}^{cand} \tag{4.11}$$

$$\overline{f_{i,j}^{cand}} = x_{ij} \overline{f_{i,j,1}^{cand}} \tag{4.12}$$

Ideally, the total power flows through a candidate AC corridor *(i,j)* should obey the following constraints:

$$f_{ij,s}^{cand} = x_{ij} S_B Y_{ij,1}^{cand} \left(\theta_{i,s} - \theta_{j,s}\right) \tag{4.13}$$

$$-x_{ij} \overline{f_{i,j,1}^{cand}} \leq f_{ij,s}^{cand} \leq x_{ij} \overline{f_{i,j,1}^{cand}} \tag{4.14}$$

where s corresponds to a specific snapshot, S_B is the base power, and $\theta_{i,s}$ is the voltage angle at bus i in snapshot s. Equation (4.13) ensures that the flow through the corridor *(i,j)* satisfies the DCLF model, whereas Eq. (4.14) ensures that the capacity constraint of the corridor is respected. Eqs. (4.13) and (4.14), together with the other equations of the problem, define a TEP problem with an unbounded number of candidate lines per corridor (TEPUNCL).

Similar to the two previous TEP reduction techniques, the goal is to relax the TEPUNCL problem in order to find a solution that provides relevant information about ways to reduce the search space. In order to achieve this, we need to linearize the integer variables x_{ij} and find a feasible region that is convex and encompasses the region defined by (4.13) and (4.14).

In the linear relaxation of the TEPUNCL problem, the binary constraint on investment decision variables is relaxed and they are unbounded, i.e., $x_{ij} \in [0; +\infty[$. Eq. (4.14) is linear and convex. However, Eq. (4.13) is not linear. Indeed, the right-

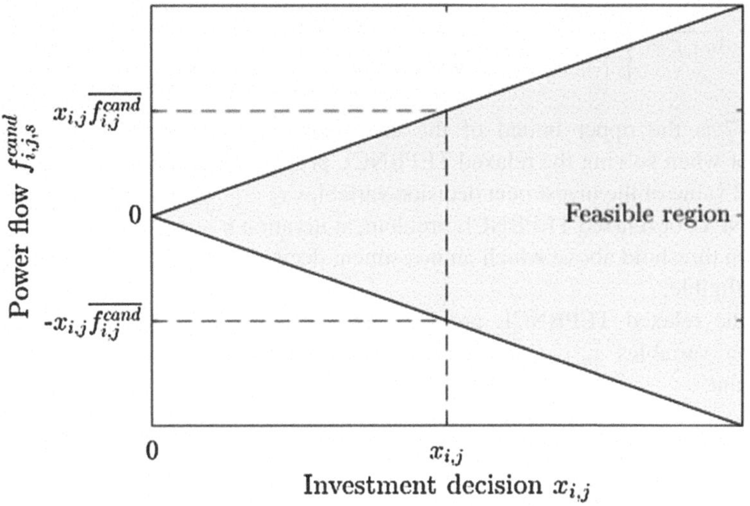

Fig. 4.3 Feasible region bounding the values of x_{ij} and $f^{cand}_{ij,s}$ in the relaxed TEPUNCL problem

hand side of the equation includes the product of an integer variable and a linear variable $x_{ij}(\theta_{i,s} - \theta_{j,s})$. The smallest feasible region that is linear and convex and includes the space defined by (4.13) and (4.14) is the space defined by Eq. (4.14). This space is illustrated in Fig. 4.3 above.

Therefore, in the linear relaxation of the TEPUNCL problem, power flows through candidate AC corridors only obey constraint (4.14) and not (4.13).

4.3.2 Relaxed TEP Problem with Bounded Number of Candidate Lines per Corridor

The relaxed TEPUNCL problem is linear and, thus, relatively easy to solve. Solving the relaxed TEPUNCL is a quick way to have an approximation of the best possible tradeoff between the benefits produced by these investments and their cost. The optimal investment decisions of this problem reflect this tradeoff. The value of these optimal investment decisions can help define an upper bound for the number of candidate lines that would be economically viable in each corridor. First, it can be assumed that the number of candidate lines that are economically viable in a given corridor is lower than the optimal investment decision value of this corridor according to the relaxed TEPUNCL problem.

As a result, a new relaxed TEP problem with a bounded number of candidate lines per corridor (TEPBNCL) can be defined. In this relaxed model, investment decision variables are bounded according to Equation (4.15):

$$\overline{x_{n,i,j}} = \begin{cases} +\infty & \text{if } x^*_{k,i,j} \leq \epsilon_{\text{inv}} \forall k \in \{0, \cdots, n-1\} \\ \left\lceil x^*_{n-1,i,j} \right\rceil & \text{otherwise} \end{cases} \tag{4.15}$$

$\overline{x_{n,i,j}}$ is the upper bound of the investment decision variable $x_{n,\,i,\,j}$ of the corridor when solving the relaxed TEPBNCL problem at iteration n. $x^*_{n-1,i,j}$ is the optimal value of the investment decision variables $x_{n\,-\,1,\,i,\,j}$ after solving the relaxed TEPUNCL, or relaxed TEPBNCL problem, at iteration $n-1$. ϵ_{inv} is the investment decision threshold above which an investment decision variable is considered to be nonnegligible.

In the relaxed TEPBNCL problem, the corridors (i,j) that have investment decision variables $x_{n,\,i,\,j}$ with finite upper bounds $\overline{x_{n,i,j}}$ satisfy the following constraint:

$$-M_{i,j}\left(\overline{x_{n,i,j}} - x_{n,i,j}\right) \leq f^{cand}_{i,j,s} - S_B \overline{x_{n,i,j}} Y^{cand}_{i,j}\left(\theta_{i,s} - \theta_{j,s}\right) \leq M_{i,j}\left(\overline{x_{n,i,j}} - x_{n,i,j}\right) \tag{4.16}$$

Equations (4.16) and (4.14) are based on the McCormick envelope [14] and represent the tightest linear and convex envelope of the feasible region defined by Eqs. (4.13) and (4.14). This envelope assumes that the voltage angle difference is bounded by $\left(\theta_{i,s} - \theta_{j,s}\right) \in \left[-M_{i,j}/S_B Y^{cand}_{i,j};\, M_{i,j}/S_B Y^{cand}_{i,j}\right]$ and that the investment decision variable is bounded by $x_{n,i,j} \in \left[0; \overline{x_{n,i,j}}\right]$. The value of the parameter $M_{i,j}$ must be large enough so the voltage angle difference between buses i and j is not unnecessarily bounded when no candidate line is installed in the corridor (i,j), i.e., $x_{n,\,i,\,j} = 0$.

An illustration of the McCormick envelope delimiting the values of the variables $x_{i,\,j}$, $f^{cand}_{i,j,s}$, and $(\theta_{i,\,s} - \theta_{j,\,s})$ is depicted in Fig. 4.4.

In the linear relaxation of the TEPBNCL problem:

- The power flows of the candidate AC corridors with unbounded investment decisions variables only obey constraint (4.14).
- The power flows of the candidate AC corridors with investment decision variables bounded by $\overline{x_{n,i,j}}$ obey both constraints (4.14) and (4.16).

The relaxed TEPBNCL problem is solved in the second and subsequent iterations of this method ($n = 1, 2, \ldots, n_{final}$). At the end of each iteration n, the optimal network investment decisions $x^*_{n,i,j}$ are updated. At the beginning of the following iteration $n+1$, the upper bounds $\overline{x_{n+1,i,j}}$ of the investment decisions variables are updated using Eq. (4.15).

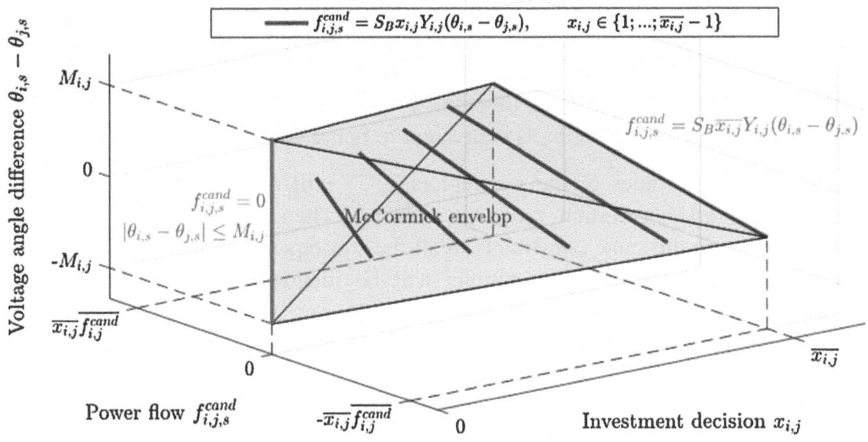

Fig. 4.4 Convex envelope delimiting the values of the variables $x_{i,j}$, $f^{cand}_{i,j,s}$, and $(\theta_{i,s} - \theta_{j,s})$ in the relaxed TEPBNCL problem. The McCormick envelope is illustrated by the orange tetrahedron. This is the smallest region that is convex and covers simultaneously all the regions for which 0, 1, ..., and $\overline{x_{i,j}}$ candidate lines are installed. The blue lines depict the feasible regions enforcing the DCLF model for a number of candidate lines between 1 and $\overline{x_{i,j}} - 1$

4.3.3 *Convergence of the Method*

The iterative process stops at iteration n_{final} when the convergence criterion (4.17) is met:

$$\left| x^*_{n_{final},i,j} - x^*_{n_{final}-1,i,j} \right| \leq \epsilon_{\text{conv}}, \quad \forall (i, j) \tag{4.17}$$

To meet this convergence criterion, the difference in optimal decisions values between this iteration and the previous one must be lower than a certain threshold for all corridors simultaneously. In other words, the optimal expansion of the network in the two subsequent iterations must be similar enough. Next, we can prove that the iterative process always stops after a finite number of iterations.

Proof. For each corridor (i,j):

- If the optimal investment decision value $x^*_{n,i,j}$ is lower than the threshold ϵ_{inv} for all iteration n, then the convergence has been achieved in that corridor. Indeed, $\overline{x_{n,i,j}} = +\infty$ for any iteration n.
- If the optimal decision variable $x^*_{n,i,j}$ is greater than the threshold ϵ_{inv} in at least one iteration N, then the investment decision variable $x_{N+1,i,j}$ will be bounded by $\left\lceil x^*_{N,i,j} \right\rceil$ in the next iteration $N + 1$. As a result, its optimal value $x^*_{N+1,i,j}$ will be bounded by the same value: $x^*_{N+1,i,j} \leq \left\lceil x^*_{N,i,j} \right\rceil$. Therefore, $\left\lceil x^*_{N+1,i,j} \right\rceil \leq$

$\left\lceil x^*_{N,i,j} \right\rceil$, and $\overline{x_{N+2,i,j}} \leq \overline{x_{N+1,i,j}}$. By recurrence, it can be deduced that the sequence $\left(\overline{x_{n,i,j}}\right)_{n\in\mathbb{N}}$ is a decreasing sequence for $n \geq N+1$. In addition, zero is a lower bound for this sequence. As a result, the sequence converges according to the monotone convergence theorem. Furthermore, because $\left(\overline{x_{n,i,j}}\right)_{n\in\mathbb{N}}$ is an integer sequence, converging is equivalent to being constant from a certain rank.

Therefore, the values of the parameters $\overline{x_{n,i,j}}$ will remain unchanged for all corridor (i,j) after a certain rank, or iteration. Then, the TEPBNCL will be formulated using the same constraints in all the subsequent iterations. The optimal solution of these subsequent iterations will be identical, and the convergence criterion (4.17) will be met.

At the end of the iterative process, the reduced search space can be finally calculated. The reduced search space of the TEP problem is defined by the maximum number of candidate lines that should be considered in each corridor and is computed with Eq. (4.18).

$$\overline{x_{i,j}} = \begin{cases} \left\lceil x^*_{n_{final},i,j} \right\rceil & \text{if } x^*_{n_{final},i,j} > \epsilon_{\text{inv}} \\ 0 & \text{otherwise} \end{cases}, \forall (i,j) \tag{4.18}$$

4.4 Conclusion

4.4.1 Summary of the Presented Techniques

The recent energy policies and advances in power system technologies have significantly increased the complexity of planning the transmission network expansion. As a result, solving the TEP problem is now more computationally intensive. This issue can be addressed by adequately reducing the size of the TEP problem in its three main dimensions:

1. The representation made of the temporal variability (temporal dimension)
2. The spatial representation
3. The set of candidate grid elements to consider

This chapter introduces three efficient reduction techniques, one reduction technique for each problem dimension , that are based on a linear relaxation of the TEP problem.

A summary of the reduction techniques used in this chapter is presented in Table 4.1.

The dimension of the TEP problem that is the most difficult to reduce is certainly the network representation. Indeed, the accuracy of the network representation is critical when identifying the optimal lines to invest in. This is because these lines must be represented as new components that belong to the reduced network. In

Table 4.1 Summary of the reduction methods described in this chapter

Dimension reduced	Temporal variability	Spatial structure	Candidate grid elements to consider
Information captured	Reinforcement benefits	Congested lines and candidate lines likely to be built	Potential reinforcement capacity
Method used	k-means clustering of the potential lines' benefits	Network partition with multicut problem and bus elimination with Kron reduction	Iterative refinement of the flow representation in candidate AC corridors

addition, to accurately identify expansion needs, it is critical to accurately represent the congestions that occur in the reduced network.

On the contrary, the set of operating situations and the set of candidate lines can be significantly reduced without dramatically impacting the quality of the TEP solution.

4.4.2 Combination and Order of the Reduction Techniques

This chapter describes three separate reduction techniques that can be applied independently to each one of the three TEP problem's dimensions. It is desirable to apply these three reduction techniques in conjunction to further decrease the computation time by reducing the size of the TEP problem. However, the combination of these different reduction methods should be performed with caution.

Both the snapshot selection and network reduction methods presented in this chapter require a finite, and preferably small, set of candidate grid elements. Therefore, the search space reduction technique should be applied first, assuming that the power system planner has not already selected a reduced set of candidate grid elements.

Furthermore, the proposed network reduction technique is more conservative in comparison to the snapshot selection technique. In fact, the procedure used in the network reduction technique partitions the network into areas where congestions are unlikely to occur. Assuming no congestion within these areas, power flows in (existing and candidate) inter-area lines are identical in the complete and reduced networks. As a result, the network reduction technique largely preserves relevant information about the TEP problem. In contrast, the snapshot selection method clusters snapshots that are comparable from the TEP standpoint but can be dissimilar from the operation perspective. Therefore, the loss of information when performing the network reduction procedure should be smaller than when performing the snapshot selection technique. This is why the snapshot selection should be last to be applied so this larger loss of information loss occurs in the last step. This way, the network reduction step is not misled.

In summary, the presented reduction techniques should be ideally applied to the TEP problem in the following order:

1. Search space reduction method
2. Network reduction method utilizing the reduced set of candidate lines from step 1
3. Snapshot selection method utilizing the reduced network from step 2 and the reduced set of candidate lines from step 1

References

1. D.Z. Fitiwi, F. de Cuadra, L. Olmos, M. Rivier, A new approach of clustering operational states for power network expansion planning problems dealing with RES (renewable energy source) generation operational variability and uncertainty. Energy **90**(2), 1360–1376 (2015)
2. Báñez Chicharro, F., Olmos Camacho, L., Ramos Galán, A., Canteli, L., & María, J. (2016). A Benefit-Based Methodology to Rank Transmission Expansion Projects
3. ENTSOE. (2015). ENTSO-E Guideline for Cost Benefit Analysis of Grid Development Projects (Guideline)
4. C.C. Aggarwal, J.L. Wolf, P.S. Yu, C. Procopiuc, J.S. Park, Fast algorithms for projected clustering, in *Proceedings of the 1999 ACM SIGMOD International Conference on Management of Data*, (ACM, New York, 1999), pp. 61–72
5. L. Kaufman, P.J. Rousseeuw, *Finding Groups in Data: An Introduction to Cluster Analysis* (Wiley, Hoboken, 2009)
6. F. Dorfler, F. Bullo, Kron reduction of graphs with applications to electrical networks. IEEE Trans. Circuits Syst. Regul. Pap. **60**(1), 150–163 (2013)
7. L.R. Ford Jr., D.R. Fulkerson, Maximal flow through a network. Can. J. Math. **8**, 399–404 (1956)
8. C. Chekuri, V. Madan, Approximating multicut and the demand graph, in Proceedings of the Twenty-Eighth Annual ACM-SIAM Symposium on Discrete Algorithms, vols. 1–0 (Society for Industrial and Applied Mathematics, 2017), pp. 855–874
9. N. Garg, V.V. Vazirani, M. Yannakakis, Approximate max-flow min-(multi)cut theorems and their applications. SIAM J. Comput. **25**(2), 235–251 (1996)
10. R. Tarjan, Depth-first search and linear graph algorithms, in *12th Annual Symposium on Switching and Automata Theory (swat 1971)*, (1971), pp. 114–121
11. W.F. Tinney, J.W. Walker, Direct solutions of sparse network equations by optimally ordered triangular factorization. Proc. IEEE **55**(11), 1801–1809 (1967)
12. W. Jang, S. Mohapatra, T.J. Overbye, H. Zhu, Line limit preserving power system equivalent, in *2013 IEEE Power and Energy Conference at Illinois (PECI)*, (2013), pp. 206–212
13. R. Villasana, L.L. Garver, S.J. Salon, Transmission network planning using linear programming. IEEE Trans. Power Syst. **PAS-104**(2), 349–356 (1985)
14. G.P. McCormick, Computability of global solutions to factorable nonconvex programs: Part I — Convex underestimating problems. Math. Program. **10**(1), 147–175 (1976)

Chapter 5
Offshore Grid Development as a Particular Case of TEP

João Gorenstein Dedecca

5.1 The Emergence of Offshore Grids

The objective of this chapter is thus to provide an overview of the aspects affecting the transmission expansion planning (TEP) of offshore grids. It focuses on the differences for TEP of a conventional offshore grid (where each offshore transmission asset serves solely to connect onshore wind farms to shore, or to interconnect onshore power systems) and an integrated offshore grid (where assets perform these functions simultaneously). The analysis is structured into five governance building blocks, and for each block the most important aspects are discussed, with some best practices identified.

The realization of an integrated offshore grid entails additional challenges compared to a conventional one, for example, concerning the increased importance of maritime spatial planning, of coordination with generation expansion, and the more complex cooperation between multiple countries. To address these challenges, it is necessary to consider not only aspects of transmission expansion planning such as cost-benefit analysis methodologies, but also broader ones such as the ownership and operation of transmission assets, or the coordination with other economic activities and marine ecosystems.

Here offshore TEP is addressed in an international context where neighbouring countries each have the jurisdiction on offshore expansion planning in their territory or exclusive economic zone (EEZ). The analysis and conclusions can be adapted to a single country where subnational government levels have similar (partial) authority, such as the US transmission planning regions. A case where decision-making is centralized into a single authority (e.g. a national planning organization) will of course circumvent many of the challenges discussed here.

J. Gorenstein Dedecca (✉)
Trinomics, Rotterdam, The Netherlands

S. Lumbreras et al. (eds.), *Transmission Expansion Planning: The Network Challenges of the Energy Transition*, https://doi.org/10.1007/978-3-030-49428-5_5

In this section, first, the drivers for the expansion of offshore grids are presented. Integrated offshore grids are then introduced, including their characteristics and potential benefits over conventional offshore grids. Next, the major risks for the expansion of integrated offshore grids are analysed, and the current situation is presented. Section 5.2 discusses the challenges for the expansion planning of offshore grids, organized around the building blocks of governance, planning, ownership, pricing and finance, and operation. Finally, Section 5.3 presents the chapter conclusions.

5.1.1 Drivers of Offshore Transmission Expansion

Three main drivers exist for the expansion of offshore grids. First, electricity interconnectors are being increasingly developed to integrate electricity markets, rather than only for security of supply purposes. Recent trends of market integration and regional planning noted by Lumbreras and Ramos [55] also drive the development of offshore interconnections, especially in the European North Sea. There, 13 different interconnectors were commissioned in the 2010–2018 period, totalling over 9 GW in transmission capacity, and at least 10 more should be commissioned in the 2019–2022 period (including some offshore wind farm connectors). There is an increase in both the average transmission capacity and the deployment rate of offshore interconnectors [27].

Second, connecting offshore wind farms to onshore systems is a strong driver for offshore transmission. In recent years, the costs of offshore wind have fallen significantly, led by innovation in turbines, scale effects, competitive site development models with reduced risk to project developers and improved financing conditions [44]. The levelized cost of energy for offshore wind has fallen by up to 49% since 2010, to less than 100 USD/MWh [15]. This is leading to the deployment of offshore wind farms in Europe, Asia and other regions. In 2018 the installed capacity amounted to 23.1 GW, with the largest capacities in the UK (8.0 GW), Germany (6.4 GW) and China (4.6 GW) [38]. Offshore wind could represent 12.1% of the world's electricity supply by 2050, with the largest markets in China, North America and Europe [22].

Third, the deployment of offshore transmission is facilitated by (and enables) innovations and cost reductions in multiple transmission components, especially high-voltage direct current (HVDC) cables and voltage-source converters (VSC). These converters have multiple operational advantages over current-source converters, including better active power control and the provision of reactive power control, besides enabling multiterminal HVDC grids as seen in Chap. 7. The HVDC transmission maximum capacity and voltage have more than doubled in the last decade and are posed to reach 13 GW and 1100 kV in 2050, respectively. Further innovations are foreseen in converter losses, maximum cable length, DC breakers and DC/DC transformers [10, 70].

5.1.2 *Integrated Offshore Grids*

Following Dedecca [20], an offshore grid can be defined as a power system combining offshore power generation (particularly from renewable sources) and transmission lines of different technologies, possibly also with the presence of offshore loads. Grid assets serve to connect offshore wind power to their national onshore systems, or to interconnect these national systems.

Until recently, these two functions were performed by separate assets (here called conventional lines). However, innovation opens the possibility of establishing an integrated offshore grid, where assets perform both functions simultaneously, leveraging multiterminal HVDC, point-to-point HVDC and high-voltage alternating current (HVAC) transmission technologies. An integrated line can be defined as a line connecting two offshore wind farms, or an offshore wind farm directly to an onshore node belonging to another country (integrated lines can be also called hybrid in the literature). Their deployment will lead to an integrated offshore grid, where the generation and transmission expansion planning considers both conventional and integrated lines. Figure 5.1 illustrates the different configurations for conventional and integrated lines, in conjunction with the offshore platforms. Note that while the topmost connector to country A could have been classified as a

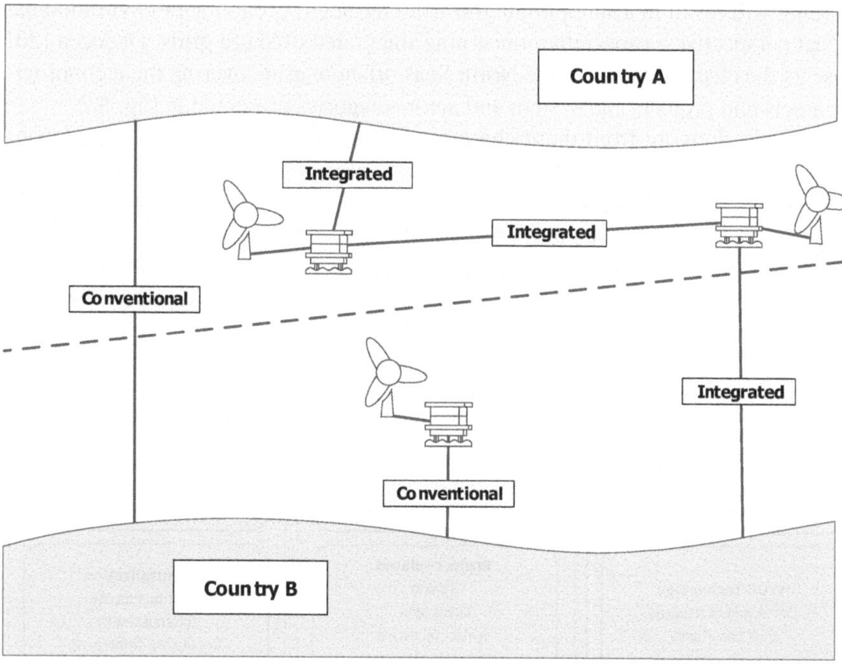

Fig. 5.1 Offshore grid typologies

conventional line, the other integrated lines ensure that it performs both connection and interconnection functions.

An integrated offshore grid can increase the socio-economic welfare compared to a non-integrated one. It can also provide important co-benefits which are, however, hard to quantify or monetize, such as technological innovation, industrial development, regional cooperation and reduced environmental impacts. Further research is needed, and these benefits are still uncertain and conditional on multiple factors, such as the international cooperation. As such, offshore expansion planning must consider the benefits of an integrated approach but also the costs, in coordination with other marine uses and the impact on the environment. Therefore, while there are large benefits to developing an integrated offshore grid, it should not be the goal per se. Moreover, given the complexity and uncertainty of the integrated offshore grid, each individual offshore transmission project should also be evaluated on its own merits [18].

5.1.3 Risk of Suboptimal Offshore Expansion

Offshore grids share many characteristics with onshore power systems. They nonetheless also have specific ones, which increase the risk that offshore expansion planning will result in a suboptimal grid from the socio-economic or environmental impact perspectives, especially concerning integrated offshore grids. Dedecca [20] presents the characteristics of the North Seas offshore grid covering the technological, assets and projects and system and actor categories presented in Fig. 5.2.

The risks deriving from these characteristics comprise first that of path dependence, where the offshore grid expansion locks into a certain (suboptimal) pathway, given initial reinforcing characteristics and in the absence of external influences [79]. Dedecca [20] develops the Offshore Grid Exploratory Model (OGEM), using myopic optimization in order to study these expansion pathways of the North Seas

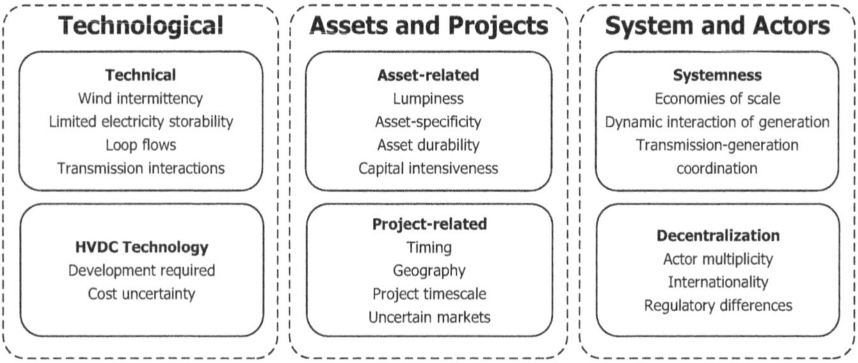

Fig. 5.2 Characteristics of offshore grids

offshore grid. It shows that the grid expansion is characterized by strong path dependence, where previous expansion periods influence following ones. This is because it may be more advantageous to connect offshore transmission lines to pre-existing ones rather than develop new lines which do not make use of the existing grid. Therefore, the topology and technology of the existing grid will favour certain investments over others. This dependency can lead to a broad range of possible final outcomes as the number of periods grow, given differences in the start of the pathways. The deployed transmission technologies affect subsequent expansions, especially for multiterminal HVDC grids (as savings in converters do not apply to point-to-point HVDC lines). Nonetheless, the offshore grid path dependence is not absolute, with factors such as the location and costs of offshore wind or the supply structure and demand levels of onshore systems still influencing the long-term expansion pathways.

The second main risk is that of countries behaving strategically and blocking expansion candidates which are beneficial from a regional socio-economic welfare point of view. Several recent studies highlight its importance and the need for reallocation of costs in order to deploy optimal expansions [18, 21, 37, 49, 51]. Interestingly, the impact of cost allocation approaches on the grid topology and deployed technologies may be more important than the impact on the regional socio-economic welfare (though this depends on other factors such as the level of offshore wind deployment).

A third important risk is that the costs and complexity of cooperation for an integrated grid steer the expansion towards conventional grid pathways, or that these costs impact the final net benefit of an integrated grid. While multiple studies quantified the benefits of a European offshore grid, there is less quantitative evidence on the complexity and costs of cooperation. Indeed, as discussed in Sect. 5.2, there are various challenges to developing an integrated grid, and hence studies comprehensively addressing those inevitably employ more qualitative approaches for some of the factors [17, 18, 21]. OGEM represents this cooperation complexity through integration governance constraints, which limit the number of integrated lines that can be developed at any given expansion period. It shows that these constraints increase the path dependence of the grid.

The fourth risk in the development of an integrated offshore grid is ignoring the larger socio-economic and environmental context and its (international) stakeholders. This can lead to developing transmission expansion plans which overlook constraints and synergies with other marine activities in multiple countries and ecosystems. This will severely hinder any integrated grid deployment, given the high competition for marine resources, and lead to suboptimal results and lack of acceptance from stakeholders. Recently, there have been a number of studies addressing this aspect, but further research and practice is needed [24, 58, 72, 78].

5.1.4 Current Research and Practice for Offshore Grids

Integrated offshore grids are still nascent: although offshore hubs have been used to simultaneously connect multiple wind farms to shore (e.g. in Belgium, Germany and the Netherlands), wind farms connected to more than one onshore point are practically inexistent, except for Kriegers Flak (as discussed below). Nonetheless, there are multiple developments concerning research, offshore projects, policy and regulation.

In the last decade, the research on offshore grids grew significantly in Europe and later to other regions. Dedecca [20] and Henneaux et al. [41] provide reviews of such research with a focus on the European North Seas, concerning planning models, operation of the offshore system and components (converters, wind turbines, DC grid protection), ownership, pricing and finance. Past modelling studies have demonstrated the important additional benefits of integrated grids, which however depend especially on the level of offshore wind development, the supply and demand structure of onshore power systems and the relative costs of different offshore transmission technologies. Offshore grids will gradually evolve to combine conventional and possibly integrated assets, employing different technologies.

While the literature on a supergrid in Asia is growing, few studies specifically address an offshore grid combined with offshore wind in the region. Ahmed et al. [6] reviews electricity supply, demand and trade in ASEAN (Association of Southeast Asian Nations) countries as well as transmission planning practices, with the view of forming a supergrid in the region. The deployment and operation of an offshore grid are highlighted as two of the major barriers to the formation of such an ASEAN supergrid, due to the need to define the grid topology and to develop and operate components such as voltage source converters and DC circuit breakers. In a follow-up study, some of the authors develop an optimal power flow model for the ASEAN supergrid [7]. Although it does not explicitly model offshore wind, the study does find that the supergrid will foster the development of renewable energy in the region and that VSC HVDC is the most suitable technology. Huber et al. [42] develop a generation, transmission and storage planning model for the ASEAN region (also reviewing previous energy modelling studies). Although the model does not include any separate offshore nodes, it does find significant offshore wind development in Southern Vietnam in low-emission scenarios, which drives transmission investment to export this energy to mainland Southeast Asia.

Research has thus advanced in ownership, pricing and finance of offshore grid assets in Europe, especially in the identification of the major challenges and identifying eventual solutions, as discussed in Sect. 5.2. This can lead already to the discussion and implementation of these solutions. In other regions, further research may be necessary, adapting to the legal, regulatory and financing regional context, while making use of the knowledge built in Europe.

Further review studies concerning specific offshore aspects exist, beyond those mentioned in this chapter. For the operation of the grid and its components, Henneaux et al. [41] review studies on possible steady-state and dynamic control issues for the grid, HVDC diode rectifiers, DC breakers and offshore wind farms. Inamdar and Bhole [46] review HVDC converters for offshore wind farms, while Zhang et al. [81] focus on offshore multiterminal HVDC grids utilizing modular multilevel converters, and Ramachandran et al. [64] analyse offshore wind farm AC grid issues when employing diode rectifiers, while Korompili et al. [50] address VSC HVDC technology including the control system and strategies. Chaithanya et al. [16] and Ruddy et al. [71] review the use of low-frequency AC (LFAC) in an offshore context and the component design, agreeing that a main issue is the design of the offshore transformer and platform. Jallad et al. [47] review frequency regulation strategies for offshore wind farms, while the IET [45] dedicated a journal issue to the coordinated control and protection of offshore wind power and combined AC/DC grid. The various academic publications on an integrated offshore grid is reflected in the number of ongoing or past research projects in Europe, as listed in Table 5.1.

The portfolio and commission of integrated offshore grid projects is accelerating. The Kriegers Flak Combined Grid Solution will be commissioned in 2019, connecting two wind farms in Denmark and Germany and allowing the trade of electricity between the two countries [29]. The 2018 Ten-Year Network Development Plan (TYNDP) of the ENTSO-E (European Network of Transmission System Operators for Electricity) lists for the first time more than 8 projects comprising offshore multiterminal technology. They range from conceptual designs to projects approaching commissioning such as the COBRA cable, which is designed to allow for the future connection of offshore wind [31]. An ambitious integrated project listed in the TYNDP is the North Sea Wind Power Hub (NSWPH), an initiative with growing participation and which in 2019 counted already with Dutch, German and Danish

Table 5.1 European research projects on integrated offshore grids

Project	Focus	Period/publication date
PROMOTioN	Technology and legal/regulator/financial framework	2016–2019
BalticIntegrid	Meshed grid in the Baltic Sea	2016–2019
BEAGINS	Baseline environmental impact assessment in the North Seas	2017
North Sea Offshore Network DK/DE/NO	Technical and economic analysis with focus on Denmark, Germany and Norway	2014–2019
Synergies at Sea	Technical design and economic and regulatory analysis with focus on NL-UK project	2013–2016
NorthSeaGrid	Technical and economic analysis of three project case studies	2013–2015
North Sea Transnational Grid	Technical and economic analysis with Dutch focus	2009–2013
ISLES I and II	Spatial plan, regulation and business model for a British Isles offshore grid	2013–2015 / 2010–2012

electricity and gas transmission system operators (TSOs). The hub is conducting studies for three different designs for implementation post-2030, with initial analysis focusing on the environmental impact and the offshore wind potential.

In 2017, the State Grid Corporation of China commissioned a 5-terminal VSC HVDC offshore network connecting the Zhoushan islands, with several offshore wind projects developing nearby, including the CGN Daishan 4 farm [1, 74]. In the United States, in 2011, the Atlantic Wind Connection project submitted a right-of-way application aimed at developing a 1271-km offshore VSC HVDC transmission system over 10 years, integrating up to 7 GW of offshore wind connected to land though up to 7 landfall points [11]. However, since then, the project has faced difficulties. In 2018, a right-of-way application was approved for another project, the New York/New Jersey Ocean Grid. Ocean Grid aims to integrate up to 5.9 GW of offshore wind with up to 5 onshore connection points [9].

Concerning international and cross-sectoral cooperation for an ecosystem-based governance of offshore grid expansion, recently, several European associations, project developers, system operators and non-governmental organizations signed the Marine Grid Declaration [66]. To minimize negative environmental impacts, the declaration seeks to foster environmentally sound and inclusive decision-making, encourage cross-sectoral and international cooperation, foster knowledge creation and provide guidance. This is done by ensuring the commitment of signatories to principles concerning maritime strategic and spatial planning, international and cross-sectoral cooperation and other areas.

Also political support is increasing, especially in Europe, where countries are actively cooperating within the framework of the North Seas Energy Initiative and the Baltic Energy Market Interconnection Plan [12, 23, 34]. The European Commission will furthermore release in 2020 an offshore wind strategy. In Asia, the Renewable Energy Institute pioneered in 2011 the concept of an Asian supergrid, later joining in the Global Energy Interconnection (GEI) initiative founded in 2016 [65]. While the GEI objectives in promoting electricity interconnections and renewable energy are general, offshore transmission could become a significant component of the initiative in Asia, given the region's geography, offshore wind potential and presence of highly populated coastal areas.

This section shows that research, projects and initiatives for developing offshore grids are accelerating in multiple regions, moved by the drivers of offshore transmission expansion discussed. This dynamic began in the North Seas of Europe, but more recently spread to Asia and the United States. These latter regions are leveraging the knowledge developed in Europe, not only regarding HVDC transmission technology but also on legal, regulatory and transmission expansion planning aspects.

5.2 Challenges for Offshore Grids

Given the multiple risks to the suboptimal expansion of offshore grids, expansion planning of an integrated offshore grid requires a holistic approach, with solutions coherently targeting the existing challenges. While any categorization of these challenges is not straightforward nor unique, they can be grouped in the governance building blocks of *governance, planning, ownership, pricing and financing, and operation* [77]. This section discusses each of these challenges (Fig. 5.3).

5.2.1 *Governance*

Governance concerns the management of organizations and institutions related to the other building blocks affecting offshore grid expansion planning. The arena where this governance occurs is characterized by a **multiplicity of actors, levels and sectors**.

Most offshore grids in development or consideration nowadays concern multiple countries (or subnational states) with (partial) authority to make independent decisions regarding the expansion planning. Examples include countries in the North Seas of Europe and East Asia, or states in the United States. In addition to this actor fragmentation, generally the offshore transmission expansion planning requires decision-making at several governance levels, from the supra- to the subnational. For example, in Europe, energy is a shared competence between the European Union and Member States, and the regional level is also important for the cooperation among countries. Additionally, subnational levels also influence

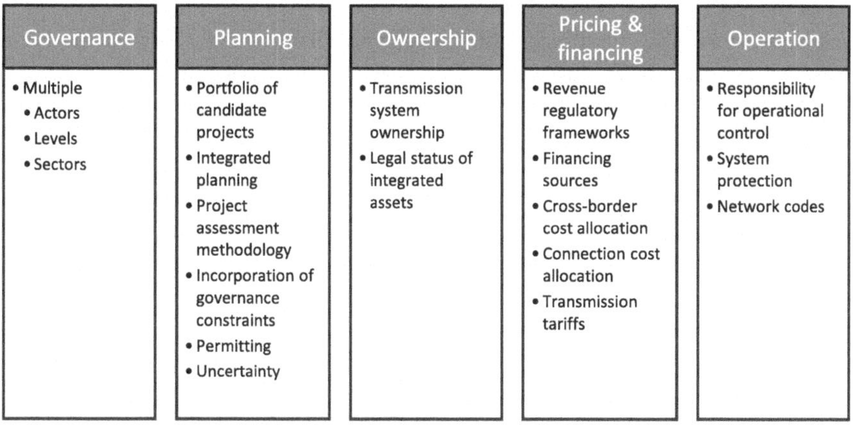

Fig. 5.3 Governance building blocks and related challenges

offshore planning, for example, regarding related onshore transmission expansions required, or nearshore wind farms.

Finally, there is a multiplicity not only of international actors and governance levels, but also of marine sectors (and the environment). An important example are the synergies between the North Sea offshore oil & gas and wind sectors regarding overlapping supply chain competences (i.e. construction and O&M), electrification of oil & gas operations and new uses for oil & gas infrastructure poised for decommissioning [43].

To address this complexity, several governance dimensions should be addressed, with Dedecca [19] focusing on three: the obligation, discretion and level dimensions. The first concerns the regulatory obligations of actors in the different levels (and the monitoring mechanisms in place to enforce obligations). The second regards the discretion that actors have to implement obligations, that is, how detailed these obligations are and how much implementation room is left for actors. Finally, the last dimension refers to the level where regulations are implemented and where the affected actors are located, from the supranational to national and subnational levels.

As an example of these dimensions, the Federal Energy Regulatory Commission (FERC) order 1000 has obligations at both the regional and interregional levels. The order requires public transmission providers to cooperate in developing regional transmission plans and to determine cost allocation methods for interregional projects. Despite these obligations, Benjamin [13] finds that the order is not sufficiently detailed (i.e. transmission providers have a high level of discretion for implementation).

The multiple marine uses coupled with the complexity of marine ecosystems call for an inclusive international and cross-sectoral cooperation in the form of ecosystem-based governance, tailored to the regional context. Such governance framework will require continuous discussion and decision-making by involved actors, guided by a long-term vision for the offshore grid and following principles such as accountability, legitimacy and transparency [72].

As an example, in Europe, the North Seas Energy Cooperation constitutes a voluntary initiative, while the obligations of the ENTSO-E and national system operators are defined in EU and national regulation. However, higher centralization for the governance, planning and/or operation of the European power system or North Sea offshore grid is also advocated [30, 73]. In the United States, the development of Ocean Grid is based on an unsolicited application to the Bureau of Ocean Energy Management and to the affected system operators, to develop an offshore grid with third-party access to offshore wind farms, relying on the market revenues from transmission services [9].

Hence, governance frameworks to each offshore region involve different approaches to the obligation, discretion and level dimensions. However, the central point is that the governance framework must evolve, and the relevant stakeholders cooperate to address the remaining challenges in the other building blocks.

5.2.2 Planning

The characteristics of the offshore grid affect the entire planning process, from the first identification of candidates to the commissioning of the projects.

An **integrated planning** is necessary in order to coordinate the long-term offshore transmission expansion planning with other marine uses, the ecosystems, offshore generation and storage and onshore transmission expansion, especially considering the multiplicity of actors, levels and sectors indicated in the governance building block above.

The first step for an integrated offshore planning is the development of a maritime spatial plan (MSP), providing the overall framework for balancing marine socio-economic uses and environmental considerations. Significant knowledge gaps on environmental impacts still exist, for example, regarding electromagnetic, noise and heat emissions [67]. A study listing several environmental, health, economic and social aspects addressed many data gaps for the European North and Irish Seas, but still others remain and may be larger for other offshore grid areas [24].

In Europe, the consideration of the impacts on ecosystems requires the development by governments of a strategic environmental assessment (SEA), where not only the impacts of the plans but also alternatives and mitigation measures are assessed. While the maritime spatial plan provides guidance to the expansion of offshore transmission and generation, multiple smaller strategic environmental assessments may be conducted according to the development of specific offshore zones, as is done in some of the countries reviewed in Ecofys and RPS [25]. By evaluating the impact of government plans and policies, the SEA complements environmental impact assessments, whose objective is to evaluate the impacts of specific offshore projects.

The planning of integrated transmission assets also requires the coordination of offshore generation and transmission expansion planning (G-TEP), leading to concurrent expansion or at least to the adoption of anticipatory measures which do not preclude asset integration later on. Otherwise, integrated plans may be infeasible or downscaled, even at the national level. This is illustrated in the case of Belgium, where offshore wind farms where not designed from the start to be connected to a single offshore wind hub [28]. In Germany, the change from overhead to underground lines delayed the implementation of the SuedLink which will transmit power from offshore wind in the North Sea to load centres in the south of the country [75].

The Renewable Energy Technology Deployment TCP discusses the centralized vs decentralized site development model for offshore wind [44]. In the latter, governments are responsible for (some of) the steps from zone identification to permitting of the offshore wind projects and construction of the offshore connection. Centralized site development reduces the risks to wind farm developers and facilitates international cooperation.

Concerning the onshore-offshore TEP coordination, the necessity arises from onshore bottlenecks often requiring simultaneous onshore transmission expansion.

In a context of long permitting and commissioning times due to public acceptance and onshore residential, recreational, economic and environmental aspects, this may be challenging. A common system operator responsible for both transmission expansion processes will increase coordination. On the other hand, in the United States, G-TEP coordination occurs by offshore transmission developers having to submit requests for onshore points of interconnection to transmission operators [60].

Also, more research is needed on the interaction of offshore storage and transmission, which could gain importance in the future with the development of power-to-gas technologies. Planning should furthermore consider the interaction of the electricity and gas systems. In Europe, the ENTSOs for electricity and gas are developing an interlinked approach which utilizes common scenarios for the planning process since 2018 and will consider several technical and market interlinkages between the electricity and gas sectors [33]. Nonetheless, this interlinked approach does not go as far as developing a single electricity-gas model, and as seen in the governance building block, sectoral interaction goes beyond the electricity and gas sectors. The need for joint planning will increase if offshore power-to-gas solutions are developed which may reuse offshore gas infrastructure poised for decommissioning.

Developing a portfolio of candidate projects is an important challenge due to the combinatorial possibilities arising from the various transmission technologies available, the 'greenfield' nature and the international dimension of the offshore grid. The combinatorial possibilities may be a significant challenge to transmission expansion models. The candidate portfolio needs to be restricted for practical purposes while considering projects deploying innovative technologies such as multiterminal HVDC and/or integrating offshore transmission and generation. Solutions for developing the portfolio include international cooperation for identifying candidates and implementing transparent and non-discriminatory planning processes which consider projects from both regulated system operators and third parties. Examples include studies conducted within the North Seas Energy Cooperation offshore grids' working group and the third-party project submission process in the European TYNDP, resulting, for example, in the iLand offshore storage island in the 2018 edition [31].

Studies and open planning processes must include and differentiate between concrete projects for short-term implementation and long-term conceptual ones. In order to support the development of a portfolio of candidate projects, financing measures must support not only permitting, final design and construction, but also planning studies in the early project phases. Thus, in the 2021–2027 period, the European Union will support studies and works for cross-border transmission as well as for renewable energy projects [35].

The project assessment methodology complements the integrated planning process, to conduct a detailed individual assessment of and compare the benefits and costs of (clusters of) projects. The assessment should provide a common, harmonized approach to compare candidate expansion plans and/or projects on an equal footing. It is thus particularly relevant for offshore transmission, which combines multiple technologies as well as conventional and integrated projects.

Olmos et al. [62] identify two approaches for regional expansion planning: either cost-benefit analysis, which monetizes all costs and benefits, or multi-criteria analysis, which weighs criteria of different dimensions to synthesize a single score. There is not a consensus on the preferred approach for either onshore or offshore expansion, as evidenced by the considerations in Bhagwat et al. [14]. In Europe, the ENTSO-E developed a second version of its cost-benefit analysis methodology, which actually constitutes a combination of the cost-benefit analysis (CBA) and multi-criteria approaches. Bhagwat et al. [14] discuss the inputs, calculation and outputs of the ENTSO-E methodology for offshore project assessment in Europe, identifying multiple improvements required. These arise especially from the uncertainty on transmission innovation, the need for a fair comparison of projects and their interaction, and the importance of the distribution of costs and benefits among actors and countries.

Expansion planning models incorporating governance constraints can help address the above challenges. One possible approach is the modelling of welfare and integration governance constraints in Dedecca [20]. Besides addressing the governance challenges, by studying the interaction of the constraints with the transmission line technologies and types (conventional or integrated), the approach can support the identification of beneficial offshore expansions in the context of complex cooperation and strategic behaviour of countries. Results show that the most visible effect of governance constraints is the change of grid topologies, the reduced participation of integrated lines and multiterminal HVDC and the increased path dependence for the latter technology.

Alternatively, Kristiansen et al. [51, 52] use the Shapley value in order to allocate costs and benefits among North Sea countries for 3 offshore expansion candidates, comparing the cost allocation to other cost allocation methods (the equal share principle of 50/50% split and the positive net benefit differential). The authors indicate the Shapley value results in a fairer allocation, which considers the ability of countries to veto candidates and the benefits these bring to the system. Moreover, other cost allocation methods such as the nucleolus are indicated for situations where countries would effectively veto projects given a Shapley value allocation.

A complementary approach is the scaling of transmission costs according to the risk of transmission projects. Li et al. [53] elaborate a transmission expansion model which accounts for the country risk (technical, economic and socio-political) and the relationship status between countries in order to scale transmission investment costs. This leads to the abandonment of transmission candidates between countries with conflictual relationships. The model could be adapted to scaling the cost of transmission offshore assets according to project risk (e.g. of asset stranding).

Exploratory and optimization models should be used complementarily and as a support to actual decision-making on the expansion of offshore grids and cost allocation. Every project should ultimately have its own evaluation of costs and benefits and cost allocation, following transparent, non-discriminatory methodologies for CBA and principles for cost allocation. Olmos et al. [62] highlight that in multilevel regional contexts, this decision should be made at the supranational level, but national planners should be able to verify the compatibility of these expansions with national transmission projects.

Permitting for offshore transmission projects involves activities intrinsically linked to the specificity of the marine environment and the competing marine uses. It should be addressed first in the maritime spatial plans, with the subsequent permitting of specific projects in alignment of the MSP. Fragmented and heterogeneous planning among countries has been identified as an important barrier to the development of offshore transmission projects. Even within a single country, the authority for issuing permits may be split among several authorities [67].

Permitting should be streamlined and harmonized internationally while still protecting marine ecosystems and being aligned with national or regional strategic environmental strategies and maritime spatial plans. In Europe, permitting solutions for trans-European energy projects include measures such as national one-stop permitting shops and regulatory limits to the permitting duration. Furthermore, countries cooperate within the North Seas Energy Cooperation to exchange best practices and develop coordinated permitting for (cross-border) offshore wind projects. The centralized site development model for offshore wind may also reduce permitting risks to transmission developers by laying the responsibility on government authorities [44].

Uncertainty over future cost and performance for offshore transmission technology further complicates the planning process and models. This concerns several components of an offshore grid, including HVDC cables, breakers and converters. The latter includes not only AC/DC converters but also DC/DC. As seen, the economic benefits of integrated offshore grid topologies are one of the main arguments for its deployment. However, these economic benefits and the specific optimal topology strongly depend on the technical possibilities and the relative costs of the components [18]. These components are at varying levels of technological maturity, with incipient technologies such as diode rectifiers showing greater uncertainty on future technical feasibility. Both market-ready and incipient technologies will still exhibit cost improvements, and the (relative) speed of these improvements is central to transmission expansion planning. The problem is compounded by uncertainty in future cost reductions for offshore wind power and particularly of floating foundations, which would open new offshore areas for development.

Possible tools for addressing the uncertainty include employing alternative scenarios and conducting sensitivity analyses on climate, load, CO_2 and fuel prices and commission delays. It is also possible to develop cost forecasts by breaking them down in components such as labour, commodity and component prices.

Advanced transmission expansion planning models such as robust optimization can help address this uncertainty. Konstantelos et al. [48] employ a min-max regret optimization for offshore transmission expansion, reinforcing the importance of the robust optimization method to address the risk of asset stranding and highlighting the strategic value of offshore-offshore lines. In order to incorporate the competition and complementarities of the offshore grid with onshore flexibility resources under uncertainty, Härtel et al. [40] developed a planning model leveraging multiple techniques such as snapshot clustering, sequential market and network optimizations and relaxation of the model to find initial feasible solutions. Offshore grid planning models may need to leverage such techniques even more than onshore ones to deal with the larger candidate expansion portfolio arising from the multiple technology combinations.

5.2.3 Ownership

Regarding the **ownership of the transmission system**, the current regulatory approaches allocate the ownership to the generator, to the transmission system operator or to a third party (merchant interconnectors or offshore transmission owners, OFTO, in the example of the UK). On the one hand, the TSO model allows for a more proactive planning of the transmission system in comparison to the others. On the other hand, the generator and third-party models allow to better provide a price signal to the generator and to introduce an element of competition to the expansion planning process [14].

In the context of difficulties in developing offshore connections due to, e.g. permitting issues and the need to coordinate offshore transmission expansion both with offshore generation and onshore transmission expansion, an approach where the TSO actively conducts this role has important coordination advantages. This has become the prevalent connection model and is employed, for example, in Belgium, Denmark, France, Germany, the Netherlands and Japan. The UK and Sweden are examples where a third-party model is applied, while in the United States, the generator model is employed [60].

The **legal status of integrated offshore assets** is an issue which only appears in integrated grids. Currently, there are open issues concerning the legal status of integrated interconnection and offshore wind connection assets at the international, European and national levels and the applicable regulation, network codes and the compatibilization between countries [59]. Several solutions are possible, from bilateral treaties to more comprehensive approaches such as tailored regional regulatory frameworks. This is an issue which is conditioned by the pre-existing regulatory framework and which will vary per region. The framework also affects the possible ownership solutions in the first place, as, for example, in Europe, where unbundling rules for interconnectors rule out their ownership by generators [73].

5.2.4 Pricing and Finance

The pricing and finance of offshore transmission assets is directly related to the previous building blocks of ownership and planning. Certainty on the legal status of integrated offshore transmission assets is a prerequisite for their development by private actors. The ownership model of transmission assets frames how their capital and operational costs can be recovered, as well as the allocation of investment and revenue risks between the transmission project developer and other actors, such as ratepayers. This will affect investment decisions in transmission assets, especially regarding the choice between conventional or integrated assets, given their different levels of risk. The most important challenges for offshore transmission pricing and finance are addressed as follows.

Regulatory frameworks for setting offshore transmission revenue should first be classified as regulating transmission revenues or not. In the first case, the revenue of the developer (usually the TSO) is set by the regulatory authority and recovered through network tariffs from users. In the non-regulated approach, merchant developers recoup their investment through congestion rents, or by selling transmission rights either in short-term markets (arbitraging between price differences between energy markets) or long-term ones.

Offshore transmission projects entail significant uncertainty regarding transport volumes, electricity price differentials and stranded costs in the case of integrated grids, if (part of) the offshore wind projects do not materialize. Offshore transmission expansion leads generally to a convergence of nodal prices, which reduces congestion rents, especially in integrated configurations. Therefore, merchant projects dependent on this revenue stream may be particularly exposed to subsequent offshore expansions. Research shows that the profitability of offshore wind farms may benefit from integrated configurations, but the exact impact is location-specific [76].

Also, even merchant investments are subject to public authorization, due to the incentive for merchant investors to undersize the transmission capacity relative to the social optimum in order to maximize congestion rents [68]. Hence, regulation of TSO revenues is the predominant approach for offshore transmission expansion. Several variants exist to determine these, such as cost of service or incentive regulation leveraging price or revenue caps, (and floors) benchmarking, menu of options and quality regulation. Additional parameters such as depreciation approaches, length of regulatory period and allowance of assets in the regulated asset base further complicate the exercise of revenue regulation.

However, these frameworks are usually determined over the entire balance sheet of TSOs rather than per specific project, while offshore projects may exhibit a higher technical, economic or financial risk than onshore ones. Therefore, TSOs may require project-specific incentives addressing risks of offshore projects, but there is the risk of regulatory complexity, information asymmetry (between the regulator and the TSO) and lack of transparency [54]. Project-specific incentives could address not only risks, but also be used to incentivize innovation by the TSOs. Moreover, merchant investors may not have the risk appetite for such an investment. Thus, hybrid models with both regulated and merchant revenue elements may be necessary for any such investment.

Possible project-specific incentives applied in Europe include the inclusion in the TSO's regulated asset base of anticipatory investments, the recognition of costs before commissioning, accelerated depreciation, shorter depreciation periods and pass-through of specific capital or operational expenditures [2]. Another solution is the separation of the regulated asset base between offshore and onshore assets for the same TSO, applying different regulatory parameters to each group, as is done in the Netherlands [5]. Also, the British regulator authorized a cap and floor mechanism limiting merchant revenue risk (positive and negative) for six offshore interconnectors due to be commissioned between 2019 and 2021 [61].

Several **financing sources** for offshore transmission are available to provide the required volumes, which in the European North Seas alone could amount to over 50 billion EUR from 2030 to 2050 in scenarios with significant offshore wind development [18]. Given the central role played by the TSO model, corporate finance, where bonds are issued and the resultant debts are included in the balance sheets of the network operators, will provide much of the necessary funding. Particularly interesting in this aspect are green bonds issued for the financing of specific low-carbon projects. They are gaining importance for the funding both of offshore transmission and wind power in Europe, for example, through the TSO TenneT in the Netherlands and especially in Germany [80].

Given the asymmetrical distribution of costs and benefits, the **strategic behaviour by parties** developing an offshore grid may be a barrier. For example, parties may oppose an integrated project if they may lead to reduced consumer surplus (due to increased energy exports) or to potential losses for offshore generators or transmission operators. This is compounded by the fact that third countries not involved in the offshore transmission projects can also be affected. Several studies have looked at the costs and benefits of different offshore grid configurations. Traber et al. [76] find that offshore generators will benefit from meshed configurations while the congestion rent of transmission projects is negatively impacted by these integrated grids, the latter finding being shared by Konstantelos et al. [49]. Dedecca et al. [20] find that strategic behaviour by welfare-losing countries such as Norway and Sweden may lead to important changes to the offshore grid topology. However, these changes may not significantly increase the welfare of the uncooperative countries, meaning those strategies are inefficient. Furthermore, while new offshore generators benefit from the offshore expansion, existing ones may lose due to the reduction in marginal prices. These reductions may lead to important welfare gains to third countries, due to increased consumer surplus, even for distant ones such as Italy. In all integrated offshore case studies of Flament et al. [37], some TSOs, generators (onshore and offshore) or consumers lose welfare in some of the North Sea countries analysed, with losses over 20 years reaching several billions of Euros.

Thus, the **cross-border cost allocation** of offshore transmission projects is central to developing an integrated grid, and cost allocation arrangements other than equal shares (50/50%) between developers may be necessary to deploy integrated projects. Research has identified certain principles which offshore cross-border cost allocation and tariff setting in general should follow. The main ones are providing short- and long-term signals to network users, allocating costs to beneficiaries and ensuring predictability and transparency [3, 37, 49]. The performance of various allocation mechanisms according to these principles varies, as research on mechanisms such as the positive net benefit differential, the Shapley value and the nucleolus shows [39, 49, 52].

Meeus and He [56] discuss the clustering of projects for cost allocation decisions. It is argued that this negotiation should occur between promoter and affected (positively and negatively) countries, with the intervention of supranational authorities (such as ACER in the European case) being solely a fallback option in case countries that are unable to reach an efficient compromise [62].

Examples of cost allocations other than equal shares are scarce. In cost allocation decisions in Europe taken since 2014, only 2 out of 11 electricity trans-European projects had alternative cost allocations, with moreover no allocation to non-hosting countries [4].

Typically, **connection cost allocation** for generators are classified as deep (where generators bear all costs, including grid reinforcements beyond the onshore coupling point), shallow (where generators bear connection costs up to the coupling point) or ultra-shallow (where all costs are borne by network users). Usually ultra-shallow cost allocation is chosen to incentivize offshore wind development, although actual approaches are more complicated, with various costs being differently allocated. When costs are to be recovered from offshore generators, this may happen with generators developing the connection themselves, or costs being recovered through lump-sum payments or transmission tariffs. If costs are allocated to all network users, they can be recovered from ratepayers or more generally from taxpayers, if there are direct government subsidies to TSOs, as in the case of the Netherlands [8]. Countries with elements of shallow or deep approaches include Denmark (shallow for nearshore projects), Belgium (shallow with partial subsidies), Norway (shallow), Sweden (deep), the UK (with parts of the cost allocated to offshore generators), Japan (shallow, with clarifications needed for new 2019 law) and the United States (deep) [14, 57, 60].

Transmission access tariffs recover capital and operational costs of transmission infrastructure as well as of losses and system services such as balancing and ancillary services (when not recovered through connection tariffs or congestion rents). Important parameters for setting access tariffs include the split of costs between generators and consumers, temporal and locational price signals, the inclusion of losses and system services, and the use of lump-sum, energy and capacity components in the transmission tariffs [14].

Beyond the tariff setting challenges common to onshore systems, three interrelated aspects are relevant to offshore grids: the support for offshore renewable energy, the provision of signals to generators and the cross-border compatibility of transmission tariffs. Tariff structures favourable to offshore renewable generators (such as low G charge components) may be justified given positive environmental externalities of offshore wind, dynamic efficiency considerations and the fact that renewable energy deployment is strongly driven by resource availability rather than by transmission access costs. Locational signals to offshore generators can be provided through connection rather than access tariffs, in order not to distort operational decisions, and the centralized offshore planning model by governments may strongly limit the freedom of developers to choose project locations anyway.

However, as important as (the lack of) locational signals within a single country is the compatibilization of national offshore transmission tariff structures. This is essential to incentivize offshore expansion where most profitable from a societal perspective and not distorting investment and operational decisions of generators which could lead to arbitrage between different regulatory environments [36].

Currently in Europe, transmission access tariffs are mostly recovered from loads, although in the North Seas region Ireland, Norway and Sweden had a generator (G) charge higher than 20% in 2018. Among the North Seas countries, most did not include a seasonal time component, while in Germany, the UK and the Netherlands, capacity components had the highest share of total transmission tariffs. It must be noted that costs unrelated to access can be included in transmission tariffs, especially of renewable energy support schemes and connection. In 2018, of the North Sea countries Germany and Belgium applied such a practice, with around 5 €/MWh added to transmission tariffs charged exclusively to loads, to finance among others the connection of offshore wind farms [32]. As a non-European example, the Japanese transmission tariff applies only to loads, with a capacity component of 27% of the total tariff, although discussions are ongoing to include a generator charge and increase the capacity component [26].

5.2.5 *Operation*

This section focuses on the operational aspects of HVDC grids which impact most directly the expansion planning of the offshore grid. The operation of the offshore grid is shaped by the characteristics of HVDC grids and transmission components such as VSC converters, diode rectifiers and DC breakers, which fundamentally alter the grid topology compared to onshore AC grids. DC/AC converters are fully controllable and therefore so are HVDC point-to-point links. However, this is not the case for meshed multiterminal HVDC offshore grids, which behave rather like AC grids, where flows are determined by Kirchhoff's laws [77].

Responsibility for control of the offshore grid operation can cover the entire onshore and offshore system, the offshore system, specific transmission assets (such as a single link) or separate national zones [77]. Asset ownership, asset operation and system control for international offshore grids are unlikely to be conducted by a single actor, but this does not impede the control to be centralized for a given national territory or even an international offshore grid.

Concerning the **protection of offshore grids**, research is ongoing on topics such as the interoperability and standardization between equipment manufacturers and technologies, development of DC breakers, fault identification and design criteria [17, 63]. For example, given the high penetration of intermittent (and thus stochastic) renewables, the absence of critical loads and the topology of HVDC grids, new stochastic security criteria and onshore security standards may be developed. These will need to be adapted for the grid characteristics and consider future evolutions such as the increase of offshore loads such as hydrogen electrolyzers.

Network codes regulate the operation of modern electricity grids, at the national level or even the international, if markets are integrated across borders. Currently, the European system counts with eight network codes in the connection, operation and market categories. From a planning perspective, the inclusion of requirements for offshore grids in network codes and (further) harmonization is needed for the integration of marine energy sources, in Europe and other regions [17, 63, 70].

5.3 Conclusions

This chapter discussed the transmission expansion planning of offshore grids, especially integrated ones. Significant practical and academic efforts are being devoted to developing such grids, particularly in Northern Europe, as they may provide additional socio-economic benefits. But since factors such as cooperation costs and the strategic behaviour of countries lead to the risk of achieving only suboptimal offshore expansion plans, it is necessary for actors to address a number of challenges.

The challenges and eventual solutions discussed here have a direct relationship with other chapters of this book. Particularly, HVDC transmission (Chap. 7), the coordination of generation and transmission expansion planning (Chap. 11) and integrated planning (Chap. 10) are covered in this book. However, the offshore grid shapes these issues so that any onshore solutions should be tailored to the specific offshore challenges. Furthermore, there is not one homogenous offshore grid concept, as the existing proposals in Europe, North America and Asia show. Factors such as the geography, renewable energy potential, onshore power systems and electricity sector regulation vary considerably. Nonetheless, many of the challenges surveyed here will be shared, especially when involving multiple areas with (partially) independent decision-making authority in transmission expansion planning.

Addressing the challenges and developing an integrated offshore grid entails costs related to, for example, the cooperation between countries, financial support for project studies or works and eventually the impacts to other marine economic activities and ecosystems. This increases the need to adopt a holistic approach to developing integrated offshore grids and requires that costs and benefits be weighed and compared to the development of a conventional offshore grid.

Moreover, not all benefits of an integrated offshore grid can be monetized, including innovation in offshore generation and transmission technologies, industrial development, reduced environmental impacts and improved cooperation between countries. These benefits can be substantial, and combined with the economic benefits, cooperating actors may well find offshore transmission expansion planning worth their while.

References

1. 4COffshore, GN Daishan 4 Offshore Wind Farm (2019)
2. ACER A for the C of ER, Recommendation on incentives for projects of common interest and on a common methodology for risk evaluation (2014)
3. ACER A for the C of ER, Scoping Towards Potential Harmonization of Electricity Transmission Tariff Structures – Conclusions and Next Steps (2015)
4. ACER A for the C of ER, Third Edition of the Agency's Summary Report on Cross-Border Cost Allocation Decisions (2018)
5. ACM AC & M, Gewijzigd methodebesluit Tenne T Net op Zee 2017–2021 (2019)
6. T. Ahmed, S. Mekhilef, R. Shah, et al., ASEAN power grid: A secure transmission infrastructure for clean and sustainable energy for South-East Asia. Renew. Sust. Energ. Rev. **67**, 1420–1435 (2017). https://doi.org/10.1016/j.rser.2016.09.055
7. T. Ahmed, S. Mekhilef, R. Shah, T. Urmee, Modelling of ASEAN power grid using publicly available data, in *2018 IEEE PES Asia-Pacific Power and Energy Engineering Conference (APPEEC)*, (2018), pp. 411–416
8. Algemene Rekenkamer, Focus op kosten windenergie op zee (2018)
9. Anbaric, Unsolicited Right-of-Way/Right-of-Use & Easement Grant Application – New York/New Jersey Ocean Grid Project (2018)
10. A.D. Andersen, J. Markard, *Innovating Incumbents and Technological Complementarities: How Recent Dynamics in the HVDC Industry Can Inform Transition Theories* (Centre for Technology, Innovation and Culture, University of Oslo, 2017)
11. AWC, Unsolicited Right-of-Way Grant Application for the Atlantic Wind Connection Project (2011)
12. I. Belet, B. Bendtsen, P. Berès, et al, Northern Seas as the Power House of North-Western Europe – Regional Cooperation in the Energy Union (2016)
13. R.M. Benjamin, Improving U.S. transmission expansion policy through order no. 1000. Contemp. Econ. Policy **34**, 614–629 (2016). https://doi.org/10.1111/coep.12158
14. P. Bhagwat, L. Meeus, T. Schittekatte, et al, PROMOTioN Intermediate Deliverable – Economic framework for offshore grid planning (2017)
15. BNEF BNEF, Battery Power's Latest Plunge in Costs Threatens Coal, Gas
16. S. Chaithanya, V.N.B. Reddy, R. Kiranmayi, A state of art review on offshore wind power transmission using low frequency AC system. Int. J. Renew. Energ. Res **8**, 141–149 (2018)
17. M. de Schepper, P. Henneaux, S. DeBoeck, et al., *PROMOTioN D12.1 Preliminary Analysis of Key Technical, Financial, Economic, Legal, Regulatory and Market Barriers and Related Portfolio of Solutions* (PROMOTioN, 2017)
18. J.G. Dedecca, *Expansion Governance of the Integrated North Seas Offshore Grid* (Delft University of Technology, 2018)
19. J.G. Dedecca, R.A. Hakvoort, P.M. Herder, The integrated offshore grid: Exploring challenges for regional energy governance. Energy Res. Soc. Sci. (2019). https://doi.org/10.1016/j.erss.2019.02.003
20. J.G. Dedecca, S. Lumbreras, A. Ramos, et al., Expansion planning of the North Sea offshore grid: Simulation of integrated governance constraints. Energy Econ. **72**, 376–392 (2018). https://doi.org/10.1016/j.eneco.2018.04.037
21. C. Delhaute, F. Gargani, G. Papaefthymiou, et al, Study on Regulatory Matters Concerning the Development of the North and Irish Sea Offshore Energy Potential – Final Report (2016)
22. DNV GL, Renewables, Power and Energy Use Forecast to 2050 – Energy Transition Outlook 2050 (2017)
23. EC EC, North Seas Countries, Political Declaration on Energy Cooperation Between the North Seas Countries (2017)
24. Ecofys, RPS, *Environmental Baseline Study for the Development of Renewable Energy Sources, Energy Storages and a Meshed Electricity Grid in the Irish and North Seas*, WP3 Final Baseline Environmental Report (2017a). https://doi.org/10.2833/720927

25. Ecofys, RPS, BEAGINS – WP3 – Appendic C – Review of Member States Plans and SEAs (2017b). https://doi.org/10.2833/720927
26. EGC E and GMSC, Electricity System and Market in Japan (2018)
27. A. Elahidoost, E. Tedeschi, Expansion of offshore HVDC grids: An overview of contributions, status, challenges and perspectives, in *2017 IEEE 58th International Scientific Conference on Power and Electrical Engineering of Riga Technical University (RTUCON)*, (2017), pp. 1–7
28. Elia, Modular Offshore Grid – Permitting challenges & Environmental concerns (2019)
29. Energinet, Kriegers Flak – Combined Grid Solution (2019)
30. ENTSO-E EN of TSO for E, Power Regions for The Energy Union: Regional Energy Forums As The Way Ahead (2017)
31. ENTSO-E EN of TSO for E, TYNDP 2018 Projects List (2018a)
32. ENTSO-E EN of TSO for E, Overview of Transmission Tariffs in Europe: Synthesis 2018 (2018b)
33. ENTSOG, ENTSO-E, Investigation on the interlinkage between gas and electricity scenarios and infrastructure projects assessment (2019)
34. EP EP, EC EC, Regulation (EU) No 347/2013 on guidelines for trans-European energy infrastructure (2013)
35. EU Council C of the EU, Proposal for a Regulation establishing the Connecting Europe Facility – progress report – ST 9951/18 + ADD 3 (2019)
36. EWEA, EWEA position paper on network tariffs and grid connection regimes (revisited) (2016)
37. A. Flament, P. Joseph, G. Gerdes, et al., NorthSeaGrid – Offshore Electricity Grid Implementation in the North Sea (2015)
38. GWEC GWEC, Global Wind Report – 2019 (2019)
39. S.Y. Hadush, C.D. Jonghe, R. Belmans, The effect of welfare distribution and cost allocation on offshore grid design. Sustain. Energy IEEE Trans. **6**, 1050–1058 (2015). https://doi.org/10.1109/TSTE.2014.2325911
40. P. Härtel, D. Mende, P. Hahn, et al., North Seas Offshore Network (NSON): Challenges and its way forward. J. Phys. Conf. Ser. **1104**, 012004 (2018). https://doi.org/10.1088/1742-6596/1104/1/012004
41. P. Henneaux, J.-B. Curis, J. Descloux, et al., *D1.3: Synthesis of Available Studies on Offshore Meshed HVDC Grids* (PROMOTioN, 2016)
42. M. Huber, A. Roger, T. Hamacher, Optimizing long-term investments for a sustainable development of the ASEAN power system. Energy **88**, 180–193 (2015). https://doi.org/10.1016/j.energy.2015.04.065
43. IEA IEA, Offshore Energy Outlook (2018)
44. IEA RETD TCP, Comparative Analysis of International Offshore Wind Energy Development (REWind Offshore) (2017)
45. IET, Guest editorial: Coordinated control and protection of offshore wind power and combined AC/DC grid. IET Renew. Power Gener. **12**, 1431–1433 (2018). https://doi.org/10.1049/iet-rpg.2018.0261
46. A. Inamdar, A. Bhole, Converters for HVDC Transmission for Offshore Wind Farms: A Review (2018)
47. J. Jallad, S. Mekhilef, H. Mokhlis, Frequency regulation strategies in grid integrated offshore wind turbines via VSC-HVDC Technology: A review. Energies **10** (2017). https://doi.org/10.3390/en10091244
48. I. Konstantelos, R. Moreno, G. Strbac, Coordination and uncertainty in strategic network investment: Case on the North Seas Grid. Energy Econ. **64**, 131–148 (2017a). https://doi.org/10.1016/j.eneco.2017.03.022
49. I. Konstantelos, D. Pudjianto, G. Strbac, et al., Integrated North Sea grids: The costs, the benefits and their distribution between countries. Energy Policy **101**, 28–41 (2017b). https://doi.org/10.1016/j.enpol.2016.11.024

50. A. Korompili, Q. Wu, H. Zhao, Review of VSC HVDC connection for offshore wind power integration. Renew. Sust. Energ. Rev. **59**, 1405–1414 (2016). https://doi.org/10.1016/j.rser.2016.01.064
51. M. Kristiansen, F.D. Munoz, S. Oren, M. Korpås, Efficient Allocation of Monetary and Environmental Benefits in Multinational Transmission Projects: North Seas Offshore Grid Case Study (2017)
52. M. Kristiansen, F.D. Muñoz, S. Oren, M. Korpås, A mechanism for allocating benefits and costs from transmission interconnections under cooperation: A case study of the North Sea offshore grid. Energy J. **39** (2018)
53. H. Li, Z. Lu, Y. Qiao, X. Guo, Transnational grid interconnection planning considering effect of country risks, in *2018 IEEE Power & Energy Society General Meeting (PESGM)*, (2018), pp. 1–5
54. L. Lind, *Investment Incentives and Tariff Design in a Meshed Offshore Grid Context*. Erasmus Mundus Joint Master in Economics and Management of Network Industries Thesis (2017)
55. S. Lumbreras, A. Ramos, The new challenges to transmission expansion planning. Survey of recent practice and literature review. Electr. Power Syst. Res. **134**, 19–29 (2016). https://doi.org/10.1016/j.epsr.2015.10.013
56. L. Meeus, X. He, *Guidance for Project Promoters and Regulators for the Cross-Border Cost Allocation of Projects of Common Interest* (Florence School of Regulation, 2014)
57. National Grid, TNUoS Tariffs in 10 minutes (2018)
58. Navigant, The North Sea as a Hub for Renewable Energy, Sustainable Economies, and Biodiversity (2017)
59. C. Nieuwenhout, M. Roggenkamp, WP7.1 Deliverable 1 Intermediate report for stakeholder review: Legal framework and legal barriers to an offshore HVDC electricity grid in the North Sea (2017)
60. M. Noonan, T. Stehly, D. Mora, et al, IEA Wind TCP Task 26: Offshore Wind Energy International Comparative Analysis (2018)
61. Ofgem, Electricity Interconnectors (2019)
62. L. Olmos, M. Rivier, I. Pérez-Arriaga, Transmission expansion benefits: The key to redesigning the regulation of electricity transmission in a regional context. Econ. Energy Environ. Policy **7** (2018). https://doi.org/10.5547/2160-5890.7.1.lolm
63. E. Pierri, O. Binder, N.G.A. Hemdan, M. Kurrat, Challenges and opportunities for a European HVDC grid. Renew. Sust. Energ. Rev. **70**, 427–456 (2017). https://doi.org/10.1016/j.rser.2016.11.233
64. R. Ramachandran, S. Poullain, A. Benchaib, et al, AC Grid Forming by Coordinated Control of Offshore Wind Farm connected to Diode Rectifier based HVDC Link – Review and Assessment of Solutions (2018)
65. REI REI, Asia International Grid Connection Study Group – Interim Report (2017)
66. RGI RGI, Marine Grid Declaration (2019a)
67. RGI RGI, Mini-workshop: Permitting, consultation and environmental protection for offshore transmission grids – Summary of key takeaways (2019b)
68. M. Rivier, J. Pérez-Arriaga Ignacio, L. Olmos, Electricity transmission, in *Regulation of the Power Sector*, ed. by I. J. Pérez-Arriaga, (Springer, London, 2013), pp. 251–340
69. E. Robles, M. Haro-Larrode, M. Santos-Mugica, et al., Comparative analysis of European grid codes relevant to offshore renewable energy installations. Renew. Sust. Energ. Rev. **102**, 171–185 (2019). https://doi.org/10.1016/j.rser.2018.12.002
70. J. Roos, P. Lundberg, Annex to D3.1 – Technology Assessment Report – Transmission Technologies: HVDC LCC, HVDC VSC, DC breakers, tapping equipment, DC/DC converters (2014)
71. J. Ruddy, R. Meere, T. O'Donnell, Low Frequency AC transmission for offshore wind power: A review. Renew. Sust. Energ. Rev. **56**, 75–86 (2016). https://doi.org/10.1016/j.rser.2015.11.033
72. K. Soma, J. van Tatenhove, J. van Leeuwen, Marine Governance in a European context: Regionalization, integration and cooperation for ecosystem-based management. Ocean Coast. Manag. **117**, 4–13 (2015). https://doi.org/10.1016/j.ocecoaman.2015.03.010

73. K. Sunila, C. Bergaentzlé, B. Martin, A. Ekroos, A supra-national TSO to enhance offshore wind power development in the Baltic Sea? A legal and regulatory analysis. Energy Policy **128**, 775–782 (2019). https://doi.org/10.1016/j.enpol.2019.01.047
74. T&D World, China Upgrades Capacity to the Zhoushan Islands (2017)
75. T. Tenne, SuedLink (2019)
76. T. Traber, H. Koduvere, M. Koivisto, Impacts of offshore grid developments in the North Sea region on market values by 2050: How will offshore wind farms and transmission lines pay? in *2017 14th International Conference on the European Energy Market (EEM)*, (2017), pp. 1–6
77. D. Van Hertem, O. Gomis-Bellmunt, J. Liang, *HVDC Grids: For Offshore and Supergrid of the Future* (Wiley, Hoboken, 2016)
78. J. van Tatenhove, J. van Leeuwen, K. Soma, Marine governance as processes of regionalization: Conclusions from this special issue. Ocean Coast. Manag. **117**, 70–74 (2015). https://doi.org/10.1016/j.ocecoaman.2015.09.009
79. J.-P. Vergne, R. Durand, The missing link between the theory and empirics of path dependence: Conceptual clarification, testability issue, and methodological implications. J. Manag. Stud. **47**, 736–759 (2010). https://doi.org/10.1111/j.1467-6486.2009.00913.x
80. WindEurope, Financing and investment trends – The European wind industry in 2017 (2018)
81. Y. Zhang, J. Ravishankar, J. Fletcher, et al., Review of modular multilevel converter based multi-terminal HVDC systems for offshore wind power transmission. Renew. Sust. Energ. Rev. **61**, 572–586 (2016). https://doi.org/10.1016/j.rser.2016.01.108

Chapter 6
HVDC in the Future Power Systems

Quanyu Zhao, Javier García-González, Aurelio García-Cerrada, Javier Renedo, and Luis Rouco

6.1 Introduction

6.1.1 A Brief History of HVDC Systems

The first commercial power plant in the USA was built at 255–257 Pearl Street in Manhattan by Edison Illuminating Company to generate direct current (DC) electricity and it started service in 1882 [50]. However, DC technology soon found a tough competitor in the alternating current (AC) technology, mainly because of the invention of the transformer, which made it possible to step up voltage easily and, therefore, reduce transmission and distribution losses. The first reliable commercial transformer was built by William Stanley in 1885 working for Westinghouse, based on the discovery of "induction" by Michael Faraday in 1831 [31, 44]. AC had clearly won the war against DC when Thomson-Houston (an AC equipment manufacturer) took control of Edison General Electric (a DC equipment manufacturer) to create General Electric in 1892 [14].

Nevertheless, DC technology has never disappeared from the battlefield. For example, in the early days after the defeat of DC, rotary converters and motor-generator sets had to be developed to connect existing DC loads to the AC supply [48], and the advantages of DC transmission over AC transmission have always been

Q. Zhao
EDF Energy Services, Beijing, China
e-mail: quanyu.zhao23@gmail.com

J. García-González (✉) · A. García-Cerrada · J. Renedo · L. Rouco
Instituto de Investigación Tecnológica (IIT), ETSI ICAI, Universidad Pontificia Comillas, Madrid, Spain
e-mail: javiergg@iit.comillas.edu; aurelio@iit.comillas.edu; javier.renedo@iit.comillas.edu; luis.rouco@iit.comillas.edu

S. Lumbreras et al. (eds.), *Transmission Expansion Planning: The Network Challenges of the Energy Transition*, https://doi.org/10.1007/978-3-030-49428-5_6

recognized in scenarios such as very-long-distance overhead lines and submarine or underground cables for moderate distances [17].

With the invention of the mercury-arc valve by an American electrical engineer, Peter Cooper Hewitt in 1901, DC transmission started to be reconsidered as an alternative. With continuous development, mercury-arc valves could be broadly classified into two categories by the mid-1930s: (1) sealed glass envelope for small ratings and (2) steel tanks with metal cooling jackets and vacuum pumps for large ratings. However, technologies back then were proven difficult to withstand high voltage (HV) and achieve low losses at the same time, which made it hard to compete with AC transmission.

In 1939, Dr. August Uno Lamm[1], who in the future was known as the father of HVDC transmission, was granted a patent for introducing grading electrodes to mercury-arc valves, which increased considerably the voltage-withstanding capabilities of the valves [53]. Soon later, in 1941, the first commercial HVDC link (± 200 kV, 115 km) was contracted to transmit 60 MW of power from the Vockerode generating station on the river Elbe to Berlin. The project was completed in 1945 but was never commissioned due to World War II [77]. As the mercury-arc valve technology matured, the first commercial HVDC link (20 MW, 100 kV) was finally built from mainland Sweden to the island of Gotland using a submarine cable and started operation in 1954. It marked the starting point of the modern era of HVDC transmission [56].

Mercury-arc valve-based systems were common until 1975 when semiconductor-based HVDC systems finally took over: firstly, thyristors were used and, more recently, insulated-gate bipolar transistors (IGBTs) have been introduced [53, 59].

The first project using thyristor-based converters was the Eel River (320 MW) back-to-back (B2B)[2] scheme commissioned in 1972. The main advantages compared to mercury-arc valves were a simpler design, less frequent maintenance, and less space needed. Since then, thyristor-based HVDC systems have been steadily developing. Currently, the Shanghaimiao-Shandong ±800-kV HVDC project will go beyond the world's record up to 6250 A [88].

The latest development for HVDC converter technology is the use of IGBT instead of thyristors. The first experimental system (3 MW, ± 10 kV) was tested in Sweden (between Hellsjön and Grängesberg) in 1997 and the first commercial project (rated at 50 MW, ± 80 kV) was commissioned by ABB again in Gotland, Sweden [8].

The substitution of thyristors by IGBTs (or similar devices) is not a mere change in the semiconductor switch: it brings a radical change of philosophy in the operation of AC-to-DC converter stations. Unlike thyristors, IGBTs can be

[1]August Uno Lamm (May 22, 1904 to June 1, 1989) was a Swedish electrical engineer and inventor. During his career, Lamm obtained 150 patents.

[2]With a back-to-back HVDC configuration, two independent neighboring systems with different and incompatible electrical parameters (frequency, voltage level, short-circuit power level) are connected via a DC link. Both rectifier (conversion from AC to DC) and inverter (conversion from DC to AC) are located in the same station.

Fig. 6.1 Scheme of an HVDC link

switched off using a control signal and, as shown later, this makes it possible to build voltage source converter (VSC) stations which are much more flexible than their thyristor-based current source converter (CSC) counterparts. Nevertheless, thyristor-based converters are still preferred when considering point-to-point bulk power transmission because it is a more mature technology when dealing with very high voltage and very high current.

In order to illustrate the functioning of a basic HVDC system, Fig. 6.1 shows a general scheme of a point-to-point (PTP) HVDC link connecting two AC grids. The link has two converter stations connected through an HVDC line. Typically, each converter is connected to its AC grid through filters and a connection transformer as it will be explained later.

6.1.2 HVDC Versus HVAC systems

Compared to conventional AC transmission, HVDC transmission is a proven technology that outperforms AC under certain conditions. The three main applications of HVDC systems are:

- Transmission of large amounts of power over long overhead lines
- Transmission of power over medium-to-long isolated cables (submarine or underground cables)
- Interconnection of asynchronous AC grids (either with the same or with different frequencies)

The advantages, for certain applications, are explained in detail as follows:

- **Investment costs:** A typical PTP link consists of two converter stations and two conductors/cables (or one in case of monopole configuration with ground return), while a three-phase AC system needs three conductors. This provides DC technology an economically competitive advantage with increasing transmission capacity, when the converter costs can be offset by the lower costs of lines. [75] provides an economic evaluation on HVDC versus HVAC.
- **Transmission losses:** In AC lines, current density is not equally distributed within a conductor due to the "skin effect." The current tends to flow near the

conductor surface. The higher the frequency, the shallower the skin depth[3]. In DC lines, the current flows uniformly through the whole conductor. This results in a smaller resistance for DC lines compared to AC lines and, thus, lower losses.

- **Long transmission distance:** AC transmission faces several challenges when the transmission distance is long. For example, reactive power compensation is needed in a long overhead AC transmission line in order to compensate its reactance, which is a limiting factor of the total bulk power one can transmit through that line. However, series compensated lines are prone to exhibit subsynchronous resonance[4] [54]. On the other hand, DC technology does not encounter limitations on distance.
- **Underground and submarine cables**: Shunt capacitance of cables is significantly higher than the one of overhead lines (OHLs), and hence, they require a much higher total current for a given power transmitted due to what could be called capacitive charging current. Limitations on transmission distance imposed by charging current of AC cables become significant with the growing need for transmitting bulk power and the trend towards higher voltages [72]. By contrast, DC cables exhibit no constraints for cable length associated with reactive current and losses can be half of those of AC counterparts. In particular, HVDC systems are of particular interest for an efficient integration of offshore wind generation [9].
- **Right of way (ROW):** As mentioned previously, DC transmission requires less conductors. Therefore, tower design of DC lines is more compact than for AC lines and less ROW is needed. This is particularly advantageous for densely populated areas. A schematic illustration can be found in [73].
- **Controllability and flexibility:** HVDC technology can easily control the active power delivered at each converter station, as long as power balance is guaranteed. This facilitates a flexible control and operation of the system and can also improve the performance (e.g., stability) of AC power systems [82 ,84].
- **Asynchronous interconnections:** DC technology is the only way of connecting AC systems of different frequencies (or otherwise asynchronous systems). An example can be found in the Eel River project commissioned in 1972.

Despite all the advantages described above, HVDC systems are expensive and their economic viability needs to be carefully examined before any large investment in transmission network reinforcement takes place.

[3]The electric current flows mainly at the "skin" of a conductor. Skin depth is a measure of how closely electric current flows along the surface of the conductor

[4]Subsynchronous resonance is a condition where the electric network has natural frequencies below the nominal frequency of the system [58].

6.2 HVDC Technology

HVDC technologies can be classified depending on the type of converters used: current source converter (CSC-HVDC), also known as line commutated converter (LCC-HVDC), and voltage source converter (VSC-HVDC). In addition, several configurations are possible, and among them, the possibility to build multiterminal HVDC grids represents a major challenge from the transmission expansion point of view. This section reviews briefly all these concepts.

6.2.1 LCC-HVDC Technology

Figure 6.2 depicts a point-to-point LCC-HVDC link. In this technology, converters are built with thyristors, after mercury-arc valves have long been surpassed. The basic building block of an LCC is the three-phase full-wave bridge circuit also referred to as a Graetz bridge (Fig. 6.3). A thyristor can stand a positive or negative voltage when in off state, until a current is fed into its gate terminal. Once this is done, the thyristor will turn on provided that the rest of the circuit is ready to impose a positive anode(A)-to-cathode(K) current through it. Once the thyristor has been turned on, it will not go off unless its A-to-K current goes to zero, even if the gate current is removed. A thyristor can be forced to switch off by applying a negative A-to-K voltage to it for a long-enough time so that its current can be reduced to zero [6].

In thyristor-based HVDC converters, voltage cannot be imposed at the AC side because thyristors cannot be switched off without the collaboration of the AC voltage. The DC-side voltage of the converter stations can be controlled using the thyristor firing angle and the DC through the link (and, therefore, the power transmitted) is imposed by controlling the DC voltages at both ends of the DC link (rectifier and inverter ends). The DC-side current is maintained through a large inductance value, and it is diverted to each of the three phases of the AC side of each converter station by the appropriate switching combination of the thyristors: the current is, therefore, imposed on the AC side. In a converter like the one in Fig. 6.2, the DC cannot be reversed because of the thyristors. However, bidirectional power flow is possible because the converter DC voltage can be made positive or negative.

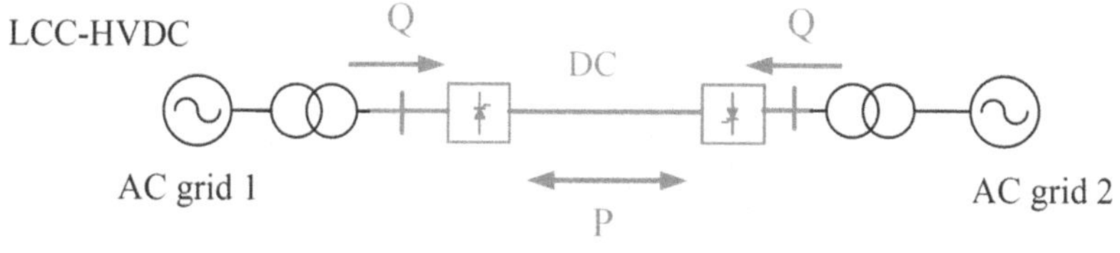

Fig. 6.2 LCC-HVDC link

Voltage rating can be raised by connecting the thyristor in series as a "stack," while current rating can be raised by connecting those stacks in parallel, and a solid experience doing this has been accumulated over the years [23]. Latest developments (the third generation) of thyristors show a blocking capability ranging from 7.2 to 8.5 kV and rating currents from 5.5 to 6.25 kA as required for the state-of-the-art ultra HVDC [80].

Nowadays, the largest CSC HVDC in operation has a rating of ±800 kV and 6.4 GW [7], and even a higher rating, i.e., ± 1100 kV and 10 GW, has been more recently planned [60].

Although CSC HVDC has been deployed largely for the past decades [65], the technology has several drawbacks [12, 69, 90]. First, CSCs rely on a strong AC network which can be expressed in terms of the short-circuit ratio (SCR)[5] at the point of common coupling (PCC). Typically, an AC network with short-circuit power at least 2.5 times of the HVDC rating is required for satisfactory operation [3]. Second, this type of converter also introduces a significant number of harmonics on both AC and DC sides which cause distortion of voltage waveforms. Therefore, filters are needed to minimize the effects. Moreover, the converters consume reactive power, which could amount to 50–60% of the converter rating. Thus, compensation is necessary, e.g., capacitor banks, static VAr compensator (SVC), a nearby generator, or a synchronous condenser. Consequently, a converter station can occupy a large space. Finally, commutation failures may occur in the inverter station often due to disturbances in the AC system such as sudden phase shifts or any voltage drop larger than 0.15 p.u. at an inverter terminal [1]. Finally, HVDC line tripping can result from a number of repeated commutation failures.

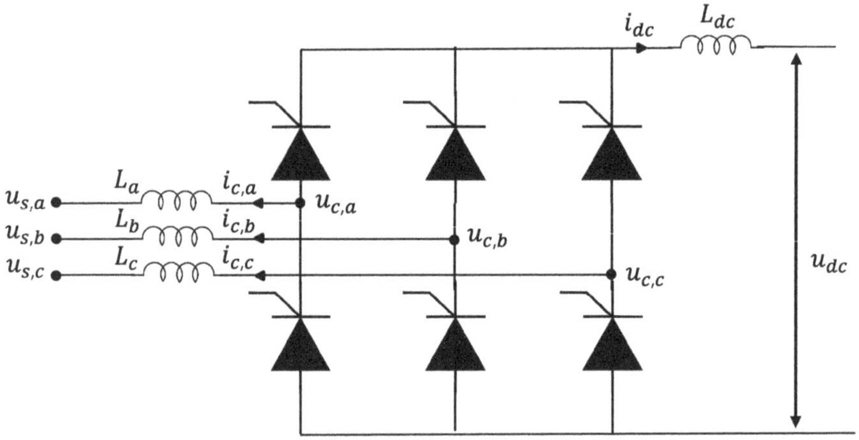

Fig. 6.3 CSC Graetz bridge

[5]SCR is defined as the ratio of system short-circuit-level MVA to the DC power MW and is often used to quantify the strength of the system [41].

6.2.2 VSC-HVDC Technology

Figure 6.4 depicts a point-to-point VSC-HVDC link. This technology is based on forced-commutation converters built with switches that can be switched off using a control terminal, such as insulated-gate bipolar transistors (IGBTs). The development of IGBTs in the 1990s opened new possibilities for HVDC technology. Currently, IGBTs have a maximum voltage rating of 6.5 kV and current rating up to 3.6 kA [78].

Since IGBTs are self-commutated, no external grid is needed for commutation. Besides, VSCs offer a number of other advantages as opposed to CSCs, such as independent and fast control of both active and reactive power, so there is no need for reactive power compensation; no commutation failure induced by disturbance from AC network; possible connection to a weak AC network or even a passive one where no generation source is available; power reversal done through current reversal instead of polarity reversal as for CSCs; and absence of lower-order harmonics and hence small filter size [39].

Figure 6.5 shows the configuration of typical VSC-HVDC converter stations. It consists of the following parts:

- **Converter transformer:** Is an essential component to match the AC grid voltage with the operating voltage of the converter. Compared to a CSC, the absence of low-order voltage harmonics allows a simpler design, similar to standard transformers.
- **Filter/phase reactor:** Since only high-order harmonics exists in VSCs due to high switching frequency, a low-pass output filter can be easily designed.

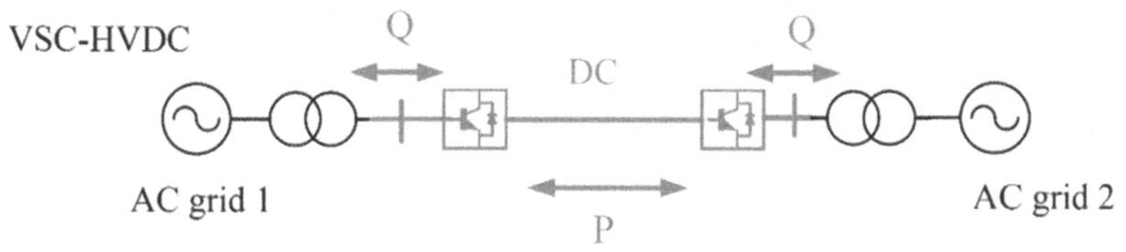

Fig. 6.4 VSC-HVDC link

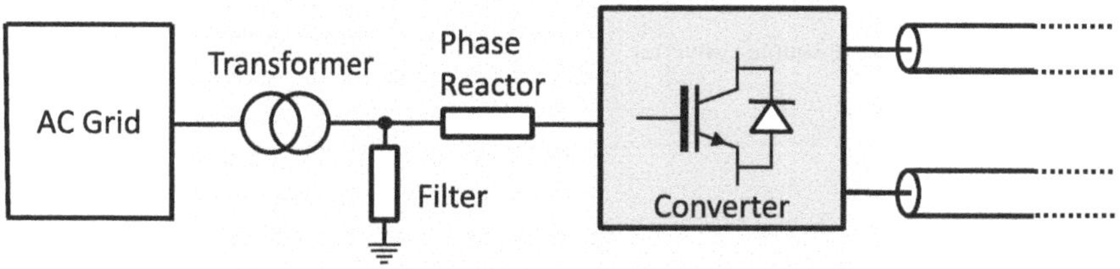

Fig. 6.5 VSC-HVDC scheme

- **Converter:** Is the most important element of the HVDC system. It converts a DC constant voltage to an AC voltage of an arbitrary size and shape by switching the IGBTs. The switching of IGBTs is controlled by the control scheme to control the complex current; thereby the active and reactive powers are controlled independently.

The topology of a VSC based on IGBTs can be roughly categorized into three types, following a chronological order of development: two-level converter, multilevel converter, and modular multilevel converter (MMC). The first generation of VSCs was based on a two-level converter topology, which was commercially developed by ABB and was known as "HVDC Light." Figure 6.6 provides a simplified schematic illustration. Each valve is represented by only one switching element. The well-known pulse width modulation (PWM) technique is used to synthesize the AC voltage waveform. Although conduction losses are low for IGBTs, losses resulted from switching elements are high. Losses in a two-level converter have been reported to be up to 3% of the total power going through the converter, while comparable losses for a CSC are 0.8% [19].

Although two-level and multilevel converters were used in the past, only the MMC seems to be practical in an ever-increasing DC voltage and power rating. Figure 6.7 provides a schematic illustration of a MMC where a full-bridge configuration has been used for each sub-module.

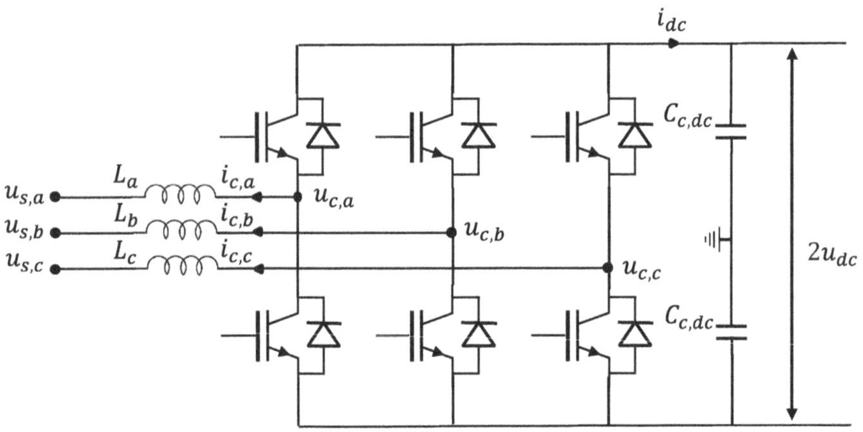

Fig. 6.6 Two-level voltage source converter

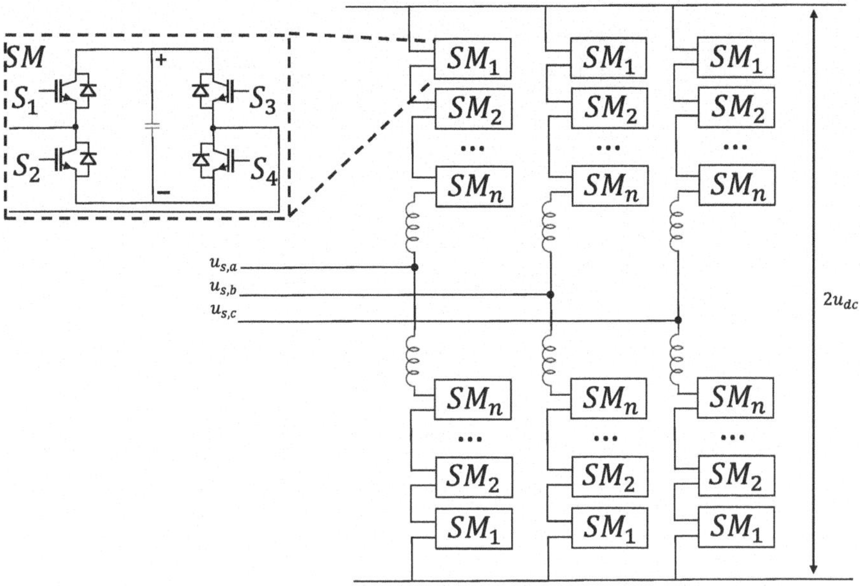

Fig. 6.7 Modular multilevel full-bridge voltage source converter

6.2.3 *HVDC Configurations*

There are several configurations available for HVDC systems which can be classified as follows: (1) monopolar, (2) homopolar, (3) bipolar, and (4) back to back (B2B) with alternative grounding strategies.

The monopolar configuration is the simplest and least expensive one [20]. For an asymmetric monopolar configuration, two converters are connected by a single pole line, a positive or a negative DC voltage. Usually a negative polarity is used due to smaller corona losses [74] and reduced radio interferences [5]. Grounding or earthing is an important part as it provides a current return path [22, 45]. In some applications, the ground or sea electrode is implemented as a conductive path for the return current. This configuration (Fig. 6.8a) can be referred to as "asymmetric monopole with ground/sea return."

The DC injection into the ground can lead to metallic corrosion of objects in the vicinity of the grounding electrodes. In the case where a sea electrode is used, DC can also cause chemical pollution in the water surrounding the electrode [42]. In addition, when the use of ground/sea return is constrained (e.g., heavily congested areas, freshwater cable crossings, or areas with high earth resistivity), a metallic neutral or a low-voltage conductor can be used as the current return path. In this case, the conductor needs to be insulated to withstand the voltage drop (along the conductor) and rise (during fault conditions) [20], resulting in higher costs. This

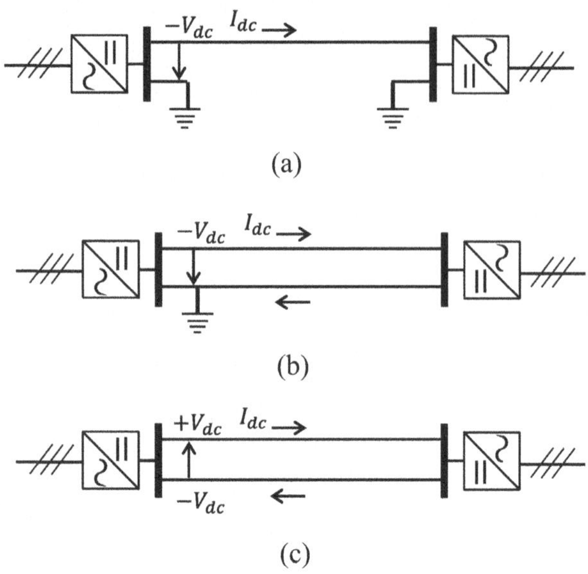

Fig. 6.8 Monopolar HVDC system configurations: (**a**) asymmetric monopole ground return, (**b**) asymmetric monopole metallic return, (**c**) symmetric monopole

configuration (Fig. 6.8b) is known as "asymmetric monopole with metallic return" [45].

Another comparable scheme (Fig. 6.8c) is the "symmetric monopole." In this case, two DC cables with full insulation are needed. The grounding can be provided through various methods, for instance, by connecting the DC capacitors' midpoint [20].

For all monopolar configurations, there is one main drawback: the failure of a converter or a cable leads to the complete loss of the entire system.

In the homopolar configuration, two conductors are operated at the same polarity (normally negative) with either ground or a metallic return. The insulation costs are reduced due to parallel operation of the two poles. However, this configuration is usually not feasible due to environmental concerns raised by large earthing current.

A bipolar configuration consists of four converters and two conductors. Unlike homopolar links, it operates with two polarities, carrying currents in opposite directions. A schematic diagram is shown in Fig. 6.9. Bipolar schemes are usually more appropriate for high transmission capacity than monopolar links due to its heavy investment costs. However, system reliability is improved due to the availability of two converters at each HVDC terminal [20, 30]. Under normal operation and a balanced load, the current rating in both poles is identical and there is no ground current. The earthing current can either flow through the ground if there are no environmental restrictions (Fig. 6.9a) or through a metallic return cable (Fig. 6.9b).

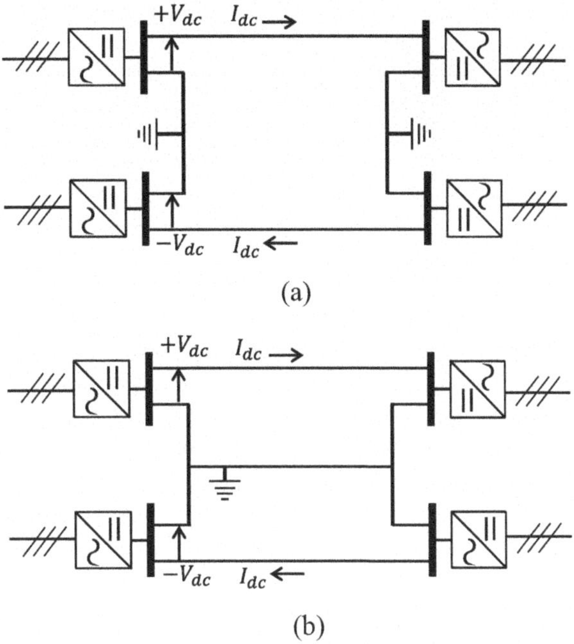

Fig. 6.9 Bipolar HVDC system configurations: (**a**) ground return, (**b**) metallic return

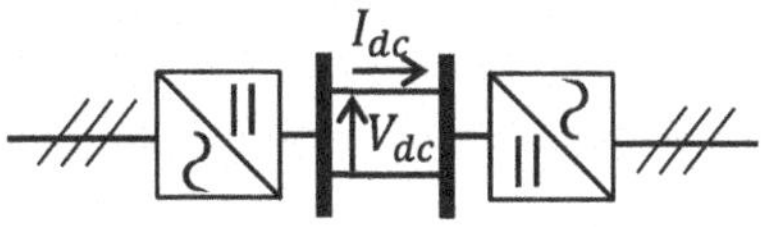

Fig. 6.10 Back-to-back HVDC system configuration

In the case of an unbalanced load, the grounding may carry a significant current. Unlike in the monopolar configuration, with a bipolar configuration, if any converter or conductor fails, power transmission can continue in the other pole with half of the original power rating.

Finally, with a back-to-back configuration, two AC systems, which may have the same or different frequencies, can be interconnected. Two converter stations are located at the same site and a long transmission line or cable is not needed [39]. A schematic illustration is shown in Fig. 6.10.

6.2.4 *Multiterminal VSC-HVDC Systems*

A multiterminal HVDC system (MTDC) consists of an HVDC system with more than two converter stations connected to the same DC network, as shown in Fig. 6.11. VSC-HVDC technology is the most appropriate technology for multi-terminal configurations since the direction of the power injections at the converter stations is determined by the direction of the current and a constant voltage level in

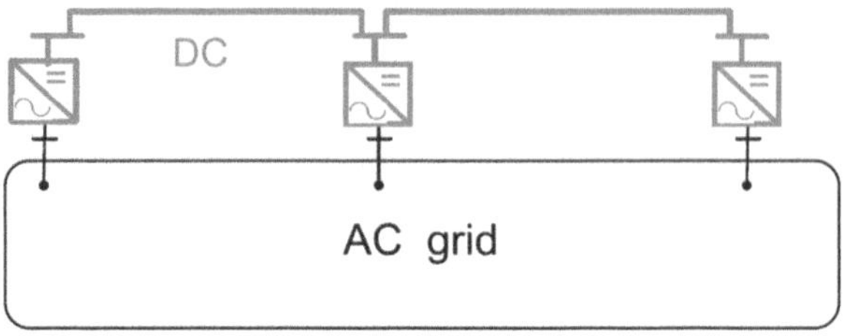

Fig. 6.11 Multiterminal VSC-HVDC system

the HVDC network can be maintained. Hence, the extension from a point-to-point to a multiterminal VSC-HVDC system is a natural evolution of HVDC systems.

One of the main advantages of multiterminal systems is its higher reliability given that VSC makes meshed MTDC systems technically feasible: in case of a cable contingency, it is possible to find an alternative pathway for the transmitted power. This increased reliability makes MTDC systems very appropriate for large offshore wind systems, and therefore, one of the most attractive applications of multiterminal VSC-HVDC systems is the integration of offshore wind energy by building submarine HVDC grids. Another advantage is that multiterminal systems require less converter stations, which are the most expensive and loss-prone components of HVDC systems.

With multiterminal VSC-HVDC technology, hybrid VSC-based AC/DC networks can be built, as shown in Fig. 6.12.

6.3 Concept of *Supergrid*

Providing low-carbon power across Europe will require a more integrated and a stronger liquid energy market, supported by upgraded and new transnational transmission networks. This will enable European citizens to benefit from a fully competitive European Internal Energy Market (IEM), and it will also make it possible to share natural resources among all Member States (MSs) and citizens, based on solidarity and market rules. In this context, the development of a *supergrid*[6] in Europe is considered one of the most promising solutions to allow MSs a large-scale integration of renewable energy sources (RES) in the system, as well as to contribute to the mitigation of global climate change [79]. Some initiatives that have studied the *supergrid* concept are the European Projects *Desertec* and *Medgrid* [24]. In addition, *Friends of the Supergrid* is a "group of companies and

[6]The *supergrid* is defined as "a pan-European transmission network facilitating the integration of large-scale renewable energy and the balancing and transportation of electricity with the aim of improving the European market" [97]. The *supergrid* assumes that a large DC grid is overlaying the conventional AC grid.

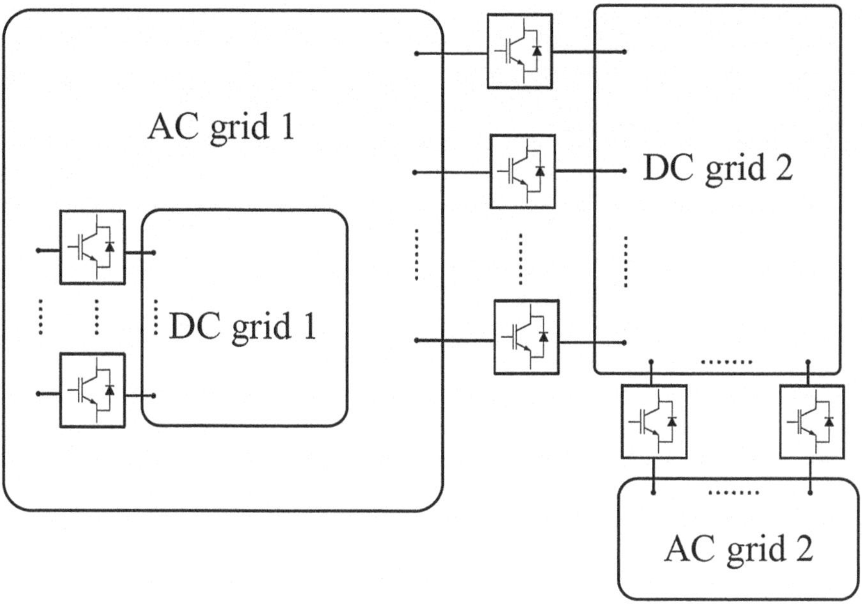

Fig. 6.12 Hybrid VSC-based AC/DC network

organizations with a mutual interest in promoting the policy agenda for a European Supergrid" [40].

Furthermore, similar concepts have also been proposed in China and North America. In China, there are a good number of HVDC systems already in operation and the first VSC-MTDC was built: Nan'ao 5-terminal system, which interconnects islands to China's onshore power system and facilitates the integration of wind energy [67]. North America is investigating potential benefits of a large-scale HVDC grid embedded into its power system, commonly known as *macrogrid* [36].

The use of DC technology would overcome the known difficulties of AC bulk power transmission over long distances [61], from generation centers, where resources are more abundant and economical, to consumption centers (above all if underground or submarine cables are required). Besides, thanks to the technical developments in power electronics, converter stations are now able to operate efficiently at high-voltage levels, and as explained earlier, several initiatives have been launched to move from point-to-point links towards multiterminal (MT) configurations in order to increase the reliability of the system and to facilitate the connection between existing AC and future DC grids.

Despite the potential economic, energy security and environmental advantages that could be brought by a *supergrid*, there remain very substantial obstacles to its development. The main challenges can be categorized as technical, regulatory, and economic.

- **Technical barriers:** Compared to conventional AC systems, new control and protection devices need to be developed to operate the *supergrid* in a reliable manner. In addition, there is a need to update current power flow and optimal power flow tools used to plan the optimal operation of the grid.
- **Economic barriers**: Although the economic opportunities associated with a *supergrid* may be promising, there are also tremendous uncertainties about the balance of expected costs and benefits. Therefore, it is crucial to demonstrate that such a large network would bring the proclaimed benefits before its deployment. Towards this end, the European Commission addressed a scalability, a replicability, and a cost-benefit analysis (CBA) in the European project *Best-Paths* [93]. One of the main conclusions of this project was that a coordinated planning of both AC and DC future networks is essential.
- **Regulatory barriers**: The *supergrid* could not only enable a large amount of renewable energy integration but could also facilitate international trade of electric power and balancing services [79]. Consequently, a harmonized European energy policy is necessary in all time horizons to improve competitiveness, to achieve the security of supply at European level, and, at the same time, to diminish external dependency. Moreover, financing such a large infrastructure needs a stable and attractive framework and a truly harmonized and collaborative European regulatory environment which are both significantly challenging. Meanwhile, it raises a nontrivial cost allocation problem that needs a careful analysis as the number of involved agents and countries can be significant leading to governance constraints. This is of particular importance for offshore HVDC grids as explained more in depth in Chap. 5.

Some additional details about the challenges related to the future evolution of HVDC grids are discussed in the following subsection.

6.4 Challenges of Future HVDC Systems

6.4.1 Technical Operation

The impact on power flows derived from point-to-point (PTP) interconnections between HVAC and HVDC systems would be even more noticeable in case of having a meshed multiterminal HVDC system overlaying an AC grid, and the control strategies of the converter stations become crucial. VSCs can provide flexible and independent active and reactive power control [39] and are ideal to interface distant wind farms as they can mitigate the propagation of voltage and frequency deviations caused by wind variations. However, the operating principles of VSCs and their potential contributions are not fully understood, yet, in power system control and power flow studies.

The problem of finding the power flow (PF) solution for the case of hybrid networks (AC with VSC-MTDC systems) is relatively new [11, 64]. Among the few papers that study the PF problem of such a hybrid network in the literature, it is possible to distinguish basically between (1) the sequential approach as in [13] where AC and DC systems are solved separately to be later linked through the losses of the converter stations and (2) the unified approach as in [10] where AC and DC systems are solved together with an explicit consideration of converter losses that are balanced with a DC slack bus per MTDC grid in the system.

Optimal power flow (OPF) was first addressed by Carpentier in 1962 [87]. The goal is to find the optimal scheduling of the generators while taking into account network constraints (capacity limits and security constraints). In order to investigate the potential impact of VSC-MTDC, traditional models that were developed for AC conventional systems need to be extended to cope with hybrid AC/DC networks. If, as mentioned before, not many research studies include VSC-MTDC systems into power flow calculations for hybrid AC/DC systems, even fewer take VSC-MTDC systems into account under an optimization context: [11] applies second-order cone programming (SOCP) techniques; [16] solves the hybrid network using the primal-dual interior point (PDIP) algorithm, with predefined control strategies for DC networks, as well as modified Jacobian and Hessian matrices; and [35] utilizes an interior point optimizer (IPOPT) to seek solutions for the nonlinear model built in the General Algebraic Modeling System (GAMS).

OPF models are not only used for short-term system operation. They can also serve as submodules to assist long-term expansion planning problems [95]. The economic assessment of considering VSC-MTDC systems as an alternative for network reinforcement of conventional AC links becomes one of the most important tasks during the overall planning stage in this emerging hybrid AC/DC scenario [35]. In this sense, the extended OPF tool for AC networks integrated with embedded VSC-MTDC systems of diverse topologies proposed in [92] could be applied for both operation and planning purposes.

Finally, there is no agreement on the way in which several technical aspects of hybrid AC/DC systems must be modelled. For example, a suitable model for converter losses is still in discussion as losses can be different depending on whether a station is working as a rectifier or as an inverter [13] and must be further investigated. In [94], it was shown that the way in which such losses are modelled can have a substantial impact on the obtained OPF solution. Therefore, the question of how converter losses must be considered when studying the optimal operation of a large AC/DC *supergrid* with many converter stations in order to have an appropriate level of accuracy without an unnecessarily high computational burden deserves additional research.

6.4.2 HVDC in TEP

Only a few works dealing with TEP consider HVDC systems. For instance, in [85], the economic benefits are split into three parts – consumer, producer, and transmission surplus – and quantified based on the market simulation program

GridView. Meanwhile, [75] assesses the economic limit based on cost minimization criteria through comparisons of costs of physical components and losses. However, only PTP HVDC interconnections are considered in those papers. [52] considers offshore wind farms of different structures and wind turbines of different topologies, combining them with either VSC-HVDC or HVAC systems. A techno-economic analysis was implemented through a discounted cash flow (DCF) approach incorporating investment costs, discounted annual costs, and revenues, to obtain cost-effectiveness. [35] proposes a model to obtain the optimal operation of a hybrid network. The model is developed in GAMS and uses the IPOPT solver. However, this model includes all the nonlinearities in the formulation, which makes it impossible to tackle a large-scale system and cannot guarantee a global optimum. Hybrid transmission expansion models have been proposed recently in [29, 34] to minimize costs, taking two-terminal HVDC links into consideration. However, [34] uses a DC equivalent model for network representation neglecting system losses completely, which could lead to underinvestment [2, 29] includes transmission losses, yet converter losses are overlooked, and this could lead to inaccurate results.

Taking into account all the issues above, in order to be able to assess the profitability of modern HVDC systems, it was clearly necessary to develop a new model able to deal with large-scale hybrid networks considering all system losses as well as a detailed technical representation of the generation and the transmission system (both AC and DC). Such a model is proposed in [92] and the main ideas will be summarized in this chapter.

6.4.3 Need of Tools for Cost-Benefit Analysis

Most studies, as well as commonly used processes and methodologies for transmission expansion planning, focus on conventional AC system, as seen in [27, 55, 62], while HVDC systems, especially VSC-MTDC, have seldom been considered. While the advantages of using innovative technologies to repower and expand EU transmission grid are potentially large, the balance between the associated costs and expected benefits is subject to uncertainty. Thus, new models capable of properly representing the particularities of VSC-MTDC systems are required. These models should be complex enough to capture all relevant features, including converter losses as discussed above, but simple enough to allow affordable computation for large-scale hybrid AC/DC systems.

Moreover, CBA intrinsically is a conceptually broad research topic, which calls for an in-depth analysis. In order to help perform CBA on European projects, the European Network of Transmission System Operators for Electricity (ENTSO-e) has approved a guideline, in which there are a number of benefit indicators [32]. Nevertheless, their implementation and calculation is not as easy and straightforward as categorizing them. Therefore, more research on assessing methodologies and on developing appropriate models to support the analyses is required.

Finally, a serious barrier is the difficulty to gather enough information and publicly available data to perform a CBA or a replicability[7] study from the perspective of the pan-European network. This data set should include transmission line parameters, the geographical location of all the buses, the generation system data (such as cost functions, technologies, etc.), and demand and RES profile time series.

6.5 Optimal Operation of a Hybrid VSC-Based AC/DC Network

6.5.1 OPF of a Hybrid VSC-Based AC/DC System

The aim of the optimal power flow (OPF) problem is to find a feasible operating point of the power system that minimizes a certain objective function and satisfies all network constraints. OPF calculations in conventional AC systems have been used successfully during the past decades. In OPF of conventional AC systems, control variables that are the active and reactive power [11] injections of generators and power flow equations of the system are included in the model. Due to the development of AC/DC hybrid VSC-based networks and the interest on this type of systems during the past years, OPF tools need to be updated so that they are capable to handle hybrid VSC-based AC/DC systems with an arbitrary topology. Several OPF tools for hybrid VSC-based AC/DC systems have been proposed during the past years [4, 11, 16, 33, 35, 49, 51, 68, 69, 70, 76, 94].

The most relevant differences of OPF tools for hybrid VSC-based AC/DC systems, in comparison to OPF tools for conventional AC systems, are:

- **Modelling**: New components (i.e., VSC stations and DC grids) need to be represented in the model.
- **Control variables**: Control variables that are manipulated by the optimization algorithm are active and reactive power injections of the generators (as in conventional OPFs) and active and reactive power injections of the VSC stations.

The next section describes in detail the OPF formulation for hybrid VSC-based AC/DC systems used in this work which is based on [94].

[7] Replicability refers to duplication of a system component at another location or time with different boundary conditions [96].

6.5.2 Modelling of VSC Stations and Their Losses

Firstly, it shall be clarified that all variables presented in this section are in [p.u.], and units of corresponding parameters are adapted accordingly.

Figure 6.13 shows the equivalent circuit of the VSC, where all parameters to be used in this section are in p.u. As described in [13, 35], the VSC converter is generally modeled as a controllable voltage source $v_c = v_c\angle\theta_c$ connected by a phase reactor $z_c = r_c + jx_c$ to an intermediate node where a lossless shunt filter is connected ($z_f = -j/b_f$). The voltage at this intermediate node is $v_f = v_f\angle\theta_f$. The transformer can be represented by its impedance: $z_{tf} = r_{tf} + jx_{tf}$. Therefore, each VSC converter station adds two AC buses to the system: the filter bus (voltage $v_f \angle \theta_f$) and the converter bus (voltage $v_c \angle \theta_c$). If the filter is not used (or when its effect can be neglected), both the phase reactor and the transformer impedance can be lumped together, eliminating bus f from the equations. In any case, the power flow within the VSC station between the nodes c, f, and s is described by standard AC power flow equations and the shunt susceptance would affect the diagonal terms of the susceptance matrix **B** at the position of the filter buses.

Assuming the most detailed representation where the intermediate filter bus is maintained, the power injected by the converter v must satisfy that $p_v^{ac} + jq_v^{ac} = p_c + jq_c = S_c$.

Regarding the power balance between the AC and DC sides of the converter, Fig. 6.14 depicts the criterion adopted to describe the power balance between the AC and DC sides of the converter. Arrows indicate positive injected power. The corresponding active power balance equation is established in (6.1), where power losses at the converter can take only positive values, i.e., $\rho_v \geq 0$. Notice that during the conversion process, the available real power at one side of the converter will be lower than the active power injected at the other side due to the converter losses, ρ_v:

$$0 = p_v^{dc} + p_v^{ac} + \rho_v, \forall v \in \mathcal{V} \tag{6.1}$$

where $\mathcal{V}$ is the set of all converter stations. There are several sources of losses within a VSC station, such as semiconductor losses (conduction and switching losses),

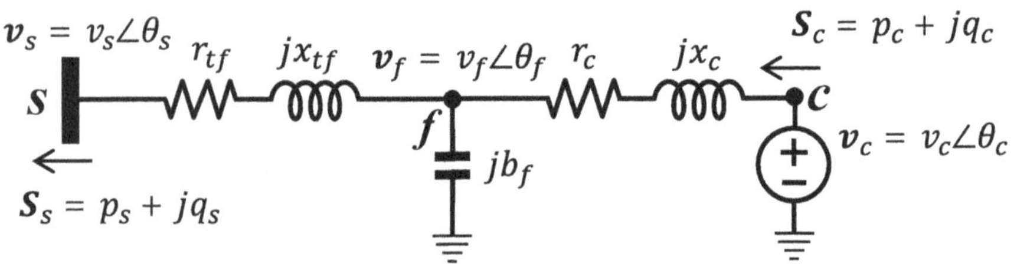

Fig. 6.13 Equivalent circuit of the VSC

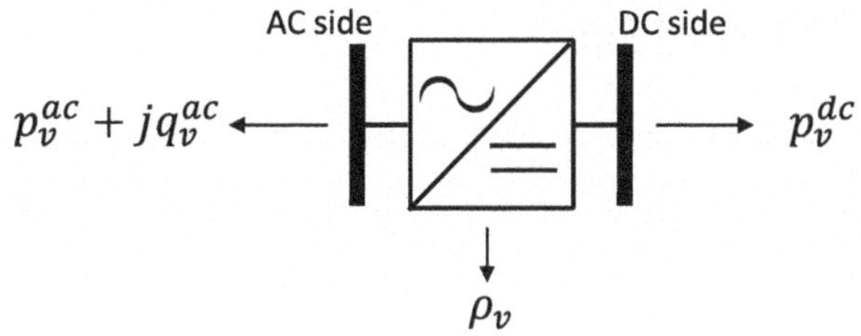

Fig. 6.14 Power balance at the VSC station

phase reactor losses, or transformer losses [71]. Different methods to address losses in VSC stations are described in [43, 63, 71, 91]. However, a very complex and detailed procedure cannot be included directly in an optimization model. Therefore, in favor of a polynomial quadratic expression can be adopted, as it is conventionally used in state-of-the-art PF calculations [13, 21], where the authors differentiate between the operation of the converter as rectifier and inverter converter modes. Since an inverter uses the IGBTs more frequently, while a rectifier uses the diodes more frequently, the coefficients of the polynomials of loss models can be different depending on the direction of the active power transferred [26].

In a PF calculation, the converter operation mode (inverter or rectifier) must be known in advance. By contrast, under the optimization context, the operation mode is a decision variable, and the optimization problem must first determine the optimal operation of every converter. The importance of the power flow direction in the calculation of converter losses was not recognized in previous OPF studies of hybrid networks such as [16, 83].

Similarly to the sign criteria for active power, the phase current of the converter (i_v) will be considered positive when the converter behaves as an inverter and negative when it behaves as a rectifier. Therefore, it is possible to write $i_v = i_v^{inv} - i_v^{rec}$, where i_v^{inv} and i_v^{rec} are auxiliary positive variables and converter losses can be computed in the OPF model as follows:

$$\rho_v = A_v + B_v \cdot |i_v| + C_v^{inv} \cdot \left(i_v^{inv}\right)^2 + C_v^{rec} \cdot \left(i_v^{rec}\right)^2 \tag{6.2}$$

where A_v, B_v, C_v^{rec}, and $C_v^{inv}$are the appropriate converter loss coefficients. Adding as an extra condition that either i_v^{inv} or i_v^{rec} *will always be zero (rectifier or inverter operation, respectively)*. Obviously, only one of the quadratic terms $C_v^{inv} \cdot \left(i_v^{inv}\right)^2$ or $C_v^{rec} \cdot \left(i_v^{rec}\right)^2$ will be activated (rectifier or inverter mode, respectively). In addition, the absolute value used in the linear term could be computed as $|i_v| = i_v^{inv} + i_v^{rec}$.

6.5.3 VSC Operation Limits

The OPF model must take into account the IGBT current limit. This can be achieved by Eq. (6.3) where $\overline{I}_v$ is the maximum current that can circulate through the arms:

$$-\bar{I}_v \leq i_v \leq \bar{I}_v, \forall v \in \mathcal{V} \tag{6.3}$$

In addition, apart from the voltage limits imposed at both sides of the converter, it is necessary to take into account that the voltage level on the DC side (v_i) limits the maximum voltage that can be obtained at the AC side of the converter (v_c). This can be modeled as follows:

$$v_c \leq k_v \cdot v_i, \forall v \in \mathcal{V}, c \in \mathcal{N}_v^{ac}, i \in \mathcal{N}_v^{dc} \tag{6.4}$$

where $\mathcal{N}_v^{ac}$ and $\mathcal{N}_v^{dc}$ are the set of AC and DC buses, respectively, that are connected to the converter stations. This way it is possible to identify which are the AC and DC buses of each converter v. In this chapter, k_v has been set to 1.1 for all case studies according to [35] as converters are assumed to be operated in nominal conditions. However, there are modulation techniques that give higher voltages at the AC buses, in which case this factor should be modified accordingly.

Another limit to be taken into account is the maximum current that can circulate through the DC cables. Since the DC voltage does not have much variation with respect to its nominal value, the limits imposed on currents can be established equivalently by setting the maximum power limit for each DC line; see Eq. (6.10).

Finally, it is worth mentioning the P-Q capability curves that relate the active and reactive power limits of the converter. In the proposed steady-state OPF formulation, such limits are implicitly taken into account by Eq. (6.5):

$$\left(p_v^{ac}\right)^2 + \left(q_v^{ac}\right)^2 = (v_c \cdot i_v)^2, \forall c \in \mathcal{N}_v^{ac}, v \in \mathcal{V} \tag{6.5}$$

The reactive power q_v^{ac} injected to the AC side by each converter will be considered positive in case of being capacitive and negative in case of being inductive. In this sense, a minimum of −0.5 p.u. of reactive power can be imposed to the converters as in [35], where $\overline{Q}_v$ is the maximum allowed reactive power (see Fig. 6.15). This is also illustrated with a P-Q diagram such as the one given for HVDC Light of ABB [81]. This limit can be easily formulated as follows:

$$q_v \geq \underline{q}_v = \frac{-\overline{Q}_v}{2} \forall v \in \mathcal{V} \tag{6.6}$$

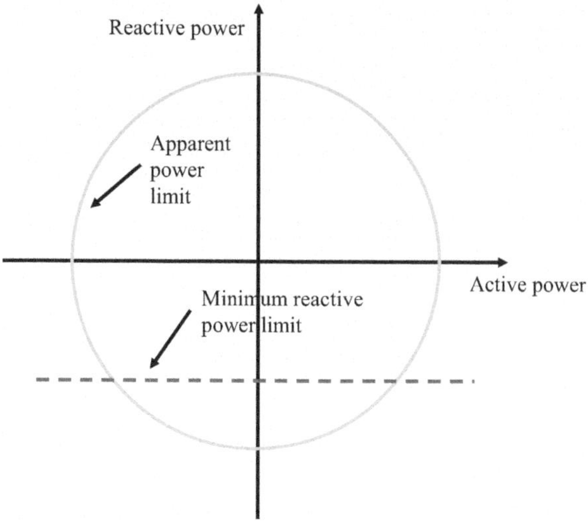

Fig. 6.15 P-Q diagram of the VSC station

6.5.4 *Modelling of Network Technical Constraints*

The voltage of every bus must be kept within certain bounds [25], which can be expressed for both the AC and DC systems as follows:

$$\underline{V}_i \leq v_i \leq \overline{V}_i, \forall i \in \mathcal{B}^{ac} \tag{6.7}$$

$$\underline{V}_i \leq v_i \leq \overline{V}_i, \forall i \in \mathcal{B}^{dc} \tag{6.8}$$

where $\mathcal{B}^{ac}$ and $\mathcal{B}^{dc}$are the sets of network buses of the AC and DC system, respectively.

In addition, the transmission lines have a maximum thermal limit that needs to be respected. For AC transmission lines, this limit can be expressed in terms of the apparent power that circulates through the line. Let $\mathcal{L}^{ac}$ be the set of all AC transmission lines of the system under study. The thermal limit can be expressed as $\sqrt{{p_l}^2 + {q_l}^2} \leq \overline{S}_l, \forall l \in \mathcal{L}^{ac}$, which is equivalent to the next quadratic constraint:

$${p_l}^2 + {q_l}^2 \leq {\overline{S}_l}^2, \forall l \in \mathcal{L}^{ac} \tag{6.9}$$

On the other hand, the power that can be transmitted through a DC cable or OHL line is bounded by its technical characteristics. Let $\mathcal{L}^{dc}$ be the set of all DC transmission lines of the system under study. The power limit of DC lines can be modelled as follows:

$$-\underline{P}_l \leq p_l \leq \overline{P}_l, \forall l \in \mathcal{L}^{dc} \tag{6.10}$$

6.5.5 Modelling of AC Network Flows

Every bus i of the AC grid is characterized by its voltage magnitude v_i and its phase angle θ_i. By denoting $\theta_{ij} = \theta_i - \theta_j$, power injections at node i satisfy the following equations [87]:

$$p_i = v_i \sum_{j \in \mathcal{B}^{ac}} v_j \left[G_{ij}^{ac} \cos\left(\theta_{ij}\right) + B_{ij} \sin\left(\theta_{ij}\right) \right], \forall i \in \mathcal{B}^{ac} \tag{6.11}$$

$$q_i = v_i \sum_{j \in \mathcal{B}^{ac}} v_j \left[G_{ij}^{ac} \sin\left(\theta_{ij}\right) - B_{ij} \cos\left(\theta_{ij}\right) \right], \forall i \in \mathcal{B}^{ac} \tag{6.12}$$

where G_{ij}^{ac} and B_{ij} are the elements (i, j) of the AC network conductance and susceptance matrices, respectively. In addition, the active and reactive power flows of every existing branch l between the pair of buses (i,j) must satisfy the following equations:

$$p_l = v_i^2 G_l - v_i v_j \left[G_l \cos\left(\theta_{ij}\right) + B_l \sin\left(\theta_{ij}\right) \right], \forall l \in \mathcal{L}^{ac} \tag{6.13}$$

$$q_l = -v_i^2 \left(B_l + B_l^{sht} \right) - v_i v_j \left[G_l \sin\left(\theta_{ij}\right) - B_l \cos\left(\theta_{ij}\right) \right], \forall l \in \mathcal{L}^{ac} \tag{6.14}$$

where G_l and B_l are the conductance and susceptance of line l, and B_l^{sht} is its half total line charging susceptance. Notice that transformers are modeled as regular lines with predefined tap ratios.

6.5.6 Modelling of DC Network Flows

In this case, every bus i of the DC grid is characterized by its voltage magnitude v_i. Every line l connecting a pair of DC buses can be represented by its resistance R_l. Assuming that the extreme nodes of such line are i and j, the real power injected at node i and power flows (from i to j) of the DC line l satisfy the following expressions:

$$p_i = n \cdot v_i \sum_{j \in \mathcal{B}^{dc}} G_{ij}^{dc} \left(v_i - v_j \right), \forall i \in \mathcal{B}^{dc} \tag{6.15}$$

$$p_l = n \cdot \left[\frac{v_i \left(v_i - v_j \right)}{R_l} \right], \forall l \in \mathcal{L}^{dc} \tag{6.16}$$

where G_{ij}^{dc} is the conductance between DC nodes i and j, and n represents the number of poles. Therefore, different configurations can be easily taken into account by its corresponding value of n. In this chapter, for all case studies, a symmetric monopole configuration has been used where $n = 2$, although different values could be established for each DC line if necessary.

6.5.7 *Power Balance at Every Bus*

Once the power flows have been established for all the AC and DC lines, it is necessary to ensure that the power balance is respected at every node of the system. Assuming without loss of generality that there is only demand for active power (P_{Di}) and reactive power (Q_{Di}) at the nodes of the AC system, the conservation of power at AC buses can be modeled by the following equations.

$$p_i = \sum_{g \in \mathcal{G}_i} (p_g) - P_{Di} + \sum_{v \in \mathcal{V}_i} p_v^{ac} + np_i, \forall i \in \mathcal{B}^{ac} \tag{6.17}$$

$$q_i = \sum_{g \in \mathcal{G}_i} (q_g) - Q_{Di} + \sum_{v \in \mathcal{V}_i} q_v^{ac} + nq_i, \forall i \in \mathcal{B}^{ac} \tag{6.18}$$

where $\mathcal{G}_i$ is the set of generation units connected to the node i, p_g and q_g are the active and reactive power generated by unit g, the terms $\sum_{v \in \mathcal{V}_i} p_v^{ac}$ and $\sum_{v \in \mathcal{V}_i} q_v^{ac}$ account for the possible power injections from a VSC converter station if the bus i is connected to it, and np_i and nq_i are positive variables that represent the possible unserved active and reactive power, respectively. It is important to highlight the dependencies among all the decision variables derived from the joint consideration of Eqs. (6.11) and (6.17) and also from (6.12) and (6.18).

For the DC buses, the power balance is guaranteed by the following expression where VSC losses ρ_v are incorporated from the DC side of the converter:

$$p_i = \sum_{v \in \mathcal{V}_i} \left(p_v^{dc} + \rho_v\right), \forall i \in \mathcal{B}^{dc} \tag{6.19}$$

In this case, the joint consideration of Eqs. (6.15) and (6.19) couples the decision variables of the DC system.

6.5.8 *Phase Current Sign*

In order to force that the sign of the phase current i_v used in the losses expression is aligned with the sign of the power injected to the corresponding AC bus, the following equation can be included for every converter v:

$$0 \leq i_v \cdot p_v^{ac}, \forall v \in \mathcal{V} \tag{6.20}$$

6.5.9 Generation Limits

In hybrid AC/DC networks, the decision variables that must be determined by the OPF model in order to optimize the objective function are the active and reactive power injections of the generators and active and reactive power injections of the VSC stations. Therefore, the limits of the output power of the generation units must be carefully taken into account when solving the OPF model.

When a generator is on, the maximum active and reactive power that can be produced is limited due to the technical characteristics of the unit. The same applies to the minimum generation. In case of active power, the minimum stable load is due to the minimum requirement of the output power to satisfy pressure and temperature conditions in thermal units, or minimum outflow requirements in hydro units. Regarding the minimum reactive power, the limits are related to the P-Q curve of the generators. Therefore, for each available generator, the following constraints must be considered in order to obtain a feasible solution:

$$\underline{P}_g \leq p_g \leq \overline{P}_g, \forall g \in \mathcal{G} \tag{6.21}$$

$$\underline{Q}_g \leq q_g \leq \overline{Q}_g, \forall g \in \mathcal{G} \tag{6.22}$$

6.5.10 Objective Function/Optimization Criterion

In this chapter, the objective function to be minimized will include the total operation costs plus penalty terms related to unserved active and reactive power:

$$\min \sum_{g \in \mathcal{G}} \left(C_g + A_g \cdot p_g + B_g \cdot p_g^2 \right) + \sum_{i \in \mathcal{B}} \left(NC_p \cdot np_i + NC_q \cdot nq_i \right) \tag{6.23}$$

where A_g, B_g, and C_g are the cost coefficients of the generators and NC_p and NC_q represent the unitary costs of unserved real and reactive power, respectively. Other alternatives, such as minimization of network losses as in [16], could also be easily adapted, if necessary.

6.6 Solution Techniques

A general OPF formulation can be expressed as in (6.24), where u and x represent state and decision (or control) variables, respectively, and $f(u, x)$ represents the objective function. Vector functions $g(u, x)$ and $h(u, x)$ characterize system equality and inequality constraints correspondingly. Depending on the f, g, and h functions, the OPF problem may be formulated as a linear, nonlinear, mixed-integer linear, or mixed-integer nonlinear problem.

$$\begin{aligned} \min \quad & f(u, x) \\ s.t \quad & \\ & g(u, x) = 0 \\ & h(u, x) \leq 0 \end{aligned} \tag{6.24}$$

6.6.1 Nonlinear OPF

The proposed nonlinear programming (NLP) OPF model can be built by minimizing the objective function formulated in (6.23), subject to all the constraints (6.1)–(6.22) presented previously. Since the (NLP) OPF includes many nonlinearities, it cannot cope with very large systems and it is not possible to guarantee that the global optimum has been reached. For that reason, a linearized version is provided in the following subsection.

6.6.2 Linearized OPF

This model is built based on the well-known DC-OPF approximation described in [87] for AC systems, and it has been extended to include MTDC networks, converters, and transmission losses. In this simplified formulation, no reactive power is considered, AC voltage magnitudes are assumed to take nominal values (1 in p.u.), and small-angle differences are assumed in order to remove the sine and cosine functions from the AC power flow equations. In the following expressions, quadratic terms might appear for the sake of simplicity, but as the model must be linear, the procedure used to linearize all these quadratic terms will be explained and clarified later. Moreover, only the expressions that differ from the nonlinear OPF formulation will be discussed hereafter:

Similarly as in (6.11), power injections at node i can be obtained as follows, where active power flows depend linearly on the difference of voltage angles:

$$p_i = \sum_{j \in \mathcal{B}^{ac}} B_{ij}\theta_{ij} + \sum_{l \subset [(i,j) \cup (j,i)]} \frac{\rho_l}{2}, \forall i \in \mathcal{B}^{ac}, (i, j) \in \mathcal{N}_l, (j, i) \in \mathcal{N}_l \tag{6.25}$$

In the previous expression, ρ_l represents AC transmission line losses that can be treated as additional loads at the two extreme nodes of the line [66], where each fictitious load represents half of the branch losses. In addition, the active power flow for every AC line satisfies the following equation:

$$p_l = B_l \theta_{ij}, \forall l \in \mathcal{L}^{ac}, (i, j) \in \mathcal{N}_l \quad (6.26)$$

Instead of expressing the losses as a function of angle differences, they can be expressed as a function of power flows on the line given that, in p.u., current is equal to the active power. Thus, losses can be readily derived as follows [38]:

$$\rho_l = f(p_l) = R_l p_l^2, \forall l \in \mathcal{L}^{ac} \quad (6.27)$$

With respect to the DC power flows, Eqs. (6.15) and (6.16) should be substituted by (6.28) and (6.29):

$$p_i = \sum_{j \in \mathcal{B}^{dc}} G_{ij}^{dc} \left(v_i - v_j\right) + \sum_{l \subset [(i,j) \cup (j,i)]} \frac{\rho_l}{2}, \forall i \in \mathcal{B}^{dc}, (i, j) \in \mathcal{N}_l, (j, i) \in \mathcal{N}_l \quad (6.28)$$

$$p_l = n \cdot \left[\left(v_i - v_j\right) / R_l\right], \forall l \in \mathcal{L}^{dc}, (i, j) \in \mathcal{N}_l \quad (6.29)$$

where the product by the voltage level v_i has been removed under the assumption that it is close to the nominal value 1 in p.u.

DC transmission losses are treated in a similar way as for AC lines:

$$\rho_l = f(p_l) = R_l p_l^2, \forall l \in \mathcal{L}^{dc} \quad (6.30)$$

Another expression that needs to be adapted is Eq. (6.5) that can be further simplified to the following expression provided that the entire formulation is made in p.u.:

$$0 = i_v - p_c, \forall v \in \mathcal{V}, c \in \mathcal{N}_v^{ac} \quad (6.31)$$

Finally, the objective function also needs to be formulated as a linear expression, where each generator can be characterized by a linear input-output cost curve and where only unserved active power is penalized:

$$\min \sum_{g \in \mathcal{G}} \left(C_g + A_g \cdot p_g\right) + \sum_{i \in \mathcal{B}} \left(NC_p \cdot np_i\right) \quad (6.32)$$

6.6.3 Linearization of Quadratic Terms

As previously mentioned, neglecting transmission losses to reduce the computational burden in case of a large-scale system can jeopardize the accuracy of TEP solutions leading to underinvestment. Transmission losses modelling in relation with TEP problems is thoroughly reviewed in Chapter 3 of [37]. There are mainly two issues that one has to pay close attention to: (1) accuracy (including the capability to limit *artificial losses*[8] [38] or fictitious losses as in [66]) and (2) computational complexity resulting from linearization. Thus, the representation should be accurate enough to address key problems, but simple enough to allow affordable computation for large-scale systems.

Existing linear loss models are reviewed in [38]. Among the three methods, the piecewise linear approximation model can achieve a certain level of accuracy while limiting artificial losses at a price that additional variables are needed to represent power flow segments. This model is described in [2, 28, 66, 89] with further details.

In addition, [38] has also proposed some other loss models to improve accuracy. Results show that although binary variables improve precision, they are very intensive computationally, which reduces the applicability for large-scale systems. With an additional constraint (compared to piecewise linear approximation) forcing segments to be filled in a sequential way, losses can be represented in a more accurate manner given a reasonable extra amount of computation time.

Thus in this formulation, piecewise linear approximation with "filled up" constraint is adopted for the LP-OPF model and, therefore, all quadratic terms are approximated by a set of linear constraints. In order to illustrate the linearization, the loss ρ_l in Equation (6.27) is used as an example hereafter.

The first step is to discretize the maximum allowed line flow P_l into a desired number of segments N. Each segment is of a positive step size ΔP_l and corresponds to a term in the final linear expression. In this chapter, $N = 4$ as suggested in [38, 66]. Consequently, a quadratic loss function can be approximated by pieces of linear functions. In this way, every segment has a flow variable Δp^n associated with the nth partition. By adding losses calculated for each segment Δp^n, losses for every line can be approximated. The following constraints are included:

$$\rho_l = R_l \sum_{n=1}^{N} (2n-1)\, \Delta \overline{P}_l \Delta p_l^n, \forall l \in \mathcal{L}^{ac} \tag{6.33}$$

$$0 \leq \Delta p_l^n \leq \Delta \overline{P}_l \tag{6.34}$$

[8] Artificial losses refer to losses that do not really exist. However, by computation of such losses, the solution leads to an artificial increase in cheap power generation, yet a reduction of the overall operation cost in the system.

$$\sum_{n=1}^{N} \Delta p_l^n = |p_l| = p_l^+ + p_l^- \tag{6.35}$$

$$\Delta p_l^{n+1} \leq \Delta p_l^n \tag{6.36}$$

where $p_l = p_l^+ - p_l^-$, and p_l^+ and p_l^- are two nonnegative auxiliary variables (i.e., $p_l^+, p_l^- \geq 0$), representing the flow through the line in the opposite directions. In addition, only one of them (p_l^+ and p_l^-) could be different from zero.

6.7 Case Study

The performance of the OPF models for hybrid VSC-based AC/DC systems will be tested and compared now in a case study. The test system used is the IEEE 14-bus test system with an embedded 5-terminal VSC-MTDC system, as shown in Fig. 6.16, and it is based on the one used in [86]. The AC network was originally proposed in [18], and the complete data of the test system can be found in [92]. The OPF models were implemented in GAMS software.

The following OPF models will be compared:

- Nonlinear OPF (NLP)
- Linear OPF (LP)

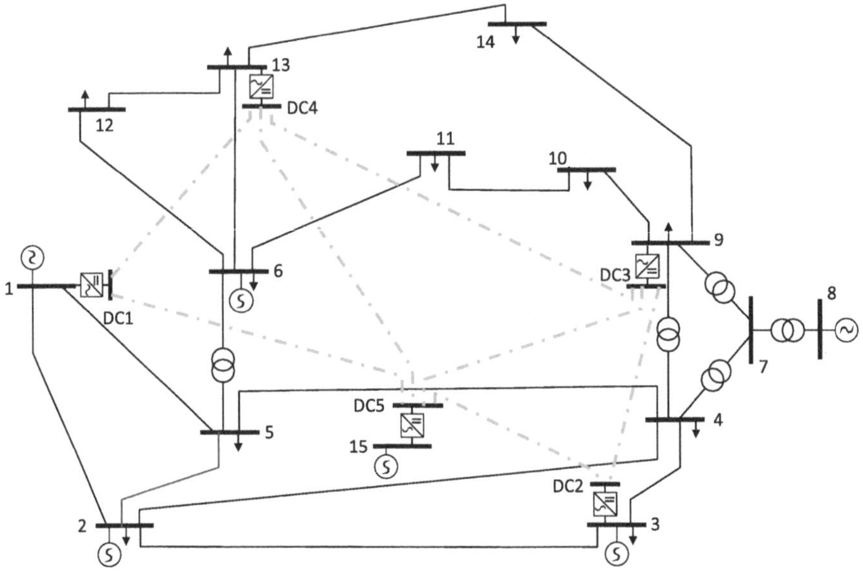

Fig. 6.16 IEEE 14-bus system with an embedded 5-bus VSC-MTDC system

Table 6.1 OPF results. Objective function

	NLP	LP	Error (%)
Objective function (€)	172.24	172.66	0.06

Table 6.2 OPF results. Active power injections of the generators

P injections of the generators (MW)	NLP	LP
P_{G1}	2.56	0.00
P_{G2}	0.00	0.00
P_{G3}	99.42	100.00
P_{G6}	69.83	71.86
P_{G8}	48.95	50.00
P_{G15}	45.06	44.23

Table 6.3 OPF results. Power flows in the AC network

Power flow (MW)	NLP	LP	Error (%)	Power flow (MW)	NLP	LP	Error (%)
P_{1-2}	10.94	14.76	3.82	P_{6-11}	10.29	12.50	4.42
P_{1-5}	−2.33	0.00	2.33	P_{6-12}	6.26	6.30	0.08
P_{2-3}	−0.99	−0.96	0.03	P_{6-13}	12.81	12.50	0.62
P_{2-4}	−3.33	−1.03	2.30	P_{7-8}	−48.95	−50.00	2.10
P_{2-5}	−6.47	−5.02	1.45	P_{7-9}	17.24	23.16	11.84
P_{3-4}	−1.79	0.00	1.79	P_{9-10}	2.37	0.17	4.40
P_{4-5}	−12.82	−16.51	7.38	P_{9-14}	6.26	4.58	3.36
P_{4-7}	−31.70	−26.84	9.72	P_{10-11}	−6.65	−8.88	4.46
P_{4-9}	−8.41	−5.51	5.80	P_{12-13}	0.11	0.15	0.08
P_{5-6}	−29.71	−29.19	0.16	P_{13-14}	8.85	10.46	3.22

Table 6.1 shows the value of the objective function obtained with both approaches. Results are very close and an error of only 0.05% is obtained (normalized with respect to the objective function obtained with NLP that is considered as the benchmark). Table 6.2 provides the active power injections of the generators. Again, very similar results have been obtained with the nonlinear OPF and with the linear OPF.

The power flows through the AC lines are compared in Table 6.3. The last column reports the normalized error, which is defined as:

$$\text{error} = \frac{\left|P_{ij}^{LP-OPF} - P_{ij}^{NLP-OPF}\right|}{\text{RatedPower}} \tag{6.37}$$

Similar results have been obtained with both models. Error values obtained are in the range 0.03% (P_{2-3}) and 11.84% (P_{7-9}).

Finally, Table 6.4 compares the power flow through the DC lines. The normalized error is also provided in the last column of the table. In this case, error values are in the range of 5.14% ($P_{DC2-DC5}$) and 11.30% ($P_{DC1-DC5}$).

Table 6.4 OPF results. Power flows in the DC network

Power flow (MW)	NLP	LP	Error (%)
$P_{DC1-DC4}$	5.03	1.88	6.30
$P_{DC1-DC5}$	−12.22	−17.87	11.30
$P_{DC2-DC5}$	−7.83	−5.26	5.14
$P_{DC2-DC3}$	12.70	8.95	7.50
$P_{DC2-DC3}$	−4.07	0.63	6.88
$P_{DC3-DC4}$	−13.80	−9.53	8.54
$P_{DC4-DC5}$	−9.74	−10.15	0.82

As a conclusion, similar results can be obtained with the nonlinear OPF (NLP) and with the linear OPF (LP). Naturally, the latter introduces errors in the results. However, the errors are comparable with the linear OPF used for AC systems. The advantage of using the linear OPF is that it presents a lower computational burden and it is suitable to be applied to OPF calculations of large-scale systems. For instance, in [92] the linear OPF (LP) has been applied to the whole pan-European network with satisfactory results for different HVDC offshore grid scenarios in 2030 with more than 7700 nodes and 8900 branches. In addition, such formulation could be implemented in large-scale transmission expansion models that need to take into account hybrid AC/DC networks.

6.8 Conclusions

This chapter has presented the fundamentals of HVDC systems and the role of this type of systems in future power systems. Special emphasis has been given to the use of hybrid VSC-based AC/DC systems to improve the operation of the systems by means of solving an optimal power flow (OPF). The formulation of an OPF of a hybrid VSC-based AC/DC system has been described in detail. Two formulations have been presented – (a) nonlinear OPF and (b) linear OPF – and results using the two approaches have been compared in a case study.

Results show that the objective function of a power system (such as generation costs or total losses) can be minimized while satisfying network constraints, by operating the VSC stations in an optimal way. Results obtained with the nonlinear OPF model are more accurate than those obtained with the linear OPF model, but the latter presents a much lower computational burden. The linear OPF model could be useful when analyzing large-scale systems in which only active power flows are of interest.

References

1. R. Adapa, High-wire act: HVdc technology: The state of the Art. IEEE Power Energ. Mag. **10**(6), 18–29 (2012). https://doi.org/10.1109/MPE.2012.2213011
2. N. Alguacil, A. Motto, A. Conejo, Transmission expansion planning: A mixed-integer LP approach. IEEE Trans. Power Syst. **18**(3), 1070–1077 (2003). https://doi.org/10.1109/TPWRS.2003.814891
3. B. Andersen, HVDC transmission-opportunities and challenges, in *8th IEEE International Conference on AC and DC Power Transmission*, (2006). https://doi.org/10.1049/cp:20060006
4. M. Aragues-Penalba, J. Beerten, J. Rimez, D. Van Hertem, O. Gomis-Bellmunt, Optimal power flow tool for hybrid AC/DC systems, in *Proc. 11th IET International Conference on AC and DC Power Transmission*, (Birmingham, 2015), pp. 1–7
5. J. Arrillaga, Y. Liu, N. Watson, *Flexible Power Transmission: The HVDC Options* (Wiley, Chichester, 2007)
6. M.J. Asghar, *Power Electronics*, 8th edn. (PHI Learning Private Limted, Delhi, 2011)
7. U. Astrom, V. Lescale, D. Menzies, W. Ma, Z. Liu, The Xiangjiaba-Shanghai 800 kv UHVDC project, Status and special aspects, in *International Conference on Power System Technology*, (2010), pp. 1–6. https://doi.org/10.1109/POWERCON.2010/5666671
8. U. Axelsson, A. Holm, C. Liljegren, K. Ericksson, L. Weimers, Gotland HVDC light transmission – World's first commercial small scale DC transmission, in *15th International Conference and Exhibition on Electricity Distribution*, (1999)
9. M. Bahrman, B. Johnson, The ABC of HVDC Transmission Technologies. IEEE Power Energ. Mag., 1–7 (2007)
10. M. Baradar, M. Ghandhari, A multi-option unified power flow approach for hybrid AC/DC grids incorporating multi-terminal VSC-HVDC. IEEE Trans. Power Syst. **28**(3), 2376–2383 (2013). https://doi.org/10.1109/TPWRS.2012.2236366
11. M. Baradar, M. Hesamzadeh, M. Ghandhari, Second-order cone programming for optimal power flow in VSC-type AC-DC grids. IEEE Trans. Power Syst. **28**(4), 4282–4291 (2013)
12. J. Beerten, *Modeling and Control of DC Grids, PhD Thesis*, Katholieke University Leuven, Belgium (2013)
13. J. Beerten, S. Cole, R. Belmans, Generalized steady-state VSC MTDC model for sequential AC/DC power flow algorithms. IEEE Trans. Power Syst. **27**(2), 821–829 (2012)
14. R.L. Bradley, *Edison to Enron: Markets and Political Strategies* (Wiley, New York, 2011)
15. C. Cagigas, M. Madrigal, Centralized vs. competitive transmission expansion planning: The need for new tools. IEEE Power Eng. Soc. Gen. Meet. **2**, 10–17 (2003). https://doi.org/10.1109/PES.2003.1270450
16. J. Cao, W. Du, W. HF, S.Q. Bu, Minimization of transmission loss in meshed AC/DC grids with VSC-MTDC networks. IEEE Trans. Power Syst. **28**(3), 3047–3055 (2013)
17. N.R. Chaudhuri, B. Chaudhuri, R. Majumder, A. Yazdani, *Multi-terminal Direct-Current Grids. Modeling, Analysis and Control* (Wiley & IEEE Press, New York, 2014)
18. R. D. Christie. *Power Systems Test Case Archive.* University of Washington, USA (1993). URL: www2.ee.washington.edu/research/pstca. Accessed 01-06-2019
19. CIGRE. (2005). *VSC TRANSMISSION Technical Report Ref. 269.* CIGRE Working Group B4.37
20. CIGRE, *HVDC Grid Feasibility Study.* CIGRE Working Group B4.52. (2013)
21. CIGRE. (2014). Guide for the Development of Models for HVDC Converters in a HVDC Grid. CIGRE Working Group B4.57.
22. CIGRE. (2017). TB 675 General Guidelines for HVDC Electrode Design. CIGRE Working Group B4.61
23. S. Cole, *PhD Thesis: Steady-State and Dynamic Model Ling of VSC HVDC Systems for Power System* (KU Leuven, Belgium, 2010)

24. S. Cole, K. Karoui, T.K. Vrana, O. Fosso, J. Curis, A. Denis, C.C. Liu, A European supergrid: Present state and future challenges, in *17th Power System Computation Conference (PSCC)*, (Stockholm, 2011), pp. 1–7
25. B. Cui, X. Sun, A new voltage stability-constrained optimal power-flow model: Sufficient condition, SOCP representation, and relaxation. IEEE Trans. Power Syst. **33**(5), 5092–5102 (2018). https://doi.org/10.1109/TPWRS.2018.2801286
26. G. Daelemans, *Master Thesis: VSC HVDC in Meshed Networks* (KU Leuven, Belgium, 2008)
27. R. de Dios, F. Soto, A. Conejo, Planning to expand? IEEE Power Energ. Mag. **5**(5), 64–70 (2007). https://doi.org/10.1109/MPE.2007.904764
28. S. de la Torre, A. Conejo, J. Contreras, Transmission expansion planning in electricity markets. IEEE Trans. Power Syst. **23**(1), 238–248 (2008). https://doi.org/10.1109/TPWRS.2007.913717
29. A. Dominguez, A. Zuluaga, L. Macedo, R. Romero, Transmission network expansion planning considering HVAC/HVDC lines and technical losses, in *IEEE PES Transmission Distribution Conference and Exposition-Latin America*, (2016), pp. 1–6. https://doi.org/10.1109/TDC-LA.2016.7805606
30. R. Dorf, *The Electrical Engineering Handbook* (CRC Press, Boca Raton, 2000)
31. Edison Tech Center, (n.d.). Retrieved July 27, 2019, from https://edisontechcenter.org/Transformers.html
32. ENTSO-e, *ENTSO-E Guideline for Cost Benefit Analysis of Grid Development Projects* (ENTSO-e, 2015)
33. H. Ergun, J. Dave, D. Van Hertem, F. Geth, Optimal power flow for AC–DC grids: Formulation, convex relaxation, linear approximation, and implementation. IEEE Trans. Power Syst. **34**(4), 2980–2990 (2019)
34. V. Escobar, L. Romero, Z. Escobar, R. Gallego, Long term transmission expansion planning considering generation-demand scenarios and HVDC lines, in *IEEE PES Transmission Distribution Conference and Exposition-Latin America*, (2016), pp. 1–6. https://doi.org/10.1109/TDC-LA.2016.7805617
35. W. Feng, L.A. Tuan, L.B. Tjernberg, A. Mannikoff, A. Bergman, A new approach for benefit evaluation of multiterminal VSC-HVDC using a proposed mixed AC/DC optimal power flow. IEEE Trans. Power Syst. **29**(1), 432–443 (2014)
36. A.L. Figueroa-Acevedo, A. Jahanbani-Ardakani, H. Nosair, A. Venkatraman, J.D. McCalley, A. Bloom, et al., Design and valuation of high-capacity HVDC macrogrid transmission for the continental US. IEEE Trans. Power Syst., 1–10 (2020). https://doi.org/10.1109/TPWRS.2020.2970865
37. D. Fitiwi, *PhD Thesis: Strategies, Methods and Tools for Solving Long-term Transmission Expansion Planning in Large-scale Power Systems* (Comillas Pontifical University, KTH Royal Institue of Technology, Delft University of Technology, 2016)
38. D. Fitiwi, L. Olmo, M. Rivier, F. de Cuadra, J. Perez-Arriaga, Finding a representative network losses model for large-scale transmission expansion planning with renewable energy sources. Energy (2015)
39. N. Flourentzou, V. Agelidis, G. Demetriades, VSC-based HVDC power transmission systems: An overview. IEEE Trans. Power Electron. **24**(3) (2009). https://doi.org/10.1109/TPE:/2008/2008441
40. FOSG, *Friends of the Supergrid* (2018). Retrieved from https://www.friendsofthesupergrid.eu/
41. A. Gavrilovic, AC/DC system strength as indicated by short circuit ratios, in *International Conference on AC and DC Power Transmission*, (1991), pp. 27–32
42. P. Girdinio, P. Molfino, M. Nervi, M. Rossi, A. Bertani, S. Malgarotti, Technical and compatibility issues in the design of HVDC sea electrodes, in *International Symposium on Electromagnetic Compatibility – EMC EUROPE*, (2012), pp. 1–5. https://doi.org/10.1109/EMCEurope.2012.6396841

43. U. Gnanarathna, A. Gole, A. Rajapakse, S. Chaudhary, Loss estimation of modular multi-level converters using electro-magnetic transients simulation, in *International Conference Power System Transient*, (2011)
44. M. Guarnieri, Who invented the transformer? IEEE Ind. Electron. Mag. **7**, 56–59 (2013)
45. H. Hamzehbahmani, H. Griffiths, A. Haddad, D. Guo, Earthing requirements for HVDC systems, in *50th International Universities Power Engineering Conference (UPEC)*, (2015), pp. 1–7. https://doi.org/10.1109/UPEC.2015.7339768
46. R. Hemmati, R. Hooshmand, A. Khodabakhshian, State-of-the-art of transmission expansion planning: Comprehensive review. Renew. Sustain. Energ. Rev., 312–319 (2013). https://doi.org/10.1016/j.rser.2013.03.015
47. M. Hesamzadeh, N. Hosseinzadeh, P. Wolfs, Economic assessment of transmission expansion projects in competitive electricity markets – an analytical review, in *43rd Internal Universities Power Engineering Conference*, (2008), pp. 1–10. https://doi.org/10.1109/UPEC.2008.4651531
48. T.P. Hughes, *Networks of Power: Electrification in Western Society, 1880–1930* (Johns Hopkins University Press, Baltimore, 1993)
49. E. Iggland, R. Wiget, S. Chatzivasileiadis, G. Anderson, Multi-area DC-OPF for HVAC and HVDC grids. IEEE Trans. Power Syst. **30**(5), 2450–2459 (2015)
50. M. Josephson, *Edison* (McGraw Hill, New York, 1959)
51. B. Kazemtabrizi, *Mathematical Modelling of Multi-Terminal VSC-HVDC Links in Power Systems Using Optimal Power Flows* (University of Glasgow, Scotland, 2011)
52. X. Kong, H. Jia, Techno-Economic Analysis of SVC-HVDC Transmission System for Offshore Wind, in *Power and Energy Engineering Conference*, (2011), pp. 1–5. https://doi.org/10.1109/APPEEC.2011.5748839
53. M. Korytowski, Uno Lamm: The Father of HVdc Transmission [History]. IEEE Power Energ. Mag. **15**(5), 92–102 (2017, September). https://doi.org/10.1109/MPE.2017.2711759
54. P. Kundur, *Power System Stability and Control* (McGraw-Hill Education, New York, 1994)
55. A. L'abbate, G. Migliavacca, G. Fulli, C. Vergine, A. Sallati, The European research project REALISEGRID: Transmission planning issues and methodological approach towards the optimal development of the pan-European system, in *IEEE Power and Energy Society General Meeting*, (2012), pp. 1–8
56. A. Lamm, The peculiarities of high-voltage dc power transmission. IEEE Spectr. **3**(8), 76–84 (1966). https://doi.org/10.1109/MSPEC.1966.5217652
57. G. Latorre, R. Cruz, J. Areiza, A. Villegas, Classification of publications and models on transmission expansion planning. IEEE Trans. Power Syst. **18**(2), 938–946 (2003). https://doi.org/10.1109/TPWRS.2003.811168
58. X. Lei, B. Buchholz, D. Poyh, Analysing subsynchronous resonance phenomena in the time- and frequency domain. Trans. Electr. Power **10**(4), 203–211 (2000)
59. V. Lescale, Modern HVDC: State of the art and development trends, in *International Conference on Power System Technology, POWERCON'98*, (1998), pp. 446–450. https://doi.org/10.1109/ICPST.1998.729003
60. Z. Liu, L. Gao, Z. Wang, J. Yu, J. Zhang, L. Lu, R&D progress of +/− 1100kv UHVDC technology. CIGRE B4-201-2012 (2012)
61. K. Meah, S. Ula, Comparative evaluation of HVDC and HVAC transmission systems, in *IEEE Power Engineering Society General Meeting*, (2007), pp. 1–5. https://doi.org/10.1109/PES.2007.385993
62. G. Migliavacca, A. L'abbate, I. Losa, E. Carlini, A. Sallati, C. Vergine, *The REALISEGRID Cost-Benefit Methodology to Rank Pan-European Infrastructure Investments* (IEEE PowerTech, Trondheim, 2011), pp. 1–7. https://doi.org/10.1109/PTC.2011.6019150
63. C. Oates, C. Davidson, A comparison of two methods of estimating losses in the modular multi-level converter, in *14th European Conference on Power Electronics and Applications*, (2011), pp. 1–10

64. M. Okba, M. Saied, M. Mostafa, T. Abdel-Moneim, *High Voltage Direct Current Transmission – A Review, Part I* (IEEE Energytech, 2012), pp. 1–7. https://doi.org/10.1109/EnergyTech.2012.6304650
65. O. Oni, I. Davidson, K. Mbangula, A review of LCC-HVDC and VSC- HVDC technologies and applications, in *16th International Conference on Environment and Electrical Engineering*, (2016), pp. 1–7. https://doi.org/10.1109/EEEIC.2016.7555677
66. A. Ramos, P. Sanchez-Martin, *Modeling Transmission Ohmic Losses in a Stochastic Bulk Production Cost Model* (Institute for Research in Technology, Madrid, 1997)
67. H. Rao, Architecture of nan'ao multi-terminal VSC-HVDC system and its multi-functional control. CSEE J. Power Energ. Syst. **1**(1), 9–17 (2015)
68. J. Renedo, A.A. Ibrahim, B. Kazemtabrizi, A. García-Cerrada, L. Rouco, Q. Zhao, J. García-González, A simplified algorithm to solve optimal power flows in hybrid VSC-based AC/DC systems. Int. J. Electr. Power Energy Syst. **110**, 781–794 (2019)
69. J. Rimez, *Optimal Operation of Hybrid AC/DC Meshed Grids, PhD Thesis*, TU Eindhoven, Eindhoven (2014)
70. J. Rimez, R. Belmans, A combined AC/DC optimal power flow algorithm for meshed AC and DC networks linked by VSC converters. Int. Trans. Electr. Energy Syst. **25**, 2024–2035 (2015)
71. S. Rohner, S. Bernet, M. Hiller, R. Sommer, Modulation, losses, and semiconductor requirements of modular multilevel converters. IEEE Trans. Ind. Electron. **57**(8), 2633–2642 (2010). https://doi.org/10.1109/TIE.2009.2031187
72. C. Schifreen, W. Marble, Charging current limitations in operation or high- voltage cable lines [includes discussion]. Transactions of the American Institute of Electrical Engineers. Part III: Power Apparatus and Systems **75**(3) (1956)
73. A.G. Siemens, *Power Engineering Guide Edition 7.1. Technical report.* (2014)
74. V. Sood, *HVDC and FACTS Controllers: Applications of Static Converters in Power Systems* (Kluwer Academic Publishers, Boston, 2004)
75. T. Sousa, M. dos Santos, J. Jardini, R. Casolari, G. Nicola, An evaluation of the HVDC and HVAC transmission economic, in *Transmission and Distribution: Latin America Conference and Exposition (TD-LA)*, (2012), pp. 1–6. https://doi.org/10.1109/TDC-LA.2012.6401828
76. R. Teixeira Pinto, M.G.-B. Aragues-Penalba, A. Sumper, Optimal operation of DC networks to support power system outage management. IEEE Trans. Smart Grid **7**(6), 2953–2961 (2016)
77. D. Tiku, Dc power transmission: Mercury-arc to thyristor HVdc valves [History]. IEEE Power Energ. Mag. **12**(2), 76–96 (2014). https://doi.org/10.1109/MPE.2013.2293398
78. A. Trzynadlowski, *Introduction to Modern Power Electronics* (Wiley, Hoboken, 2015)
79. D. van Hertem, M. Ghandhari, Multi-terminal VSC HVDC for the European supergrid: Obstacles. Renew. Sustain. Energ. Rev. **14**(9), 3156–3163 (2010)
80. J. Vobecky, K. Stiegler, M. Bellini, U. Meier, New generation large area thyristor for UHVDC transmission, in *International Exhibition and Conference for Power Electronics, Intelligent Motion, Renewnable Energy and Energy Management*, (2017), pp. 1–4
81. P. Vormedal, *Master Thesis: Voltage Source Converter Technology for Offshore Grids: Interconnection of Offshore Installations in a Multiterminal HVDC Grid using VSC.* (2010)
82. F. Wang, *PhD Thesis: On Techno-economic Assessment of a Multi-terminal VSC-HVDC in AC Transmission Systems* (Chalmers University of Technology, Gothenburg, 2013)
83. F. Wang, T. Anh Le, L. Tiernberg, A. Mannikoff, A. Bergman, A new approach for benefit evaluation of multiterminal VSC-HVDC using a proposed mixed ac/dc optimal power flow. IEEE Trans. Power Delivery **29**(1), 432–443 (2014). https://doi.org/10.1109/TPWRD.2013.2267056
84. H. Wang, M. Redfern, The advantages and disadvantages of using HVDC to interconnect AC networks, in *45th Universities Power Engineering Conference (UPEC)*, (2010), pp. 1–5
85. S. Wang, J. Zhu, L. Trinh, J. Pan, Economic assessment of HVDC project in deregulated energy markets, in *3rd Internal Conference on Electric Utility Deregulation and Restructuring and Power Technologies*, (2008), pp. 18–23
86. R. Wiget, *Combined AC and Multi-Terminal HVDC Grids–Optimal Power.* PhD thesis, ETH Zurich, Switzerland (2015)
87. A. Wood, B. Wollenberg, *Power Generation, Operation, and Control* (Wiley, Hoboken, 2012)

88. K. Zha, J. Cao, W. Ouyang, B. Sun, C. Gao, H. Luan, Design of 6250 A/+− 800 kV UHVDC converter valve, in *13th IET International Conference on AC and DC Power Transmission (ACDC2017)*, (2017), pp. 1–6. https://doi.org/10.1049/cp.2017.0048
89. H. Zhang, *PhD Thesis: Transmission Expansion Planning for Large Power Systems* (Arizona State University, Tempe, 2013)
90. L. Zhang, *PhD Thesis: Modeling and Control of VSC-HVDC Links Connected to Weak AC Systems* (Royal University of Technology, Stockholm, 2010)
91. Z. Zhang, Z. Xu, Y. Xue, Valve losses evaluation based on piecewise analytical method for MMC-HVDC links. IEEE Trans. Power Delivery **29**(3), 1354–1362 (2014). https://doi.org/10.1109/TPWRD.2014.2304724
92. Q. Zhao, *PhD Thesis: Technical and Economic Impact of the Deployment of a VSC-MTDC Supergrid with Large-Scale Penetration of Offshore Wind* (Universidad Pontifica Comillas, Madrid, 2019)
93. Q. Zhao, J. García-González, *D13.3 Identified barriers for replicability. Best Paths EU Project: Beyond State-of-the-art Technologies for rePowering AC corridors and multi-Terminal HVDC Systems* (2018)
94. Q. Zhao, J. García-González, O. Gomis-Bellmunt, E. Prieto-Araujo, F.M. Echavarren, Impact of converter losses on the optimal power flow solution of hybrid networks based on VSC-MTDC. Electr. Power Syst. Res **151**, 395–403 (2017)
95. J. Zhu, *Optimization of Power System Operation*, 2nd edn. (Wiley-IEEE, Chichester, 2015)
96. L. Sigrist, K. May, A. Morch, P. Verboven, P. Vingerhoets, L. Rouco, On scalability and replicability of smart grid projects - a case study. Energies. **9**(3), 1–19 (2016)
97. Z. Casey, Build a European supergrid for jobs and energy. The European Wind Energy Association (EWEA), (2012) Online: http://www.ewea.org/blog/2012/12/build-a-european-supergrid-for-jobs-and-energy. Accessed 1-9-2020

Chapter 7
Transmission Expansion Planning Outside the Box: A Bilevel Approach

Sonja Wogrin, Salvador Pineda, Diego A. Tejada-Arango, and Isaac C. Gonzalez-Romero

Nomenclature

Indices and Sets

g	Generating unit index
t	Time period index
n, n'	Node index
$\mathcal{GN}(g, n)$	Set of generating units at a given node
$\mathcal{GO}(g)$	Set of generating units owned by a Genco
$\mathcal{LN}(n, n')$	Dynamic set of possible transmission lines

Parameters

C_g	Linear cost parameter of generating unit g (€/MWh)
I_g	Annualized investment cost of generating unit g (€/MW)
$IT_{nn'}$	Annualized transmission investment cost of line between n and n' (€/MW)
$X_{nn'}$	Reactance of line between n and n' (p.u.)
SB	Base power (MW)

S. Wogrin (✉) · D. A. Tejada-Arango · I. C. Gonzalez-Romero
Comillas Pontifical University, Madrid, Spain
e-mail: Sonja.Wogrin@comillas.edu

S. Pineda
University of Malaga, Malaga, Spain
e-mail: spineda@uma.es

S. Lumbreras et al. (eds.), *Transmission Expansion Planning: The Network Challenges of the Energy Transition*, https://doi.org/10.1007/978-3-030-49428-5_7

ρ_{gt} Capacity factor of generating unit g and time t (p.u.)
S_{tn}^0 Demand slope at node n and time t (MW^2/€)
D_{tn}^0 Demand intercept at node n and time t (MW)
Φ_g Conjectured-price response of generating unit g (€/MW^2)
$\underline{M}_{gt}^p, \underline{M}_{gt}^{\beta}, \overline{M}_{gt}^p, \overline{M}_{gt}^{\overline{\beta}}, \underline{M}_{tnn'}^f, \underline{M}_{tnn'}^{\underline{\gamma}}, \overline{M}_{tnn'}^f, \overline{M}_{tnn'}^{\overline{\gamma}}, \underline{M}_{tn}^d, \underline{M}_{tn}^{\underline{\alpha}}$ Auxiliary large constants used for linearization

Primal Variables

p_{gt} Output of generating unit g in time t (MW)
$\overline{p}_g$ Capacity investment of generating unit g (MW)
$f_{tnn'}$ Power flow at time t from node n to node n' (MW)
θ_{tn} Voltage angle at time t and node n (rad)
$l_{nn'}$ Capacity investment in line from node n to node n' (MW)
d_{tn} Satisfied demand at time t and node n (MW)
λ_{tn} Electricity price at time t and node n (€/MW)

Lagrange Multipliers (Dual Variables)

$\underline{\beta}_{gt}$ Dual of lower bound on production (MW)
$\overline{\beta}_{gt}$ Dual of upper bound on production (MW)
$\underline{\alpha}_{tn}$ Dual of lower bound on demand (MW)
$\underline{\gamma}_{tnn'}$ Dual of lower bound on power flow (MW)
$\overline{\gamma}_{tnn'}$ Dual of upper bound on power flow (MW)
$\kappa_{tnn'}$ Dual of definition of power flow (MW)
ξ_t Dual of slack bus definition (rad)

Auxiliary Binary Variables

$\underline{b}_{gt}^p, \overline{b}_{gt}^p, \underline{b}_{tnn'}^f, \overline{b}_{tnn'}^f, \underline{b}_{tn}^d$ Auxiliary binary variables used for linearization ($\{0,1\}$)

7.1 Introduction

The liberalization of the electricity sector and the introduction of electricity markets first emerged in the 1980s in countries like Chile, the UK, and New Zealand. Nowadays, the vast majority of all developed countries have undergone such a liberalization process, which has complicated the organization of the electricity and energy sector greatly. In regulated electricity systems, many tasks – such as expansion planning, for example – are usually carried out by a centralized planner who minimizes total cost while meeting a future demand forecast, reliability constraints, and environmental requirements identified by the government. Planning the investment and operation of such a regulated system can be regarded as stable, relatively predictable, and essentially risk-free for all entities involved.

However, under a liberalized framework, many decisions are no longer regulated but to a large extent, up to the responsibility of individual entities or private companies that act strategically and might have opposing objectives. Within the realm of strategic decision-making, the hierarchy in which decisions are taken becomes extremely important.

A strategic generation company (GENCO) is not going to invest the same capacity as a centralized planner that maximizes social welfare nor will the built capacity be operated in the same way. Therefore, even regulated entities such as a transmission system planner should take into account strategic investment decisions of GENCOs and market feedback from a potentially non-perfectly competitive market when taking transmission expansion decisions. Another reason for considering sequential transmission and generation expansion planning decisions is the time horizon in which these decisions take place. While construction times of power plants can range from 1 up to a couple of years, it can take between 10 to 15 years to build a transmission line, including planning, permitting, land acquisition, and construction. In order to characterize strategic behavior or considerations of different time horizons, bilevel models inspired by Stackelberg [52] games become necessary and important decision-support tools in electricity and energy markets.

Bilevel models – which have first been used in the electricity sector to formulate electricity markets, for example, by Cardell et al. [8], Berry et al. [4], Weber and Overbye [53], Hobbs et al. [18] or Ramos et al. [40] just to name a few – allow us to represent a sequential decision-making process as opposed to single-level models where all decisions are considered to be taken simultaneously, which can be a gross simplification of reality and distort model outcomes.

In energy and electricity markets, there are a plethora of different applications of interesting bilevel optimization and bilevel equilibrium problems. The purpose of this chapter is to provide an introduction of how bilevel programming and Stackelberg-type games can be applied to the problem of transmission expansion planning (TEP). Hence, the remainder of this chapter is organized as follows. First of all, in Sect. 7.2, we present some basic mathematical concepts of bilevel programming in order for the reader to follow the subsequent analysis and terminology used in the remainder of the paper. Then, Sect. 7.3 contains a detailed literature review of bilevel applications in TEP and gives a brief outlook on existing solution techniques and pending challenges in bilevel models for TEP. In Sect. 7.4, we formulate a bilevel model that takes TEP decisions while accounting for strategic market feedback and compare it to a traditional model in a numerical case study in Sect. 7.5. Finally, Sect. 7.6 concludes the chapter.

7.2 Basics of Bilevel Programming

Generally, an optimization problem is mathematically formulated as follows:

$$\min_{x} \quad f(x) \tag{7.1a}$$

$$\text{s.t.} \quad g(x) \leq 0, \tag{7.1b}$$

where $x \in \mathbb{R}^n$ is the *decision variable*, $f(x) : \mathbb{R}^n \to \mathbb{R}$ is the *objective function*, and $g(x) \leq 0$ represents the *constraints* that determine the *feasible region*. Note that $g : \mathbb{R}^n \to \mathbb{R}^m$ and thus we have m constraints. The objective of this problem is to find the point that belongs to the feasible region with the lowest value of $f(x)$.

A Stackelberg game is a strategic game that involves two participants who have different objectives, act in sequence, and whose objectives rely on the other's decision. In this sequential game, a *leader* moves. First, a *follower* observes the leader's action and then moves. Furthermore, both the leader and the follower have knowledge of the other's objective function and decision, making it a game of perfect information. The Stackelberg game is illustrated in Fig. 7.1.

A Stackelberg game cannot be mathematically modeled using single-level optimization models as (7.1) but needs to be formulated as the following generic bilevel programming problem [1]:

$$\min_{x,y} \quad f(x, y) \tag{7.2a}$$

$$\text{s.t.} \quad g_i(x, y) \leq 0, \quad \forall i \tag{7.2b}$$

$$y \in \arg\min_{y} \ h(x, y) \tag{7.2c}$$

$$\text{s.t. } k_j(x, y) \leq 0, \quad \forall j \tag{7.2d}$$

In the bilevel problem (7.2), there are two different decision variables: x belonging to the leader and y belonging to the follower. Functions $f(x, y)$ and $h(x, y)$ are, respectively, the leader's and follower's objective functions, and $g_i(x, y)$ and $k_j(x, y)$ are the leader's and follower's constraint functions, in that order. Note

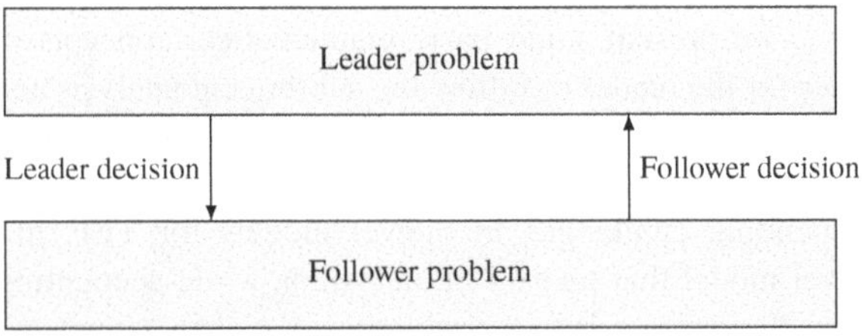

Fig. 7.1 Stackelberg game

that the follower's optimization problem is embedded in the leader's optimization problem in order to account for the sequence of decisions. Since the leader acts first, the follower considers the leader's decision as a parameter in its own problem. The leader, however, can anticipate the follower's action, and therefore, its problem now is to decide x such that the follower chooses y in the best way possible for the leader.

Let us next illustrate the differences between single-level and bilevel optimization using the following small problem:

$$\max_{x,y} \quad y \tag{7.3a}$$

$$\text{s.t.} \quad -x \leq -2 \tag{7.3b}$$

$$y \in \arg\min_{y} \quad y \tag{7.3c}$$

$$\text{s.t.} \quad -x - y \leq -7 \tag{7.3d}$$

$$y \leq 9 \tag{7.3e}$$

$$x - 2y \leq 1 \tag{7.3f}$$

$$6x + y \leq 45 \tag{7.3g}$$

Note that the leader aims at maximizing y, while the follower's target is to minimize y. The problem is illustrated in Fig. 7.2.

The constraint region formed by all feasible points is the gray shaded area. If the leader could ignore the follower's objective function, then the leader would choose a value of y equal to 9 (maximum feasible value). On the other hand, if the follower

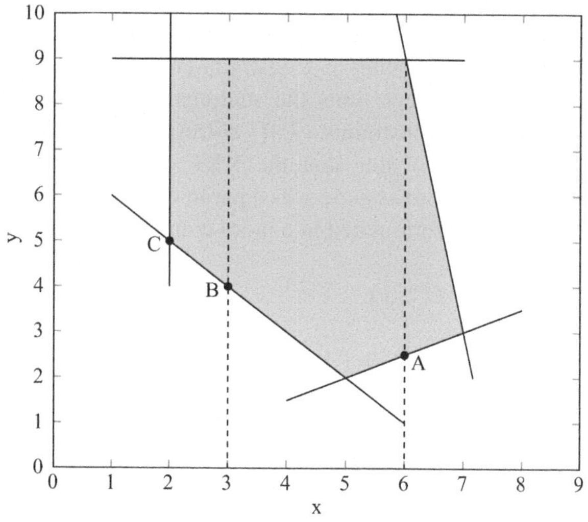

Fig. 7.2 Illustrative example of bilevel optimization

could choose the value of y without having to abide by the leader's decision, the follower would make y equal to 2 (minimum feasible value). We see next that the solution of the bilevel problem (7.2) is neither of these two values. Let us assume that the leader decides $x = 6$. Then the follower would choose the minimum value of y for all feasible points with $x = 6$ (point A). If the leader decides $x = 3$, then the follower would choose the minimum y (point B) again. Therefore, since the leader wants to maximize y, the optimal decision is to make $x = 2$ so that $y = 5$ (point C). As the reader has probably realized, determining the optimal solution of a bilevel problem graphically is more challenging than doing the same for single-level optimization problems with a unique objective function.

If the lower-level problems are convex and satisfy some constraint qualifications, then it can be replaced by its necessary and sufficient KKT optimality conditions. Therefore, the usual strategy to solve a bilevel problem of type (7.2) is to reformulate it as the following single-level optimization problem equivalently: sss

$$\min_{x,y} \quad f(x, y) \tag{7.4a}$$

$$\text{s.t.} \quad g_i(x, y) \leq 0, \quad \forall i \tag{7.4b}$$

$$k_j(x, y) \leq 0, \quad \forall j \tag{7.4c}$$

$$\nabla_y h(x, y) + \sum_j \lambda_j \nabla_y k_j(x, y) = \mathbf{0} \tag{7.4d}$$

$$\lambda_j \geq 0, \quad \forall j \tag{7.4e}$$

$$\lambda_j k_j(x, y) = 0, \quad \forall j \tag{7.4f}$$

where λ_j are the Lagrange multipliers (or in some cases also referred to as dual variables or shadow prices) of lower-level constraints, and ∇_y denotes the gradient with respect to lower-level variables y. Constraint (7.4c) represents all lower-level constraints. Equation (7.4d) represents the stationarity condition, dual feasibility is imposed by (7.4e), and constraints (7.4f) correspond to the complementarity conditions. It is important to note that the KKT conditions of the lower-level problem must be determined to assume x as a parameter.

Problem (7.4) is usually formulated in a more compact way as follows:

$$\min_{x,y} \quad f(x, y) \tag{7.5a}$$

$$\text{s.t.} \quad g_i(x, y) \leq 0, \quad \forall i \tag{7.5b}$$

$$\nabla_y h(x, y) + \sum_j \lambda_j \nabla_y k_j(x, y) = \mathbf{0} \tag{7.5c}$$

$$0 \leq \lambda_j \perp k_j(x, y) \leq 0, \quad \forall j \tag{7.5d}$$

where "perp" operator $\perp$ denotes the inner product of two variables equal to zero. Problem (7.5) is also referred to as mathematical programming with equilibrium constraints (MPEC). While all bilevel problems with a convex lower-level problem can be reformulated as MPECs, the converse is not true. In that sense, MPEC problems are more general than bilevel problems [26]. Problem (7.5) is nonconvex and nonregular, and, therefore, finding its optimal global solution is hard even if all functions f, g_i, h, k_j are linear. The interested reader can find a comparison of different methods to solve bilevel linear problems using off-the-shelf optimization software in [33]. In Sect. 7.4, we present a TEP example problem that we solve using one of these methods, in particular, mixed-integer programming (MIP). Let us now explore the different applications of bilevel programming and sequential decision-making with respect to the transmission expansion planning problem.

7.3 Literature Review on Bilevel Transmission Expansion Planning

At the core of every multilevel optimization or equilibrium problem, there lies a sequential decision-making process. When it comes to transmission expansion planning (TEP) in liberalized electricity markets, the main factor of decision hierarchy stems from the generation expansion planning (GEP). Does the transmission planner take its decisions after the generation has been decided and sited, or do generation companies plan their investments after transmission assets have been decided? What comes first, the chicken or the egg? The sequence of TEP versus GEP defines two different philosophies: proactive TEP and reactive TEP.

Under proactive TEP approaches, the transmission company (TRANSCO) is the Stackelberg leader, whereas generation companies (GENCOs) are the Stackelberg followers. This means that TRANSCO has the first-mover advantage and decides the best possible TEP taking into account the feedback from generation companies. In other words, the network planner has the ability to influence generation investment and, furthermore, the spot market behavior. Such an assumption can also closely be related to the time frame of TEP versus GEP decisions. While building a power plant takes a couple of years, building a new transmission line can take up to several decades. It is, therefore, a reasonable assumption to assume that the more long-term decision, i.e., TEP, is taken before the GEP decision, which can be considered short term (in comparison to the TEP time horizon). Note that while the market operation is an important part of TEP and GEP problems, in terms of decision-making sequence, market decisions clearly happen after both GEP and TEP. Hence, the market is not the main focus in the bilevel TEP-GEP discussion.

On the other hand, under a reactive TEP approach, the network planner assumes that generation capacities are given (GENCOs are Stackelberg leaders), and then the network planner optimizes transmission expansion based only on the subsequent market operation. Reactive planning is thus represented by a model with multi-leader GENCOs and one or several TRANSCOs as followers.

Additionally, there exist some alternative TEP approaches like the one presented in [21] where the upper level represents both GEP and TEP simultaneously, while the lower level represents the market operation (MO). In [3], the authors decide optimal wind and transmission expansion in the upper level, subject to a market-clearing. Such an approach is more suitable for a centralized power system where the network planner and the generation companies belong to the same decision-making agent. This is usually not the case in liberalized electricity markets. The work of [41] explores the interactions between an ISO with security of supply objectives with private profit-maximizing entities deciding GEP and TEP investment simultaneously. The arising problem can be interpreted as some kind of hierarchical TEP-GEP. The entire problem, however, is never explicitly formulated as a bilevel problem. Instead, an iterative solution procedure is proposed. Since our main focus lies in the sequence between TEP and GEP, we continue our detailed literature review on bilevel TEP problems distinguishing between proactive and reactive approaches.

While, according to [47] in practice, most of the TRANSCO companies in the world follow a reactive TEP approach, most of the TEP-GEP literature on multilevel TEP-GEP models employ a proactive planning approach. This is partly explained by the different time frames of TEP and GEP models. While generation power plants can be built in 1–3 years, the construction of new transmission lines may take up to 10 years or more. Therefore, it makes more sense to assume that transmission expansion is decided first, anticipating the reaction of generating companies. Moreover, as mentioned in [38], there are some other approaches that are close to proactive planning. For example, a regulation was approved in the USA that includes the concept of anticipative (proactive) transmission planning to obtain higher social welfare [10]. Additionally, in the current European context, ENTSOE plays the role of a centralized agent that proposes future planning pathways in which regional coordination takes place, and then nationally, generation companies can react to its decisions. Finally, it is worth mentioning that transmission investments are usually less expensive than generation investments, and therefore, the goal is that the network should not impose any constraints on market results. Under this regulatory context, a proactive planning approach would make more sense, and then we begin our analysis focusing on proactive planning.

7.3.1 Proactive Transmission Expansion Planning

Before discussing bilevel TEP approaches in the literature, we first want to focus on contributions of multilevel (and in particular more than two levels) proactive TEP works in the literature. For a comprehensive review of GEPTEP co-planning models under a market environment, please refer to [16].

The first work on proactive TEP we want to discuss is presented by Sauma and Oren [43]. In [43], the authors extend their work done in [42] and explore optimal transmission expansion given different objectives, considering policy implications and anticipating responses of strategic players such as GENCOs. They also consider a spot market where the distinctive ownership structures are reflected (one GENCO can own several units), as proposed in [59]. In particular, the authors consider a kind of three-level structure: first TEP, then GEP, and, finally, the market. The market equilibrium by itself is a linear complementarity problem. When adding strategic GEP, the problem already becomes an equilibrium problem with equilibrium constraints (EPEC). An additional layer of TEP is set on top of that. The authors attempt to solve this very complex type of nonconvex problem as follows: they formulate a bilevel problem and in particular, as MPECs, where a strategic GENCO takes investment decisions in the upper level, while the market equilibrium is the lower level. This MPEC is solved iteratively, eliminating dominated GEP strategies, which yields an equilibrium solution to the GEP EPEC problem. The authors solve the TEP problem by some kind of enumeration of TEP strategies and repeating the iterative process that yields the EPEC solution. It is fair to say that while [43] takes the first ambitious step into modeling proactive TEP with bilevel programming techniques, it does not achieve to solve the full (all possible TEP options) three-level problem exactly and to global optimality.

Pozo, Contreras, and Sauma [36] extend the groundwork of [42, 43] and propose a first complete model formulation of the three-level TEP problem. The three levels are still: first TEP, then GEP, and, finally, the market-clearing. In the second level (the GEP stage), all GENCOs' investment strategies are enumerated, expressing Nash equilibria as a set of inequalities. The advantage of doing so is that it avoids having to pass to a complicated MPEC-EPEC formulation of the GEP market problem and emulates this bilevel equilibrium in an MIP setting. The shortcoming of this method is that it only works when second-stage (GEP) strategies are discrete and finite. In other words, if there are a lot of different investment options and combinations of strategies, the problem size could blow up and yield an intractable model. They do, however, find all possible equilibria. Note that, in the market, perfect competition is assumed as opposed to [42, 43], where Cournot competition is considered at the lower level. From a formulation point of view, the market-clearing problem and the GEP problem are converted into MIP constraints, and then the TEP level is put on top of that, yielding an MIP. Nonlinear terms (stemming from complementarity conditions, and nonlinear terms in the objective function) are linearized using binary expansion and Fortuny-Amat [13]. While this work represents an important step forward in terms of actually formulating a three-level problem in closed form, the proposed model does not seem to be computationally efficient.

Pozo, Sauma, and Contreras [37] extend their work of [36] by including uncertainty in demand and apply the model to a realistic power system in Chile. The same authors overcome the computational shortcoming of their previously proposed models in [38]. First of all, the market-clearing is converted into MIP constraints by using the primal-dual formulation (instead of the KKT conditions) and linearizing nonlinear terms in the strong duality constraint using Fortuny-Amat. This yields an MIP with fewer binary variables than if all complementarities would have been linearized. Then, they propose a novel column-and-row decomposition technique for solving the arising EPECs (GEP and market-clearing), which ultimately yields the globally optimal solution, i.e., Nash equilibrium of the EPEC, by iteratively solving two problems. This greatly increases computational efficiency with respect to previous works, while not compromising the rigorousness of the mathematical formulation. Furthermore, the authors propose a pessimistic and optimistic network planner (the pessimistic case is used to eliminate multiple equilibria by considering the worst generation expansion case) to describe all possible outcomes of the EPEC in the lower level. The authors conclude that, in practice, if multiple generation expansion exists in the equilibrium, proactive planning does not always yield the best welfare results, and it can even reduce social welfare.

Jin and Ryan [22] also propose a three-level TEP problem, but they extend previous approaches by considering Cournot's strategic decisions in the market. Note that they do not replace the GEP level by Nash equilibrium constraints but consider the actual optimization problems, yielding an EPEC for the GEP and market problem alone. In the end, they propose a hybrid approach to solve the three-level problem, which takes advantage of two different ways of solving the EPEC: the diagonalization method (iteratively solving MPECs until convergence) and formulating the EPEC as a complementarity problem (which in case of the nonconvex MPEC only yields stationary points). The proposed algorithm proves computationally efficient; however, it does not guarantee to find the global optimum.

Another multilevel approach to TEP is presented by Motamedi et al. [29]. The authors characterize their problem as a four-level problem; however, since two of those levels refer to problems that are considered simultaneously, mathematically speaking, the framework boils down to a three-level problem: TEP, GEP, and market (pairs of price and quantity bids). The authors propose an iterative algorithm using search-based techniques and agent-based modeling to solve the arising problem.

In Taheri et al. [48], the authors also tackle a three-level TEP problem: TEP, GEP, and market-clearing through price and offer quantities. The temporal representation of this work resorts to a load duration curve, which does not allow for modeling intertemporal constraints such as ramping or commitment constraints. In order to solve this problem, the authors do the following: the market-clearing is replaced by its KKT conditions. Then, the GEP problem is represented by an EPEC, where each GENCO faces an MPEC. The authors then write the KKT conditions of each MPEC, and the resulting nonlinear and nonconvex system of equations represents the stationarity conditions of the EPEC. This nonlinear EPEC system is then linearized and transformed into mixed-integer constraints. The TEP objective function is put on top, which yields the three-level problem represented as an MIP. Since the EPEC is a nonconvex problem, the obtained point might just be a saddle point. Hence, the authors have to carry out an ex post validation technique to check if the obtained point is actually an equilibrium. The advantage of the approach is that the all three levels are represented fully; however, if the obtained point does not represent an equilibrium to the GEP problem, then it gets messy in terms of finding the *next best point that is actually an equilibrium*. Computationally speaking, this method is inefficient due to the high amount of binary variables necessary to linearize the EPEC constraints.

Apart from the three-level proactive approaches mentioned above, there are two-level approaches where transmission investment decisions are taken first, and then generation investment and operation decisions are taken simultaneously. On the one hand, the work in [20] models the market-clearing in the lower level, which includes GEP investment variables as well. Authors in [20] consider perfect competition in the lower level and define different objective functions in the upper level that are compared with a vertically integrated one-level approach. Additionally, they consider a network fee so the TRANSCO can recover investments in the case of a flow-based fee regulation typically used in the USA. In terms of formulation, the lower level is replaced by its KKT conditions, which are then linearized using [13]. The resulting MPEC is then solved as a mixed-integer linear program (MILP).

Pisciella et al. [35] also present a proactive TEP approach where the TSO takes investment decisions subject to a lower-level market equilibrium of GENCOs and a market operator that exchange price and quantity bids and a market-clearing. In this problem, there appear binary variables both in the upper and in the lower level. The authors apply a progressive penalization algorithm to solve the arising problem.

Later, Weibelzahl and Märtz [54] extend the work of Jenabi and Fatemi Ghomi [20] and choose a pessimistic TRANSCO. The authors prove some subsequent uniqueness properties. Since their model uses chronological time steps, they additionally consider battery expansion in their framework. The resulting MPEC is reformulated as an MIP and solved as such. In [50], the authors present various types of TEP-GEP models, one of which represents a proactive approach with a pessimistic TEP in the upper level and strategic GENCOs in the lower level.

On the other hand, Maurovich et al. [27] consider a stochastic bilevel model with a merchant investor of transmission in the upper level and GEP including wind expansion and market operation in the lower level, where Cournot competition is considered. The resulting bilevel problems are converted into MIPs.

Finally, Gonzalez-Romero et al. [17] apply the same structure, but they consider storage expansion and Cournot competition in the lower level. Both works [17, 27] find counterintuitive results when considering Cournot competition in the lower level compared to a perfect competition case. The lower level replaced by its KKT conditions is linearized. The resulting MPEC is solved as an MIP.

7.3.2 Reactive Transmission Expansion Planning

One of the first reactive TEP planning approaches was proposed in [43], where some theoretical properties were shown, and some other practical results were presented for fixed transmission plans. Unfortunately, the subsequent research on this regard is limited. In general, under this approach several GENCOs are considered as the leaders and a single TRANSCO as the follower. However, it could also be the case that only one GENCO is the leader and the remaining GENCOs and TRANSCO(s) are followers, yielding a *one leader multiple followers* structure, which is easier to solve numerically speaking.

Therefore, there are only three subsequent studies on reactive planning. In [50], Tohidi and Hesamzadeh propose a new comparison between the proactive and the reactive approach. In contrast to the work done by Pozo et al. [37], authors in [50] do not consider anticipation of market outcomes by GENCOs and propose the elimination of the multiple Nash equilibria by considering a pessimistic or optimistic TRANSCO.

We now discuss a type of three-level approach of reactive TEP by Dvorkin et al. [12]. While the authors differentiate three levels: upper level representing a merchant storage owner (some kind of GEP); the middle level that carries out centralized TEP; and a lower level simulating the market-clearing, to our best knowledge mathematically speaking, this is not a three-level structure as the middle (TEP), and the lower (market) levels are solved simultaneously. Mathematically speaking, this belongs to the bilevel structures when merchant investors are in the upper level, and then TEP and market decisions are taken simultaneously (each problem has been replaced by its primal-dual formulation) in the lower level. The arising problem is solved numerically efficiently by means of a column generation algorithm, which is applied to a real-size case study. Note that the conventional GEP is omitted from this framework as the main focus of this work lies in the energy storage owner and not on the transmission expansion. The authors conclude that the co-planning of storage and transmission lead to greater cost savings than independent storage planning.

Table 7.1 provides a summary of the different system sizes of TEP problems mentioned in this literature review in order to see real applications.

Table 7.1 Summary of system sizes (* Expansion of existing units, + Suppliers, ** Supercomputer, – No data available)

Reference	Years	Periods	System	Gen.	Buses	Lines	Cand. Gen.	Cand. Lines	CPU (h)
[43]	1	1 Period	Chilean-32	32	32	37	32*	3,3*	–
[42]	1	1 Period	Cornell-32	6	30	39	3	4,4*	–
[20]	1	1 Period	IEEE-21	13	21	29	4	7	22.6
[54]	1	1 Day (24h)	Garver-6	3	6	8	3*	4	–
[21]	1	1 Period	6-bus	5, 4+	6	7	5*	5	–
[41]	10	4 Load Lev.	30-bus	7	30	41	11	8	–
[3]	1	5 Load Lev.	IEEE-118	54	118	186	3	5	12.5
[29]	5	5 Load Lev.	5-bus	3	5	6	3*	5	–
[37]	1	Load Lev.	4-bus	3	4	3	3*	3,1*	0.6
[36]	1	4 Load Lev.	Chilean 34	8	34	38	7*	6	22.15
[38]	1	1 Period	IEEE-24	32	24		10	6*	0.05
[35]	5	1 Period	Garver-6	10	6	9	5*	2	0.17
[22]	1	1 Period	IEEE-118	54	118	179	54*	4	1.5
[12]	1	5 days (24h)	WEEC-240	157	240	448	240*	179	12**
[27]	1	2 hours	3-bus	1	3	0	2	3	–
[17]	1	4 days (24h)	4-bus	2	4	2	2	3	0.28

7.3.3 Overview of Solution Methods for Bilevel Problems in Energy

As shown in the previous section, bilevel programming allows us to model a wide variety of situations that involve two decision-makers with their own objectives and constraints. However, the computational complexity of bilevel programming problems is extremely high. In fact, the authors of [6] theoretically show that the bilevel knapsack problem is contained in the complexity class Σ_2^p. Simulation results supporting the high computational burden of this bilevel problem are reported in [7]. Bilevel problems are hard to solve even under linearity assumptions, and, hence, most research efforts are currently focused on solving linear bilevel problems (LBP) [11].

Current methods to solve LBP highly depend on whether upper-level variables are discrete or continuous. If upper-level variables are discrete and lower-level variables continuous, then an equivalent single-level mixed-integer reformulation of the original bilevel can be obtained using the strong duality theorem [61]. Although such equivalent problems belong to the NP-complete class, current commercial optimization software can be used to compute globally optimal solutions.

If both upper- and lower-level variables are continuous, the solution procedures proposed in the technical literature are not as straightforward. From a practical point of view, methods to solve LBP can be divided into two main categories. The first category includes those methods that make use of dedicated solution algorithms to solve bilevel problems [5, 25, 45]. While these methods are usually efficient and ensure global optimality, they involve substantial additional and ad hoc coding work to be implemented in commercially available off-the-shelf optimization software. The second category includes the methods that can be implemented in or in combination with general-purpose optimization software without any further ado. Most of the existing applications of bilevel programming to energy-related problems rely on solution methods that belong to this second category, and, therefore, we explain next the most commonly used ones in more detail. There also exist bilevel problems that can be solved analytically, such as [28, 58]; however, such models are usually toy problems in terms of model size.

All the solution methods described next are based on reformulating the original LBP as an equivalent single-level problem that can be solved using off-the-shelf optimization software. To do so, the lower-level optimization problem is replaced by its necessary and sufficient Karush-Kuhn-Tucker (KKT) optimality conditions. Unfortunately, the KKT equations include nonlinear complementarity conditions, thus yielding a nonconvex single-level problem. Additionally, such a problem violates the Mangasarian-Fromovitz constraint qualification at every feasible point, and, therefore, the computation of globally optimal solutions becomes extraordinarily difficult. Existing methods to solve these nonconvex nonregular optimization problems differ regarding their strategy to deal with the nonlinear complementary conditions.

In energy-related problems, the most commonly used method first proposed in [13] consists of reformulating the complementary constraints by an equivalent set of linear inequalities to obtain a single-level mixed-integer problem. This strategy has, however, two main drawbacks. Firstly, it requires a large number of extra binary variables and big M parameters that increase the computational burden of the problem. Secondly, the equivalence between the single-level reformulation and the original BLP is only guaranteed if valid bounds on lower-level dual variables can be found. It is worth mentioning that most research works in the technical literature do not pay much attention to this aspect, which may lead to highly suboptimal solutions as proven in [34]. Another drawback is the fact that the big M method scales badly when extending the underlying MPECs to stochastic problems as in [55].

A related method, also based on the combinatorial nature of complementarity conditions, has been proposed in [46]. Such a method uses special order sets variables, and its computational advantages are problem-dependent. Alternatively, a regularization approach to solving the nonconvex single-level reformulation of LBP was first introduced in [44] and further investigated in [39]. This method solves a set of regular nonconvex optimization problems in which the complementarity conditions are iteratively imposed in a smooth manner.

Methods based on the combinatorial aspect of the complementarity conditions as those proposed in [13, 46] are computationally intensive but provide optimal solutions for valid bounds on the lower-level dual variables. On the other hand, solving the arising nonlinear MPEC directly, as done in [9], or other nonlinear-based methods investigated in [39, 44] are fast but only guarantee locally optimal solutions. There exist other heuristics [56] in order to approximate bilevel equilibrium problems with single-level equilibrium problems. Finally, the authors of [33] propose a method that combines the advantages of the two aforementioned approaches. Numerical simulations show that the method significantly improves the computational tractability of LBP.

7.3.4 Challenges of Bilevel Programming in TEP

The challenges for bilevel programming in energy markets are, in general, problem-dependent. However, there are certain topics that constitute gaps in the literature common to many of the works discussed in this chapter. We analyze each of these topics in the following paragraphs.

The first topic for discussion is the full optimal power flow in AC or AC-OPF. Many of the discussed articles [3, 17, 27], just to name a few, involve the representation of the transmission or the distribution network. The way that power flows in an electricity network is governed by Kirchhoff's Laws, which can be expressed mathematically as a set on nonlinear and nonconvex equations. This problem is often referred to as AC power flow, or if a specific objective is considered then AC-OPF. In terms of mathematical programming, and especially when considering bilevel programming, taking into account the full AC-OPF constitutes a serious issue. For example, the AC-OPF cannot be considered in the lower level of a bilevel problem as it cannot be replaced by its KKT conditions, for example (since the problem is nonconvex). As a solution, in the literature, the AC-OPF is replaced by its linear and thereby convex approximation, the DC-OPF. The assumption upon which this simplification is based can fail in low-voltage networks (like distribution systems) or when the network is not well meshed with a high transfer of power between areas (e.g., considering the transmission network of Spain and France which is not well-interconnected). When working with the DC-OPF formulation, important variables such as reactive power are not captured. In a future power system with increasing penetration of renewables and decreasing the amount of conventional synchronous machines that are important for system stability, not taking into account reactive power in planning models, for example, might lead to suboptimal solutions. As one possible solution, the DC-OPF could be replaced by a convex quadratic problem called second-order cone program [24] which replicates the AC-OPF exactly under certain hypotheses; however, these hypotheses need to be evaluated very carefully for each individual application.

Another shortcoming of existing bilevel models, and in particular, GEP-TEP applications, is the fact that many of them disregard storage technologies. If the focus of the bilevel model lies on TEP, then usually the GEP corresponding to storage is omitted, for example, in [55]. Taking into account that storage technologies will very likely play an important role in power systems of the future, they should be included in these models in order to correctly capture power system operation. Moreover, different types of storage technologies should be considered to cover the different range of services they can supply: for example, long-term storage technologies such as pumped hydro facilities that can carry out energy arbitrage over long time periods; and, short-term technologies such as batteries that have several charge and discharge cycles within one day. One of the practical reasons for disregarding storage is based on the fact that many planning models either use load periods [3, 55], representative days [12, 32], or just a few sequential hours [22, 27, 38] in order to reduce the temporal dimension which brings us to another challenge which is appropriate representation of time in medium- and long-term models. All previously mentioned methods have difficulties in representing storage constraints. Traditional load period models cannot formulate short-term storage constraints unless considering improvements such as system states [57]. Models that use representative days can formulate short-term storage well. However, they fail to properly represent long-term storage whose cycles go beyond the representative day. Tejada-Arango et al. [49] fix this shortcoming by introducing enhanced representative periods that are linked among each other.

An important issue that is also related to the previous one is the proper characterization of power system operations via unit commitment (UC)-type constraints, e.g., start-up, shutdown, ramping constraints. Such constraints either involve binary variables (e.g., start-up or shutdown) or require sequential time steps (e.g., ramping). The fact that UC constraints involve binary variables makes them very undesirable for lower-level problems as there are no meaningful KKT conditions. However, it is highly questionable to omit important constraints in bilevel problems just because it makes solving them even more difficult. Apart from involving binary variables, many UC constraints require sequential time steps, which makes them challenging for load-period models, for example. UC constraints have a relevant impact on expansion planning models, as shown in [31]. Since then, some attempts have been made in the literature to tackle the issue of binary variables in equilibrium problems or bilevel problems [19, 60]; however, this is still an open field of research with a lot of room for improvement.

Finally, there is a general need to develop more powerful computational methods to solve bilevel problems more efficiently. Currently, and due to the lack of adequate computational methods, the vast majority of bilevel applications in energy do not classify as large-scale problems. This becomes even more important when considering stochasticity. Many existing works in the literature, e.g., [32, 36, 48, 54], disregard the stochastic nature of many investments and operating problems. Therefore, introducing stochasticity in many bilevel applications and being able to solve them efficiently still constitute a major challenge in bilevel optimization.

7.4 Proactive Transmission Expansion Model Formulation

In this section, we present a simple proactive bilevel TEP problem, where a social TEP planner acts as a Stackelberg leader and GENCOs, consumers, and a system operator (SO) as Stackelberg followers. In Sect. 7, we present the nomenclature used in the mathematical formulations of the subsequent bilevel TEP model, followed by a step-by-step derivation of the formulation of the lower level of the bilevel problem in Sect. 7.4.1. Section 7.4.2 extends the lower-level equilibrium to a bilevel problem resulting in an MPEC. Finally, Sect. 7.4.3 transforms the previously obtained MPEC into an equivalent MIP, one of the most common ways of obtaining a globally optimal solution of bilevel problems in energy markets. Note that all of the models formulated in this section are available online.[1]

7.4.1 Lower-Level Equilibrium

In this section, we derive step-by-step the formulation of the lower level of the bilevel TEP problem that we want to solve. In particular, this lower level represents an equilibrium problem similar to the one presented in [23], and not just a single optimization problem. In order to formulate and discuss this equilibrium,

[1] https://github.com/datejada/PROTEM

we first introduce the different types of players: a social planner or transmission expansion planner, strategic generation companies (GENCOs), consumers, and the system operator (SO). The equilibrium situation that we are facing is such that each individual GENCO maximizes its total profits while deciding investment in generation capacity, i.e., GEP, and production at the same time. One might argue that investment and production decision are sequential per se, as we have seen in many three-level approaches discussed in the literature review; however, we assume that GEP and production decisions are taken simultaneously for the sake of model simplicity. The consumers simply maximize consumer surplus while deciding demand.

The system operator, on the other hand, is concerned with maximizing congestion rents arising from price differences, while maintaining power flow within the established limits. Finally, the demand balance equation is what links the optimization problem of the SO, the consumers, and the GENCOs. For the sake of simplicity, in the model below, we assume that the time steps t have a duration of 1 hour. This means that the conversion from MW to MWh is carried out by multiplying the MW amount by 1. Or alternatively, this can be interpreted as MW values and MWh values being the same (a MWh value is necessary when calculating linear variable generating costs in the GENCO's objective function for example).

7.4.1.1 GENCOS' Profit Maximization

Each generation company faces its own profit maximization problem, given by problem (7.6), deciding generation capacity and production. In the objective function, each GENCO maximizes its total profits, subject to lower and upper bounds on production. The dual variable of each constraint is indicated after the colon. Even though these dual variables are not explicitly used in the primal optimization problem (7.6) at hand, they will appear when taking KKT conditions of this problem later on. Please note that, in order to achieve an equilibrium among GENCOs, problem (7.6) has to be considered as many times as there are companies in the market. In order to indicate this in the formulation, we have made use of the dynamic set $\mathcal{GO}(g)$.

$$\max_{p_{gt},\overline{p}_g} \quad \sum_{t,g\in\mathcal{GO}(g)} p_{gt}\left(-C_g + \sum_{n\in\mathcal{GN}(g,n)} \lambda_{tn}\right) - \sum_{g\in\mathcal{GO}(g)} I_g\overline{p}_g \tag{7.6a}$$

$$\text{s.t.} \quad 0 \leq p_{gt}, \quad \forall g \in \mathcal{GO}(g), t \quad : \underline{\beta}_{gt} \geq 0 \tag{7.6b}$$

$$p_{gt} \leq \overline{p}_g \rho_{gt}, \quad \forall g \in \mathcal{GO}(g), t \quad : \overline{\beta}_{gt} \geq 0 \tag{7.6c}$$

The profit maximization[2] problem given by (7.6) is a convex[3] optimization problem, and can hence be replaced by its KKT conditions given in (7.7). Note that the nonlinear system of equations and inequalities given below represents the KKT conditions of all GENCOs (not only of a single one) as can be seen by the fact that we formulate equations $\forall g$ instead of $\forall g \in \mathcal{GO}(g)$. Moreover, we want to draw the reader's attention to the parameter Φ_g, which is defined as the partial derivative of market price with respect to production, i.e. $\partial\lambda/\partial p$. This conjectured-price response is a kind of conjectural variation and represents the belief that each generator has on how much it can impact price by varying its production. If Φ is zero, then generators believe they have no impact on price whatsoever – a situation which is referred to as perfect competition. Depending on the value of Φ, and in particular if it is greater than zero, then the problem below represents equilibrium conditions of an imperfect market. One particular case, which we examine later in the illustrative case study, will be the Cournot oligopoly, which is achieved when Φ is equal to $1/S^0$, the slope of the demand curve.

Constraints (7.7a) and (7.7b) represent the partial derivative of the Lagrangian with respect to the primal variables. Constraints (7.7c) and (7.7d) are the primal constraints. Constraint (7.7e) represents the nonnegativity of the Lagrange multipliers of the primal inequality constraints. Finally, constraints (7.7f) and (7.7g) are the nonlinear and nonconvex complementarity constraints.

$$\frac{\partial \mathfrak{L}}{\partial p_{gt}} = C_g - \sum_{n \in \mathcal{GN}(g,n)} \lambda_{tn} + \Phi_g p_{gt} - \underline{\beta}_{gt} + \overline{\beta}_{gt} = 0, \quad \forall g, t \tag{7.7a}$$

$$\frac{\partial \mathfrak{L}}{\partial \overline{p}_g} = I_g - \sum_t \overline{\beta}_{gt} \rho_{gt} = 0, \quad \forall g \tag{7.7b}$$

$$0 \leq p_{gt}, \quad \forall g, t \tag{7.7c}$$

$$p_{gt} \leq \overline{p}_g \rho_{gt}, \quad \forall g, t \tag{7.7d}$$

$$0 \leq \underline{\beta}_{gt}, \quad 0 \leq \overline{\beta}_{gt}, \quad \forall g, t \tag{7.7e}$$

$$p_{gt} \underline{\beta}_{gt} = 0, \quad \forall g, t \tag{7.7f}$$

$$(\overline{p}_g \rho_{gt} - p_{gt}) \overline{\beta}_{gt} = 0, \quad \forall g, t \tag{7.7g}$$

[2]If we convert the maximization into a minimization problem by multiplying the objective function with minus one, then (7.6) is a convex problem, which is what we need to derive meaningful KKT conditions. However, currently the problem is a maximization problem, so actually what we have is a concave objective function with convex constraints. In conclusion, (7.6) satisfies all the requirements for the KKT conditions to be sufficient.

[3]Taking the first derivative of the objective function (7.6a) with respect to $\overline{p}_g$ is linear an hence both convex and concave, and taking it with respect to p_{gt} yields: $\lambda_{tn} - \Phi_g p_{gt} - C_g$. Then, the second derivative is $-2\Phi_g$, which is smaller or equal to zero for each value of $\Phi_g \in [0, 1/\alpha]$, which yields concavity (convexity) of the maximization (minimization) objective function.

7.4.1.2 Consumer Surplus Maximization

At each of the nodes in the network, there are consumers that decide the demand d_{tn} for each time step by maximizing consumer surplus. For the sake of simplicity, however, we assume that the consumers of all nodes act as one single player. We make this assumption because it means that there is only one optimization problem for the consumers, instead of having as many optimization problems as nodes in the network. Moreover, the subsequent KKT conditions of both approaches lead to the exact same set of equations and inequalities.

That being said, let us derive the expression of consumer surplus. We maximize $\sum_{tn} \int_0^{d_{tn}} (D^0_{tn} - x)/S^0_{tn} dx - \sum_{tn} \lambda_{tn} d_{tn}$. In the first term, we integrate over the linear demand function, in which D^0 corresponds to the demand intercept and S^0 corresponds to the demand slope. Simplifying the integral in the first term becomes $D^0_{tn} d_{tn}/S^0_{tn} - d^2_{tn}/(2S^0_{tn})$. The second term represents the cost of energy the consumers have to pay, i.e., market prices times demand.

Therefore, the consumers' optimization problem is given as follows.

$$\max_{d_{tn}} \quad \sum_{tn} \left(\frac{D^0_{tn} d_{tn}}{S^0_{tn}} - \frac{d^2_{tn}}{2S^0_{tn}} \right) - \sum_{tn} \lambda_{tn} d_{tn} \tag{7.8a}$$

$$\text{s.t.} \quad 0 \le d_{tn}, \quad \forall t, n \quad : \underline{\alpha}_{tn} \ge 0 \tag{7.8b}$$

The previous optimization problem is convex for a known set of prices (since the consumers do not act strategically they are considered as price-takers) and can hence be replaced by its KKT conditions, which are indicated below.

$$\frac{\partial \mathcal{L}}{\partial d_{tn}} = \lambda_{tn} - \frac{D^0_{tn} - d_{tn}}{S^0_{tn}} - \underline{\alpha}_{tn} = 0, \quad \forall t, n \tag{7.9a}$$

$$0 \le d_{tn}, \quad \forall t, n \tag{7.9b}$$

$$0 \le \underline{\alpha}_{tn}, \quad \forall t, n \tag{7.9c}$$

$$d_{tn} \underline{\alpha}_{tn} = 0, \quad \forall t, n \tag{7.9d}$$

7.4.1.3 System Operator's Congestion Rent Maximization

The system operator maximizes congestion rents while ensuring correct power flow and voltage angles. This optimization problem is given by (7.10). Please note that even though $l_{nn'}$ has been defined as a variable in the nomenclature, it is not a lower-level variable but an upper-level variable. In particular, at this lower-level stage, the system operator does not choose transmission line capacities. They are considered a parameter in this particular optimization problem.

$$\max_{\theta_{tn}, f_{tnn'}} \sum_{t,n,n' \in \mathcal{LN}(n,n')} (\lambda_{tn'} - \lambda_{tn}) f_{tnn'} \tag{7.10a}$$

$$\text{s.t.} \quad -l_{nn'} \leq f_{tnn'}, \quad \forall t, n, n' \in \mathcal{LN}(n, n') \quad : \underline{\gamma}_{tnn'} \geq 0 \tag{7.10b}$$

$$f_{tnn'} \leq l_{nn'}, \quad \forall t, n, n' \in \mathcal{LN}(n, n') \quad : \overline{\gamma}_{tnn'} \geq 0 \tag{7.10c}$$

$$f_{tnn'} = (\theta_{tn} - \theta_{tn'}) \frac{SB}{X_{nn'}}, \quad \forall t, n, n' \in \mathcal{LN}(n, n') \quad : \kappa_{tnn'} \tag{7.10d}$$

$$\theta_{t,n=1} = 0, \quad \forall t \quad : \xi_t \tag{7.10e}$$

Since problem (7.10) is also convex for a known set of price values,[4] we now derive the KKT conditions of problem (7.10). Please note that since constraints (7.10d) and (7.10e) are equality constraints, the corresponding Lagrange multipliers are free variables. Hence, in the KKT conditions, the complementarity between the constraint and the variable is always satisfied and therefore omitted.

$$\frac{\partial \mathfrak{L}}{\partial \theta_{tn}} = \sum_{n' \in \mathcal{LN}(n',n)} \frac{SB}{X_{n'n}} \kappa_{tn'n} - \sum_{n' \in \mathcal{LN}(n,n')} \frac{SB}{X_{nn'}} \kappa_{tnn'} + (\xi_t)_{n=1} = 0, \quad \forall t, n \tag{7.11a}$$

$$\frac{\partial \mathfrak{L}}{\partial f_{tnn'}} = \lambda_{tn} - \lambda_{tn'} + \kappa_{tnn'} - \underline{\gamma}_{tnn'} + \overline{\gamma}_{tnn'} = 0, \quad \forall t, n, n' \in \mathcal{LN}(n, n') \tag{7.11b}$$

$$-l_{nn'} \leq f_{tnn'}, \quad \forall t, n, n' \in \mathcal{LN}(n, n') \tag{7.11c}$$

$$f_{tnn'} \leq l_{nn'}, \quad \forall t, n, n' \in \mathcal{LN}(n, n') \tag{7.11d}$$

$$f_{tnn'} = (\theta_{tn} - \theta_{tn'}) \frac{SB}{X_{nn'}}, \quad \forall t, n, n' \in \mathcal{LN}(n, n') \tag{7.11e}$$

$$\theta_{t,n=1} = 0, \quad \forall t \quad : \xi_t \tag{7.11f}$$

$$0 \leq \underline{\gamma}_{tnn'}, \quad 0 \leq \overline{\gamma}_{tnn'}, \quad \forall t, n, n' \in \mathcal{LN}(n, n') \tag{7.11g}$$

$$(l_{nn'} + f_{tnn'}) \underline{\gamma}_{tnn'} = 0, \quad \forall t, n, n' \in \mathcal{LN}(n, n') \tag{7.11h}$$

$$(l_{nn'} - f_{tnn'}) \overline{\gamma}_{tnn'} = 0, \quad \forall t, n, n' \in \mathcal{LN}(n, n') \tag{7.11i}$$

[4]Again, since we are maximizing the objective function is actually concave. But multiplying the objective function with minus one makes it convex. Actually, (7.10) is a linear problem, and therefore the resulting KKT conditions are sufficient. Note that the SO is also considered a price taker.

7.4.1.4 Demand Balance and Market-Clearing

Finally, all optimization problems (the GENCOs' given by (7.6), the consumers' given by (7.8) and the SO's given by (7.10)) are linked by the nodal demand balance constraint (or market-clearing) in (7.12).

$$d_{tn} = \sum_{g \in GN(g,n)} p_{gt} + \sum_{n' \in \mathcal{L}N(n'n)} f_{tn'n} - \sum_{n' \in \mathcal{L}N(nn')} f_{tnn'}, \quad \forall t, n \tag{7.12}$$

7.4.1.5 Formulation of Lower-Level Equilibrium

With all of the above, we can finally define the lower-level equilibrium problem as the following system of equations and inequalities:

$$\text{Constraints: (7.7), (7.9), (7.11), (7.12)} \tag{7.13}$$

Mathematically speaking, one can view (7.13) simply as a nonlinear, nonconvex (due to the complementarities) system of equalities and inequalities. And one could proceed to solve it as such with a standard NLP solver. However, this approach does – in general – not work very well as it does not take advantage of the fact that the nonlinearity of (7.13) is a very specific one with an *orthogonal* kind of nature. Therefore, the most common approach to solving (7.13) would be to implement is as a complementarity problem [14] and use a standard solver such as PATH to solve it. In the remainder of this section, we will also propose an alternative method via mixed-integer programming.

7.4.2 Upper Level and Complete Bilevel TEP Problem

In this section, we start out by formulating the upper level of the proactive TEP problem that we set out to solve. In its simplest form, the upper-level problem only consists of an objective function given in (7.14a). In particular, the social planner that decides transmission expansion (which we sometimes refer to as TSO in this chapter) decides investments in transmission lines $l_{nn'}$ while minimizing total system costs being total transmission investment cost, total generation capacity investment cost, and total operating cost.

The upper-level objective function is minimized subject to the lower-level equilibrium constraints that have been derived in the previous section. The complete formulation of the proactive bilevel TEP is therefore given by (7.14), resulting in a mathematical problem with equilibrium constraints [26] (MPEC).

$$\min_{l_{nn'}} \quad \sum_{n,n' \in \mathcal{LN}(n,n')} IT_{nn'} l_{nn'} + \sum_{g} I_g \overline{p}_g + \sum_{t,g} C_g p_{gt} \tag{7.14a}$$

$$\text{s.t.} \quad \text{Lower-Level Equilibrium: (7.7), (7.9), (7.11), (7.12)} \tag{7.14b}$$

The MPEC given in (7.14) can be solved as such using standard MPEC solvers that exploit the complementarity structure of the lower level. However, problem (7.14) is a nonconvex optimization problem due to the complementarity conditions in the lower level. Therefore, standard solvers will at most yield a local optimum, and in general not a global one, even though a global optimum is the desired outcome. Moreover, MPECs are by definition nonregular problems without any strictly feasible point, which makes nonlinear optimization solvers fail very frequently.

In order to achieve global optimality, we propose to convert this MPEC into an *equivalent* MIP, and we elaborate on how to do so in the following section.

7.4.3 Linearization of TEP MPEC

In this section, we transform the nonconvex MPEC (7.14) into an MIP. In this particular MPEC, almost all constraints are linear in primal and dual variables.[5] The only source of nonlinearity and nonconvexity are the complementarity conditions indicated in (7.7f), (7.7g), (7.9c), (7.11h), and (7.11i).

We now apply a linearization of these conditions a la Fortuny-Amat [13] by introducing binary variables and big M constants. Let us start with complementarity (7.7f), which corresponds to the lower bound on production p_{gt}. Let us repeat constraint (7.7f) here for demonstration purposes.

$$(7.7\text{f}) : p_{gt} \underline{\beta}_{gt} = 0 \quad \forall g, t.$$

This complementarity condition only has one objective: to ensure that not both of the variables p and $\underline{\beta}$ are strictly positive at the same time. Or in other words, at least one of the two variables must be zero.

Since this complementarity is defined $\forall g, t$, in order to linearize it, we need a binary variable that depends on the same indices, i.e., $\underline{b}^p_{gt}$. We furthermore require parameters $\underline{M}^p_{gt}$ and $\underline{M}^{\underline{\beta}}_{gt}$ that in layman's terms are referred to as big M constants. These parameters can be thought of as sufficiently large numbers. We will come back to what sufficiently large means in this context. We now introduce the two following mixed-integer constraints, which will replace the nonconvex (7.7f) as proposed by [13].

[5]This is not always the case. For example, if we considered a different objective function that maximized social welfare, the arising objective function would not be linear.

$$p_{gt} \leq \underline{b}^{p}_{gt} \underline{M}^{p}_{gt} \quad \forall g, t \tag{7.15a}$$

$$\underline{\beta}_{gt} \leq (1 - \underline{b}^{p}_{gt}) \underline{M}^{\beta}_{gt} \quad \forall g, t \tag{7.15b}$$

The set of constraints (7.15) achieves exactly what the original constraint (7.7f): not both variables p and $\underline{\beta}$ can take a nonzero value. How does it work? Well, per definition, a binary variable can either be zero or one. If the binary is zero, then constraint (7.15a) forces p to go to zero, and the complementarity constraint is met. On the other hand, if the binary is one, then (7.15b) forces $\underline{\beta}$ to be zero, which also satisfies the complementarity.

Let us now come back to the big M constraint. We have already said that if the binary goes to one, then $\underline{\beta}_{gt}$ will be forced to be zero as well, but what happens to constraint (7.15a) when the binary goes to one? It simply states that the variable p_{gt} is bounded from above by this large constant $\underline{M}^{p}_{gt}$. In order for this MIP transformation of the original nonconvex complementarity constraint to actually be *equivalent*, we have to make sure that the big M constant is never actually binding. Meaning that $\underline{M}^{p}_{gt}$ has to be sufficiently large to never actually impact the natural variable space of p_{gt}. If $\underline{M}^{p}_{gt}$ is picked too small, then it will change the optimal solution of p_{gt} leading to the fact that the corresponding solution of the MIP is not really the global optimum. On the other hand and from a more practical point of view, one cannot simply set this parameter as infinity because – even though theoretically, it would be correct – that would cause numerical issues when solving the problem. One therefore has to pick the smallest possible number that will never limit p_{gt}.

How to come up with these big M values in practise? When talking about a primal variable such as production, then it is relatively easy to come up with a natural upper bound. For example, the production of each generating unit will never be larger than the maximum system demand, i.e., $\sum_n D^0_{tn}$. What is important here is to find a reasonable number as upper bound that is not infinity. One does not have to find the lowest possible number there is, because that might take considerable effort to find as well. If you find a reasonably close number fast, then that is good enough. Now, how to find these big M parameters when we have to bound dual variables, such as $\underline{\beta}_{gt}$ in (7.15b)? Often, it is difficult to get an idea of the actual variable space. Some practical piece of advice is to run several cases of the equilibrium, as a complementarity problem, for example, and observe the values these variables take. Then pick the big M constant bigger than that (maybe even ten times as big). Finally, this is important when transforming equilibrium problems into MIPs. Once the MIP has solved, you have to check ex post whether any of these linearized complementarity constraints are binding. If so, then you have to increase the value of the constant. For more detail about the pitfalls of big M constants, the reader is referred to [34].

Below we include the remaining linearized complementarity conditions that correspond to (7.7g), (7.9c), (7.11h), and (7.11i).

$$\overline{p}_g \rho_{gt} - p_{gt} \le \overline{b}^{p}_{gt} \overline{M}^{p}_{gt} \quad \forall g, t \tag{7.16a}$$

$$\overline{\beta}_{gt} \le (1 - \overline{b}^{p}_{gt}) \overline{M}^{\overline{\beta}}_{gt} \quad \forall g, t \tag{7.16b}$$

$$d_{tn} \le \underline{b}^{d}_{tn} \underline{M}^{d}_{tn} \quad \forall g, t \tag{7.16c}$$

$$\underline{\alpha}_{tn} \le (1 - \underline{b}^{d}_{tn}) \underline{M}^{\underline{\alpha}}_{tn} \quad \forall t, n \tag{7.16d}$$

$$l_{nn'} + f_{tnn'} \le \underline{b}^{f}_{tnn'} \underline{M}^{f}_{tnn'} \quad \forall t, n, n' \in \mathcal{LN}(n, n') \tag{7.16e}$$

$$\underline{\gamma}_{tnn'} \le (1 - \underline{b}^{f}_{tnn'}) \underline{M}^{\underline{\gamma}}_{tnn'} \quad \forall t, n, n' \in \mathcal{LN}(n, n') \tag{7.16f}$$

$$l_{nn'} - f_{tnn'} \le \overline{b}^{f}_{tnn'} \overline{M}^{f}_{tnn'} \quad \forall t, n, n' \in \mathcal{LN}(n, n') \tag{7.16g}$$

$$\overline{\gamma}_{tnn'} \le (1 - \overline{b}^{f}_{tnn'}) \overline{M}^{\overline{\gamma}}_{tnn'} \quad \forall t, n, n' \in \mathcal{LN}(n, n') \tag{7.16h}$$

With these linearizations in place, the TEP MPEC (7.14) can be transformed into the following MIP given by (7.17).

Obj.function : (7.14a) (7.17a)

s.t. Derivatives of Lagrangian: (7.7a), (7.7b), (7.9a), (7.11a), (7.11b) (7.17b)

Primal feasibility: (7.7c), (7.7d), (7.9b), (7.11c), (7.11d), (7.11e), (7.11f) (7.17c)

Nonnegativity of multipliers: (7.7e), (7.9c), (7.11g) (7.17d)

Demand balance: (7.12) (7.17e)

Linearized complementarities: (7.15), (7.16) (7.17f)

7.5 Illustrative Case Study

Having analyzed both proactive and reactive TEP approaches in the literature review, we now present an illustrative example of proactive TEP and demonstrate how anticipating strategic market outcomes can drastically change the obtained network plan. We present an illustrative case study to show some insightful results of why bilevel modeling yields different results with respect to single-level models (e.g., centralized planning).

7.5.1 Centralized Planning Model

The centralized planning model (CP) is the traditional approach that TSOs follow in order to plan the network. In this framework, TSOs consider the following assumptions: perfect competition in the market; perfect information about all data including variables costs; inelastic demand; and, simultaneous TEP, GEP, and market decisions made by a single centralized agent.

Therefore, this model is represented by objective function (7.14a), primal feasibility conditions (7.17c), and demand balance (7.12). In the CP model, the variable d_{tn} representing demand is considered to be a parameter. Hence, the CP model assumes perfectly inelastic demand in (7.12).

7.5.2 Comparison Framework

The objective of this case study is to analyze what are the differences in cost, welfare, and profits of considering a Bilevel Proactive model (BP) rather than a centralized planning model (CP). In particular, we want to compare how strategic operation and investment decisions affect the transmission planning results compared to a perfect and non-anticipative competition case.

Therefore, we would like to quantify what is the cost of not considering that GENCOs are strategic players that maximize their own profits, rather than players that minimize system costs. To that purpose, we solve the BP model considering a Cournot oligopoly among GENCOs, and we compare the results to the CP model. It should be pointed out that, given that the BP model considers Cournot competition and elastic demand, while the CP model considers perfect competition and perfectly inelastic demand, some considerations must be made in order to compare the models properly. Therefore, we propose the following methodology:

First, we solve the BP model considering Cournot competition. Since the BP is a model considering elastic demand, we observe the demand that is obtained at equilibrium. Second, in the CP model, which uses inelastic demand, we fix the demand to what has previously been obtained by the BP, then we solve the CP model, which obtains some TEP and GEP that might differ from the ones previously obtained by the BP. We refer to this model as the *Naïve CP*; it is "naïve" because it accounts for investment and operation results considering that the market is perfect and disregarding the market feedback. Hence, TEP and GEP obtained by the Naïve CP might be erroneous given that they assume perfect competition, which is not always the case (in fact, these results are not an actual Cournot equilibrium). Third, in order to figure out what is the actual cost of disregarding the correct market feedback of the Naïve CP, we fix the TEP solution obtained by the Naïve CP and rerun the BP model (which boils down to solving only the lower level) in order to see how much the "wrong" TEP decision is going to distort the market equilibrium and GEP decisions and ultimately how this is going to affect system costs, welfare and profits. We refer to these results as the *Actual CP* as it accounts for the "actual" cost of the potentially erroneous planning decisions made by the CP model. Finally, the regret of using a CP approach is computed as the total cost of the Actual CP minus the total cost of the BP.

7.5.3 Data

In order to test the previous models and for the sake of illustration, we assume a three-node system, a 1-hour period, and a greenfield approach, i.e., no initial generating nor transmission capacity is considered. To that purpose, we consider existing demand and candidate generators at each node as well as all possible candidate lines. Please note that we consider that each generation unit belongs to a single GENCO. Finally, we consider the capacity factor $\rho_{gt} = 1$ for every GENCO and time t.

Table 7.2 shows the capital and variable costs of candidate generators on an hourly basis. As mentioned in [51] and [30], the capital costs of generators of different technologies goes from 3000 € to 10,000 € per (kW). Considering a 3% discount rate and 25 years horizon, it implies a 7% annual recovery cost that, in turn, implies an hourly recovery value ranging from 24 € to 80 € per (MW-hour).

Table 7.3 shows the candidate transmission line characteristics. As mentioned in [51], the capital cost of transmission lines depends on a variety of factors such as geographical conditions, environmental impact, and technology implemented. Therefore the cost of construction of a transmission line can vary between 746 $ and 3318 $ per (MW-km) for long distance and between 1491 $ and 6636 $ per (MW-km) for lower voltage transmission [51]. This information is in line with the data used in other GEPTEP papers [2, 15, 20, 27, 54] that actually consider values

Table 7.2 Generator and demand characteristics

	Generator characteristics		Demand	
Nodes	Capital cost €/(MW-hour)	Variable cost €/(MW-hour)	Intercept (MW)	Slope (MW^2/€)
1	20	35	90	1.0
2	17	30	70	0.8
3	25	25	80	1.2

Table 7.3 Line characteristics

Nodes	Reactance (p.u)	Capital cost (€/MW-hour MW)
1	0.10	2.8
2	0.15	2
3	0.12	1.5

between 1200 € and 8000 € per (MW-km). We consider values between 3500 € and 1900 € per (MW-kM). Additionally, we compute the recovery factor with a 3% discount factor and considering a 25-year horizon. This implies that the annual value is around 7% of the total cost. Therefore, on an hourly basis, the capital cost of a new line lies between 2.8 € and 1.52 € per (MW-hour).

7.5.4 Results

In this section, we analyze the results of a bilevel proactive TEP and a traditional single-level approach, as described in Sect. 7.5.2. Note that all the results are given for a single representative hour.

As shown in Table 7.4, at first glance, the total cost of the Naïve CP seems lower than the total cost of the BP, as well as the welfare of the Naïve CP is greater than the welfare of BP. Please keep in mind that these seemingly better results are fictitious as they – incorrectly – assume a perfectly competitive market, while in reality, the market is, in fact, a Cournot oligopoly. Hence, the welfare and total cost obtained by the Naïve CP are fictitious as they do not represent reality. In reality, when following a traditional centralized approach, a social planner would invest in 34.2 MW of TEP (instead of the optimal 12.1 MW obtained by the BP). Fixing this TEP amount and rerunning the lower level of the original bilevel model, i.e., referred to as the Actual CP, we observe that in reality, the traditional approach actually ends up leading to a different, more expensive solution.

Regarding capacity investments, the Naïve CP invests more in TEP (than the BP) and the same total amount in GEP but only in the cheapest generator at node

Table 7.4 Investment and cost results for one representative hour

	Node	BP	Naïve CP	Actual CP
Welfare (€)		1117.8	1191.7	1032.8
Total cost (€)		2210.5	2136.7	2297.8
TEP (MW)	1	6.6	20.1	20.1
	2	0.0	10.7	10.7
	3	5.5	3.4	3.4
Total TEP (MW)		12.1	34.2	34.2
GEP (MW)	1	11.5		15.3
	2	19.5	43.7	15.6
	3	12.8		12.8
Total GEP (MW)		43.7	43.7	43.7
Demand (MW)	1	23.5	23.5	19.6
	2	12.9	12.9	16.8
	3	7.3	7.3	7.3
Total demand (MW)		43.7	43.7	43.7

two. The GENCO investments under the BP is more distributed among different nodes. After having fixed the Naïve CP TEP investments and rerunning the BP, in Table 7.4, we observe that the Cournot equilibrium forces investments to be more equally distributed within the network than what would happen under perfect competition. Given that GENCOs are not perfectly competitive, they would actually react strategically to the TEP obtained by the Naïve CP. Therefore, if TEP is planned according to the Naïve CP but there is actually a strategic competition, then GENCOs would be able to exercise even more market power compared to the BP. In fact, in the Actual CP, GENCOs actually invest in more expensive capacity at each node (as seen in Table 7.4), even though the total demand is the same. This leads to a higher total cost of 2297.8 €, which in turn leads to an hourly regret of 87.3 €, equivalent to a yearly regret of 0.7 M€, of not considering Cournot competition in the market. In other words, the suboptimal TEP plan arising from the traditional CP planning paradigm actually leads to system costs that are 4% higher than what could have been achieved when employing a bilevel planning model. The negative impact of CP planning on social welfare is even higher and amounts to almost 8%.

The regret of the Actual CP, mentioned above, is given mainly by the higher GEP and TEP costs, while the operational profits (computed as revenues minus operation cost minus investment costs) are actually lower, as seen in Table 7.5. In the Actual CP, the total GENCOs' profits are 674.8 €, while in the BP the profits account for 741.5 €. This decrease in the profits of the Actual CP is mainly due to the lower prices coming from the spare capacity in the transmission lines. This spare capacity is a consequence of fixing the TEP from Naïve CP case, which leads to much lower usage of the lines, which, in turn, results in lower nodal prices.

Additionally, it is interesting to note that the total profits in the Naïve CP are actually zero. This is a consequence of the fact that there is only one GENCO producing in the market, which means that there is no room for windfall profits for other GENCOs. In general, these zero profits (economic efficiency) are a direct consequence of the perfect competitive assumption of the Naïve CP, which implies that the GENCO should only get the necessary revenues to pay back their investments and variable costs. It is important to note that the Actual CP model

Table 7.5 Power flows, prices and profits for one representative hour

	Node	BP	Naïve CP	Actual CP
Flows (MW)	1→2	-6.6	-20.1	-0.9
	2→3	0	10,7	-2.1
	3→1	5.5	3.4	3.4
Prices (€/MW)	1	66.5	55.0	70.4
	2	71.4	47.0	66.5
	3	60.6	35.0	60.6
Profits	1	131.7	0.0	235.6
	2	474.6	0.0	303.2
	3	135.2	0.0	136.0
Total profits		741.5	0.0	674.8

results in the worse off situation overall, given that it achieves the lowest welfare among the three models analyzed. Thus, this case study demonstrates that system planning that considers market anticipation and strategic behavior leads to a better-off long-term result.

7.6 Conclusions

This chapter provides a general introduction of bilevel programming applications in transmission expansion planning. After a brief, introductory discussion on the basics of bilevel programming, we present a thorough literature review on proactive and reactive transmission expansion planning, on available solution methods for bilevel programs, and we point out the specific challenges of bilevel optimization in transmission expansion planning such as handling the nonconvexity of the full AC OPF formulation; incorporating storage technologies and intertemporal constraints, which are closely linked to the representation of time in and the computational complexity of power system models; incorporating UC constraints involving binary variables in lower levels of bilevel problems; developing computational methods to solve bilevel problems more efficiently; and extending them to include stochasticity.

In order to demonstrate the important insights that can be gained from bilevel problems, we present an application of a social transmission expansion planner that takes into account the market feedback of strategic generation companies. We formulate the arising MPEC of such a social planner and show how to linearize it. In an illustrative example, we compare the proactive bilevel transmission expansion approach to the ones of a centralized planner using a single-level optimization problem. We identify that disregarding market feedback from strategic agents can lead to suboptimal and inefficient expansion planning with higher total system cost and lower social welfare. This simple example perfectly illustrates that bilevel models are extremely useful to capture trends and decisions made by strategic agents in energy and electricity markets.

Acknowledgments This work was supported by Project Grants ENE2016-79517-R and ENE2017-83775-P, awarded by the Spanish Ministerio de Economia y Competitividad.

References

1. J. F. Bard, *Practical Bilevel Optimization: Algorithms and Applications* (Springer, 1998)
2. L. Baringo, A. Baringo, A stochastic adaptive robust optimization approach for the generation and transmission expansion planning. IEEE Trans. Power Syst. **33**(1), 792–802 (2018)
3. L. Baringo, A.J. Conejo, Transmission and wind power investment. IEEE Trans. Power Syst. **27**(2), 885–893 (2012)
4. C.A. Berry, B.F. Hobbs, W.A. Meroney, R.P. O'Neill, W.R. Stewart, Understanding how market power can arise in network competition: a game theoretic approach. Util. Policy **8**(3), 139–158 (1999)

5. H.I. Calvete, C. Galé, P.M. Mateo, A new approach for solving linear bilevel problems using genetic algorithms. Eur. J. Oper. Res. **188**(1), 14–28 (2008)
6. A. Caprara, M. Carvalho, A. Lodi, G.J. Woeginger, A complexity and approximability study of the bilevel knapsack problem, in *Lecture Notes in Computer Science (Including Subseries Lecture Notes in Artificial Intelligence and Lecture Notes in Bioinformatics).* LNCS, vol. 7801 (Springer, Berlin/Heidelberg, 2013), pp. 98–109
7. A. Caprara, M. Carvalho, A. Lodi, G.J. Woeginger, Bilevel Knapsack with interdiction constraints. INFORMS J. Comput. **28**(2), 319–333 (2016)
8. J.B. Cardell, C.C. Hitt, W.W. Hogan, Market power and strategic interaction in electricity networks. Resour. Energy Econ. **19**, 109–137 (1997)
9. E. Centeno, S. Wogrin, A. López-Pena, M. Vázquez, Analysis of investments in generation capacity: a bilevel approach. IET Gener. Transm. Distrib. **5**(8), 842 (2011)
10. Federal Energy Regulatory Commission et al. Order no. 1000 – transmission planning and cost allocation (2012)
11. S. Dempe, Annotated bibliography on bilevel programming and mathematical programs with equilibrium constraints. Optimization **52**(3), 333–359 (2003)
12. Y. Dvorkin, R. Fernández-Blanco, Y. Wang, B. Xu, D.S. Kirschen, H. Pandžić, J.P. Watson, C.A. Silva-Monroy, Co-planning of investments in transmission and merchant energy storage. IEEE Trans. Power Syst. **33**(1), 245–256 (2018)
13. J. Fortuny-Amat, B. McCarl, A representation and economic interpretation of a two-level programming problem. J. Oper. Res. Soc. **32**(9), 783–792 (1981)
14. S.A. Gabriel, A.J. Conejo, J.D. Fuller, B.F. Hobbs, C. Ruiz, *Complementarity Modeling in Energy Markets*, vol. 180 (Springer Science & Business Media, 2012). Berlin, Germany
15. L.P. Garces, A.J. Conejo, R. Garcia-Bertrand, R. Romero, A bilevel approach to transmission expansion planning within a market environment. IEEE Trans. Power Syst. **24**(3), 1513–1522 (2009)
16. I.C. Gonzalez-Romero, S. Wogrin, T. Gomez, A review on generation and transmission expansion co-planning under a market environment (Accepted). *IET GTD* (2019)
17. I.C. Gonzalez-Romero, S. Wogrin, T. Gomez, Proactive transmission expansion planning with storage considerations. Energy Strateg. Rev. **24**, 154–165 (2019)
18. B.F. Hobbs, C.B. Metzler, J.-S. Pang, Calculating equilibria in imperfectly competitive power markets: An MPEC approach. IEEE Trans. Power Syst. **15**, 638–645 (2000)
19. D. Huppmann, S. Siddiqui, An exact solution method for binary equilibrium problems with compensation and the power market uplift problem. Eur. J. Oper. Res. **266**(2), 622–638 (2018)
20. M. Jenabi, S.M.T. Fatemi Ghomi, Y. Smeers, Bi-level game approaches for coordination of generation and transmission expansion planning within a market environment. IEEE Trans. Power Syst. **28**(3), 2639–2650 (2013)
21. S. Jin, S.M. Ryan, Capacity expansion in the integrated supply network for an electricity market. IEEE Trans. Power Syst. **26**(4), 2275–2284 (2011)
22. S. Jin, S.M. Ryan, A tri-level model of centralized transmission and decentralized generation expansion planning for an electricity market – part I. IEEE Trans. Power Syst. **29**(1), 132–141 (2014)
23. V. Krebs, L. Schewe, M. Schmidt, Uniqueness and multiplicity of market equilibria on dc power flow networks. Eur. J. Oper. Res. **271**(1), 165–178 (2018)
24. H. Le Cadre, I. Mezghani, A. Papavasiliou, A game-theoretic analysis of transmission-distribution system operator coordination. Eur. J. Oper. Res. **274**(1), 317–339 (2019)
25. H. Li, L. Fang, An evolutionary algorithm for solving bilevel programming problems using duality conditions. Math. Prob. Eng. **2012**, 1–14 (2012)
26. Z.-Q. Luo, J.-S. Pang, D. Ralph, *Mathematical Programs with Equilibrium Constraints* (Cambridge University Press, Cambridge, 1996)
27. L. Maurovich-Horvat, T.K. Boomsma, A.S. Siddiqui, Transmission and wind investment in a deregulated electricity industry. IEEE Trans. Power Syst. **30**(3), 1633–1643 (2015)
28. E. Moiseeva, S. Wogrin, M.R. Hesamzadeh, Generation flexibility in ramp rates: strategic behavior and lessons for electricity market design. Eur. J. Oper. Res. **261**(2), 755–771 (2017)

29. A. Motamedi, H. Zareipour, M.O. Buygi, W.D. Rosehart, A transmission planning framework considering future generation expansions in electricity markets. IEEE Trans. Power Syst. **25**(4), 1987–1995 (2010)
30. Office of Energy Analysis, Annual energy outlook 2019 with projections to 2050. Technical report, U.S. Energy Information Administration (2019)
31. B. Palmintier, M. Webster, Impact of unit commitment constraints on generation expansion planning with renewables, in *2011 IEEE Power and Energy Society General Meeting* (2011), pp. 1–7
32. H. Pandžić, Y. Dvorkin, M. Carrión, Investments in merchant energy storage: trading-off between energy and reserve markets. Appl. Energy **230**, 277–286 (2018)
33. S. Pineda, H. Bylling, J.M. Morales, Efficiently solving linear bilevel programming problems using off-the-shelf optimization software. Optim. Eng. **19**(1), 187–211 (2018)
34. S. Pineda, J.M. Morales, Solving linear bilevel problems using big-ms: Not all that glitters is gold. IEEE Trans. Power Syst. **34**(3), 2469–2471 (2019)
35. P. Pisciella, M. Bertocchi, M.T. Vespucci, A leader-followers model of power transmission capacity expansion in a market driven environment. Comput. Manag. Sci. **10**, 87–118 (2014)
36. D. Pozo, J. Contreras, E. Sauma, If you build it, he will come: anticipative power transmission planning. Energy Econ. **36**, 135–146 (2013)
37. D. Pozo, E. Sauma, J. Contreras, A three-level static MILP model for generation and transmission expansion planning. IEEE Trans. Power Syst. **28**(1), 202–210 (2013)
38. D. Pozo, E. Sauma, J. Contreras, When doing nothing may be the best investment action: pessimistic anticipative power transmission planning. Appl. Energy **200**, 383–398 (2017)
39. D. Ralph, S.J. Wright, Some properties of regularization and penalization schemes for MPECs. Optim. Methods Softw. **19**(5), 527–556 (2004)
40. A. Ramos, M. Ventosa, M. Rivier, Modeling competition in electric energy markets by equilibrium constraints. Util. Policy **7**, 233–242 (1999)
41. J.H. Roh, M. Shahidehpour, Y. Fu, Market-based coordination of transmission and generation capacity planning. IEEE Trans. Power Syst. **22**(4), 1406–1419 (2007)
42. E.E. Sauma, S.S. Oren, Proactive planning and valuation of transmission investments in restructured electricity markets. J. Regul. Econ. **30**(3), 358–387 (2006)
43. E.E. Sauma, S.S. Oren, Economic criteria for planning transmission investment in restructured electricity markets. IEEE Trans. Power Syst. **22**(4), 1394–1405 (2007)
44. S. Scholtes, Convergence properties of a regularization scheme for mathematical programs with complementarity constraints. SIAM J. Optim. **11**(4), 918–936 (2001)
45. C. Shi, J. Lu, G. Zhang, H. Zhou, An extended branch and bound algorithm for linear bilevel programming. Appl. Math. Comput. **180**(2), 529–537 (2006)
46. S. Siddiqui, S.A. Gabriel, An SOS1-based approach for solving MPECs with a natural gas market application. Netw. Spat. Econ. **13**(2), 205–227 (2012)
47. E. Spyrou, J.L. Ho, B.F. Hobbs, R.M. Johnson, J.D. McCalley, What are the benefits of co-optimizing transmission and generation investment? Eastern interconnection case study. IEEE Trans. Power Syst. **32**(6), 4265–4277 (2017)
48. S.S. Taheri, J. Kazempour, S. Seyedshenava, Transmission expansion in an oligopoly considering generation investment equilibrium. Energy Econ. **64**, 55–62 (2017)
49. D.A. Tejada-Arango, M. Domeshek, S. Wogrin, E. Centeno, Enhanced representative days and system states modeling for energy storage investment analysis. IEEE Trans. Power Syst. **33**(6), 6534–6544 (2018)
50. Y. Tohidi, M.R. Hesamzadeh, F. Regairaz, Sequential coordination of transmission expansion planning with strategic generation investments. IEEE Trans. Power Syst. **32**(4), 2521–2534 (2017)
51. K. Vaillancourt, Electricity transmission and distribution. Technical report, IEA ETSAP (2014)
52. H. Von Stackelberg, *Marktform und Gleichgewicht* (J. Springer, 1934). Berlin, Germany
53. J.D. Weber, T.J. Overbye, A two-level optimization problem for analysis of market bidding strategies, in *IEEE Power Engineering Society Summer Meeting*, Edmonton (1999), pp. 682–687

54. M. Weibelzahl, A. Märtz, Optimal storage and transmission investments in a bilevel electricity market model. Ann. Oper. Res. **287**(2), 911–940 (2020)
55. S. Wogrin, E. Centeno, J. Barquín, Generation capacity expansion in liberalized electricity markets: a stochastic MPEC approach. IEEE Trans. Power Syst. **26**(4), 2526–2532 (2011)
56. S. Wogrin, E. Centeno, J. Barquin, Generation capacity expansion analysis: open loop approximation of closed loop equilibria. IEEE Trans. Power Syst. **28**(3), 3362–3371 (2013)
57. S. Wogrin, D. Galbally, J. Reneses, Optimizing storage operations in medium- and long-term power system models. IEEE Trans. Power Syst. **31**(4), 3129–3138 (2016)
58. S. Wogrin, B.F. Hobbs, D. Ralph, E. Centeno, J. Barquín, Open versus closed loop capacity equilibria in electricity markets under perfect and oligopolistic competition. Math. Program. **140**(2), 295–322 (2013)
59. J. Yao, I. Adler, S.S. Oren, Modeling and computing two-settlement oligopolistic equilibrium in a congested electricity network. Oper. Res. **56**(1), 34–47 (2008)
60. Y. Ye, D. Papadaskalopoulos, J. Kazempour, G. Strbac, Incorporating non-convex operating characteristics into bi-level optimization electricity market models. IEEE Trans. Power Syst. **35**(1), 163–176 (2020)
61. M.H. Zare, J.S. Borrero, B. Zeng, O.A. Prokopyev, A note on linearized reformulations for a class of bilevel linear integer problems. Ann. Oper. Res. **272**(1–2), 99–117 (2019)

Chapter 8
The Impact of Distributed Energy Resources on the Networks

Carlos Mateo, Fernando Postigo, and Álvaro Sánchez-Miralles

8.1 Introduction

The power systems supply electricity to the final consumers through the transmission and distribution networks. The transmission network covers vast areas to distribute the electricity of the generation plants, which can be located very far away from the consumers. Instead, the distribution network is fed from the transmission substations and is used to distribute the energy locally. In distribution, typically three voltage levels can be distinguished. High voltage network (also called subtransmission) uses nominal voltages in the range of about 36–200 kV and is used for higher distances within distribution. The high voltage networks connect the primary substations and the high voltage consumers to the transmission substations. The medium voltage network with nominal voltages in the range of about 1–36 kV is used for lower distances and connects the distribution transformers and medium voltage consumers to the primary substations. Finally, the low voltage network is used for the final connection between the distribution transformers and the low voltage consumers.

This traditional approach is being altered nowadays with the deployment of distributed energy resources (DER), mainly distributed generation (DG), storage, electric vehicles (EV), and demand response. These distributed energy resources are installed typically in the distribution networks. Distributed renewable generation and electric vehicles can help to reduce the environmental impact of the power system. Storage and demand response can help to soften the constraint of the generation having to be at any time equal to the consumption (plus energy losses). Moreover, the costs of DG, electric vehicles, and storage have been drastically

C. Mateo (✉) · F. Postigo · Á. Sánchez-Miralles
Comillas Pontifical University, Madrid, Spain
e-mail: Carlos.Mateo@iit.comillas.edu

S. Lumbreras et al. (eds.), *Transmission Expansion Planning: The Network Challenges of the Energy Transition*, https://doi.org/10.1007/978-3-030-49428-5_8

reduced in the latest years, making them more affordable. There is however still room for improvement. Electric vehicles will require a massive deployment of EV charging stations. Storage, although with a sharp reduction in prices, is still too expensive for certain applications and applied mainly on pilot projects or under certain specific conditions (e.g., on islands).

The incipient paradigm change caused by the penetration of DERs in the system is producing an abandonment of the traditional conception of providing electricity services. Historically, the power flows have been unidirectional, circulating from large power plants to the transmission system, from there, the electricity was delivered through the distribution system to the end customers. Commonly this system has been called a "top-down" approach [1].

However, the DERs are mostly located in the distribution system. This new conception of the problem involves a migration from the "top-down" to "bottom-up" approach. In this paradigm, the power flows from distribution feeders to the transmission system, and, in order to balance generation and demand, a reduction in traditional synchronous generation must be carried out, mainly under light load circumstances.

The new paradigm shift derived from the deployment of DER implies that the system is evolving from large generation plants located far away from the consumers to a more local balance of demand and generation, with small distributed photovoltaic (PV) units or wind generation, located in the vicinity of the consumers. As these DER are mainly deployed in the distribution network, they are expected to initially have a higher impact on these networks, as these phenomena will start being mainly local. However, as penetration levels of DER are increased, they could also have an impact on the transmission network. In some cases, conflicts may appear. For example, storage or demand response programs could require different (or even contradictory) actions in transmission, distribution, and in the electricity market. This brings the necessity of coordinating transmission and distribution and designing solutions to optimize the whole power system.

8.2 Impact in the Operation and Planning of the Networks

The two major tasks of system operators are related to operating and planning the networks. These tasks are significantly impacted by the increasing levels of DG penetration, requiring specific measures in order to achieve efficient integration.

8.2.1 Operation

The operation of the system includes several actions. On one side, in relation to the steady state of power flows and technical constraints, congestion management and voltage control are necessary. Regarding contingencies, specific actions are required to ensure reliability targets. For this purpose, in distribution networks, an adequate operation of maintenance crews and the installation of remote-controlled switches can help to improve reliability. In transmission, network is designed with N-1 criteria, ensuring that it can be operated to avoid interruptions in case of single faults; this is if two faults do not occur simultaneously.

8.2.1.1 Congestion Management and Voltage Control

Regarding congestion management, DG modifies the power flows in the system. The unidirectional power flows that are typical in the case of conventional networks with demand only scenarios turn to be bidirectional in the case of DG penetration. This challenges the operation [2]. It is necessary to consider that not only high levels of demand can create congestions, but also DG generation injections can also result in congestions, with power flowing upward.

The bidirectional nature of power flows in the presence of DG also has implications in terms of the voltage profile. With conventional demand only scenarios, the voltage profiles used to be quite straightforward, voltage decreasing when nodes get further from the transformers. The only exception used to be taps in substations, distribution transformers, and voltage regulators. In demand only scenarios transformer taps can be used to increase voltage, normally between voltage levels, except in the case of voltage regulators that help to control voltage within the medium voltage networks.

DG has the opposite effect of demand, in this case voltage increasing along the feeder. In this scenario, voltage profiles can be significantly more complex to analyze and forecast, since the dynamics of voltage along the feeder depends on the direction of power flows and therefore on the local balance between generation and demand. For example, regarding voltage regulators, the adjustment of taps in demand only scenarios is easier to design, with voltage increasing in the transformer if the upper limit is not overpassed. This increases the margin for the voltage decrease along the feeder. In the case of DG, more sophisticated strategies are required, since increasing voltage can be counterproductive if DG is driven power flows upstream. New technical solutions, like storage and demand response, can help to address the technical challenges, but they also increase the complexity of the operation, since it is necessary to determine how their flexibility is optimally managed.

8.2.1.2 Contingencies

In order to improve reliability, preventive and corrective maintenance must be carried out. Preventive maintenance anticipates to the failure of components (mainly power lines and transformers). Corrective maintenance instead takes remedial actions, once a failure occurs. In distribution, control centers supervise and direct maintenance crews in order to minimize the impact of faults. This impact is typically assessed in terms of reliability targets like SAIDI and SAIFI, which measure the duration and frequency of interruptions, respectively. Maintenance crews are optimally managed, in order to reduce the duration of faults affecting consumers, isolating the fault, and restoring the service as quickly as possible. For this purpose, remote-controlled switches can also help to accelerate the operations and reduce the duration of interruptions. In transmission, N-1 criteria are followed, ensuring that the system can be operated in the case of any single component of the network failing.

DG also affects the works of maintenance crews. For example, one aspect that is critical for the safety of maintenance crews is the disconnected state of power lines. When a power line is being repaired, one of the first steps is ensuring that the line is disconnected, verifying that there is no power flow though the power line. In the conventional case of large generation plants connected only in transmission, the operation of the upper switch allowed to ensure the disconnected state of power lines, although this was also verified in the field in order to prevent risks. With DG connected at the distribution level, the operation of the upstream switches is not enough to ensure that there is no power flow through the lines. It is necessary also that local DG is also disconnected. This results in the necessity of grid codes to ensure that DG disconnects when there is a fault in the system. This requires DG to identify the occurrence of faults in order to command disconnection automatically. However, a global verification of the disconnection state is not straightforward, having to rely on communications or in local control management. Even though the system is designed to avoid islanding, cases have been reported of DG failing to identify the fault and creating local islands with DG supplying local demand till the local balance is disrupted and automatic protections trigger the control elements, disconnecting DG.

Besides the safe operation of maintenance crews, DG also affects the precision of the calculation of the fault location, which is necessary to ensure the reliable operation of the power system. Two methods are mainly used for this purpose, which are based on (1) impedance measurement and (2) travelling waves generated by short-circuit faults [3]. Methods based on impedance measurement analyze voltage and current before the fault. The direction and the amplitude of the fault current are affected by the penetration of DG. In particular, synchronous DG injects more current during the fault than inverted-based DG and can affect more the estimation of the fault location using impedance measurements.

8.2.2 *Planning*

Planners of the networks aim at finding the most cost-efficient solution to connect consumers and distributed generators, subject to technical and reliability constraints. The planner has to locate and size new substations and feeders as well as determine new size of components that require reinforcements [4]. For guaranteeing adequate levels of reliability, system operators can also install interconnection switches and loops in the networks. In the new scenarios with active networks and with the integration of DERs, these new elements also must be considered. The possible solutions depend up to some extent on the regulation in each country. For example, in some countries, system operators can be also involved in planning DG, while in other countries, there must be a clear separation of activities, and system operators cannot locate and size new DG because this would interfere with the generation activities. Demand response and electric vehicle charging may also require controlling dynamically the demand, avoiding traditional reinforcements by improving the operation of the users and components involved. Several types of solutions can be provided, ranging from coordinated solutions to more distributed approaches or even solutions in the forms of more advanced and cost-reflective tariffs. Storage can also be a powerful tool to increase the flexibility of the system, which will become more widespread as its price continues decreasing. The location and size of voltage control devices, like capacitors and voltage regulators, are also more complex in active networks.

The planners of the networks have to take into account economic, technical, and environmental aspects [4]. If formulating mathematically the problem, some of these aspects, like reliability, can be part of the objective function or, instead, can be considered as a constraint in terms of a minimum level of reliability. The economic factors involve the minimization of investment, corrective and preventive maintenance, and energy losses costs, aiming to maximize the net profit value of the company. Guaranteeing adequate levels of energy losses could be modelled as a constraint instead, depending on the regulation of each country. Technical aspects involve designing robust networks to prevent congestions, maintain voltage profiles within the specified limits, and limit the short-circuit power of the networks. Finally, environmental objectives aim at reducing carbon emissions, and for this aspect, the integration of renewable DG can play a major role.

8.2.3 Crosscutting Topics

DER integration and network digitalization mark a turning point in the asset management strategy; an adequate use of this chance opens an infinity of research lines whose objective is to extract the maximum benefit from whole the system. However, the power system is still increasing its complexity. Electricity networks are becoming a set of integrated systems, which combine two main roles, power transmission, and communications. A reliable and secure communications channel allows optimizing the exploitation of the network and the connected assets. In this section, two interconnected crosscutting topics are going to be introduced, network resilience and cybersecurity.

8.2.3.1 Resiliency

The resilience of the network is one of the most affected factors by this new paradigm. It can be defined as the capability of a system to rapidly overcome adversities that may compromise the proper functioning of the network. These adversities can vary from been simple faults to highly destructive natural disasters such as hurricanes.

One of the most relevant events in the last few years is the Italian blackout that happened in September 2003 [5]. This event involved the loss of supply of 56 million people for 12 h and the cancellation of all of the flights in the country. Several studies analyzed this incident from the electrotechnical and complex networks point of view [6], concluding that it was produced by a cascade effect triggered by a trip of a Swiss line due to a tree flashover.

Moreover, several hurricanes have hit the US territories. On the one hand, the Hurricane Maria caused in 2017 that one-third of the residential customers in Puerto Rico lose the electricity supply during 5 months, and it was not completely restored until 11 months later [7]. On the other hand, 99% of the basic services including electricity supply were restored in less than 10 days when Hurricane Irma hit the US mainland [8]. Looking at this comparison, the necessity of hardening network resilience in those areas like geographic islands where the logistics and the availability of supplies is limited becomes of utmost important [9].

Additionally, the resilience of the electrical network is very dependent on the volt-age level. For instance, the transmission network is meshed, and commonly each node can be supplied at least from two different points. However, the distribution network is mostly radial (especially in rural networks), and therefore, an incident can lead to the loss of supply of all consumers located downstream of it. In recent years, a large number of studies have analyzed how the integration of DERs and the increase in the degree of automation of the distribution network opens a wide range of possibilities that would allow improving the resilience of the system.

As above mentioned, commonly an incidence in the distribution network leaves all the consumers located downstream of the contingency without service, until restoration actions are undertaken. Nevertheless, with DER integration, the operation of the distribution network changes radically. In future advanced configurations, faults could be isolated, and non-affected areas could be connected in electrical islands (micro-grids) supplied from DERs, while the contingency is corrected.

8.2.3.2 Cybersecurity

As this chapter remarks, the digitalization of the electricity system has brought with it a wide variety of new possibilities such as optimized operational strategies or novel maintenance approaches. Nevertheless, from the early 1990s, the digitalization of the system has led to new potential hazards which may threaten the resilience of the whole system [10]. This is the case of cyberattacks [11].

According to [12] the energy sector has suffered more than 50% of the listed cyberattacks, being the USA the preferred target for cyberattackers receiving 73% of the registered cybercrimes. Germany, Ukraine, the USA, Romania, Italy, and Poland are among the top 15 countries where cyberattacks originate [13]. Over the last decade, several cyberattacks have been detected in the energy sector [14]. One of the most relevant events occurred on December 23, 2015, in Ukraine. Based on the provided information, a foreign attacker remotely controlled the SCADA management system of three distribution companies, causing that 225,000 consumers lose their service in a period from 1 to 6 h. This kind of threat is extremely costly; thus, the entire system has to be safe and prepared to block any attack.

This is the reason why worldwide recognized institutions like the Pacific Northwest National Laboratory (PNNL) are proposing novel approaches like Blockchain [15] to overcome these new threats by creating safe environments for electricity companies and customers.

Prevention should be the fundamental objective to avoid cyberattacks; hence, authentication and authorization mechanisms play a fundamental role in obtaining a secure and reliable communications channel [16]. However, it is of utmost important to develop predefined performance protocols in case of contingency in order to guarantee the security of the system.

8.3 Impact Assessment

8.3.1 Network Model

In this chapter, in order to analyze the impact of DERs in the distribution networks, we use Reference Network Models (RNM) [17]. These models include algorithms to plan distribution networks. There are two main versions: greenfield and brownfield. The greenfield RNM can be used to model an initial network, given the location and demand of the consumers. It also has applications to design synthetic or representative networks [18] and supporting the regulators in the calculation of the efficient cost incurred by distribution companies. Instead, the brownfield RNM takes as input the initial network and determines the reinforcements required to accommodate additional demand or additional DERs. Figure 8.1 illustrates a Spanish urban network that has been built using the greenfield RNM to quantify the impact of DERs in this chapter.

Tables 8.1, 8.2, 8.3, 8.4 and 8.5 show the main parameters of this network that supplies 83,425 consumers, with 361 km of power lines, 681 distribution transformers, and 2 substations. The costs of the network are based on public reference unitary costs of the regulator in Spain [19].

Despite the network is large-scale, the model is very detailed, as Fig. 8.2 shows, superimposing the street map in the area to the low voltage and medium voltage feeders and the distribution transformers. An automatically generated street map is used to build the network, making it more realistic [17].

Fig. 8.1 Spanish urban network model: (**a**) street map and (**b**) distribution network

Table 8.1 Consumers

Consumers	Installed power (MW)	Number
Low voltage	387.7	83,425
Medium voltage	32.3	75

Table 8.2 Power lines

Power lines	Overhead (km)	Underground (km)	Investment cost (€)
Low voltage	139.06	63.89	5,984,849
Medium voltage	1.12	148.79	16,474,827
High voltage	6.86	1.81	5,425,903

Table 8.3 Distribution transformers

Capacity (kVA)	Number	Investment cost (€)
400	295	12,325,985.00
250	325	12,472,200.00
100	61	1,748,260.00

Table 8.4 Substations

Capacity (MVA)	Number	Investment cost (€)
120	1	13,318,169
80	1	10,827,482

Table 8.5 Trenches, façades, and posts

	Length (km)				Cost (€)			
Voltage level	Façade	Post	Trench	Total	Façade	Post	Trench	Total
Low voltage	111.55	1.9	37.89	151.34	167,331	19,007	2,273,346	2,459,684
Medium voltage	0	74.56	43.26	117.82	0	1,118,395	4,326,232	5,444,627
High voltage	0	0	0	0	0	0	0	0

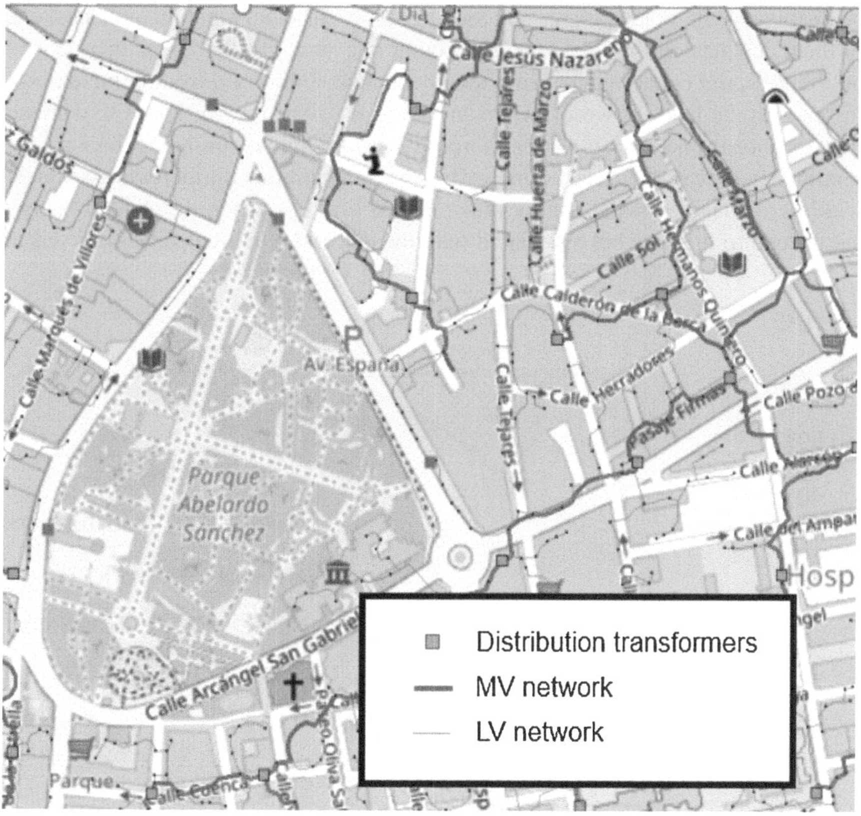

Fig. 8.2 Spanish urban distribution network model, zoom in

8.3.2 *Traditional Approach*

In the traditional approach, the consumers were connected to the distribution system, and they were supplied from large generation plants directly connected to the transmission. Under this approach, the drivers to expand the networks and reinforce them were mainly new consumers or demand growth. During many years demand steadily grew, increasing the power flow in power lines and bringing the necessity of additional reinforcements.

A sensitivity to demand growth in the distribution network described in Sect. 8.3.1 is shown in Fig. 8.3. For the LV network, MV/LV distribution transformers and MV network, there is a steady increase in the need of reinforcements. In these three components, the higher increments are expected in the distribution transformers, being the increase in low voltage network the lowest. In HV network and HV/MV substations, due to the size of the network, the components are discrete, and the trends show a sharp increase when such discrete elements are reinforced.

In the case of energy losses, in the distribution networks, two main types must be distinguished: no load losses and load losses. As its denomination indicates, one of them depends on the load, while the other does not. For the losses that depend on the load, the equations that rule the phenomenon show a quadratic increase of losses with the load. However, when analyzing the sensitivity of losses to demand growth when the network is also reinforced (as in the simulations in this chapter), the reinforcements can decrease the energy losses, compensating the increase associated with the demand growth. This is illustrated in Fig. 8.4, where in general energy losses tend to increase but associated with the discrete reinforcements observed in

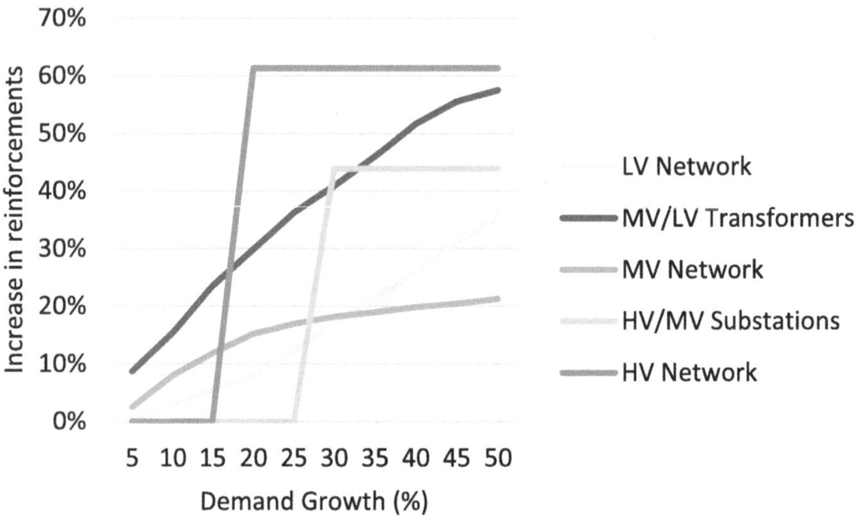

Fig. 8.3 Spanish urban distribution network model, reinforcements associated with different levels of demand growth

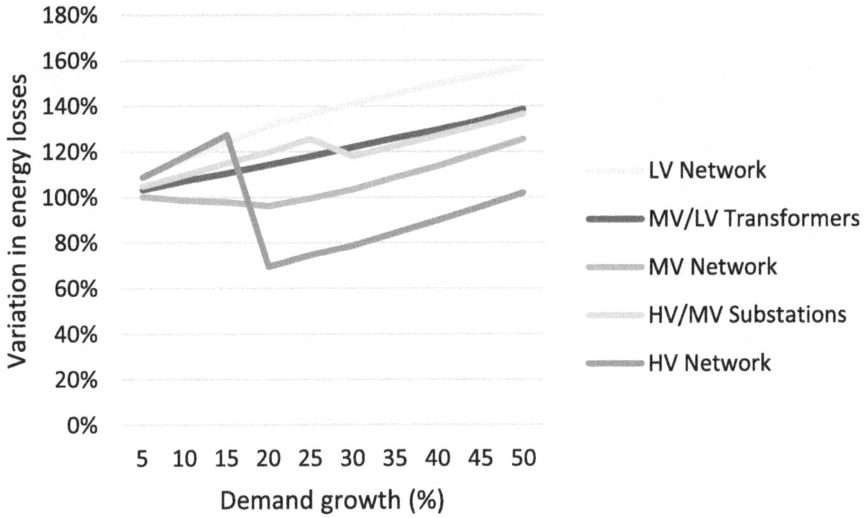

Fig. 8.4 Variation in energy losses for different levels of demand growth, after reinforcing the network

HV/MV substations and HV network in Fig. 8.3, there is also a decrease in losses in Fig. 8.4. In the LV network, MV/LV transformers and MV network energy losses increase, but less than quadratically due to the reinforcements required every time, the demand is increased.

8.3.3 *Impact of Distributed Energy Resources*

8.3.3.1 Distributed Generation

The impact that DER will have on the power system depends mainly on the penetration level. Small penetration levels of DG can be beneficial for the system, compensating the local demand and reducing power flows along with the network. However, as DG penetration levels increase, the local generation can even overpass the local demand, which can create reverse power flows. As initially the system was not designed to connect this DG, the power system must be adapted to withstand these reverse power flows, updating the protection devices accordingly. If the penetration levels of DG are increased further, the power flows can be even greater than the original ones in a demand only scenario. This results in a U-shape for energy losses that is illustrated in Fig. 8.5 [20], shown as percentage respect the base scenario (no DG).

As the networks are sized for the peak scenario, this implies that the size of transformers and power lines can limit the maximum amount of DG penetration, unless reinforcement actions are undertaken. For new installations, there is a paradigm

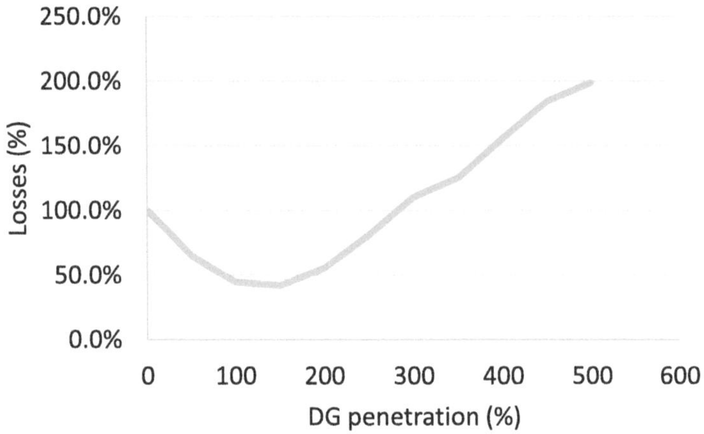

Fig. 8.5 Sensitivity of energy losses to distributed generation

shift when sizing equipment. With the traditional approach, equipment only had to be sized for the peak demand. Nowadays, with the increase penetration of DG, two scenarios should be considered: net peak demand and net peak generation. In the net peak demand scenario, the demand is maximum, and the generation is minimum. In the net peak generation scenario, the generation is maximum, and the demand is minimum. These two scenarios define the maximum power flow of network components, and under this new paradigm, both have to be taken into account for sizing network components.

Several alternatives exist to mitigate the impact of distributed generation in the networks. The most straightforward approach would be to curtail generation during certain peak hours. This, however, implies wasting renewable generation, which is only justified under certain conditions [21]. Flexibility, in the form of storage or demand response, can also be valuable to efficiently integrate it. In the case of storage, there were, for example, incentives in Germany, with the so-called KfW Bankengruppe program, to incentivize the installation of storage in PV installations, with the requirement of reducing 70% of the peak generation. In the case of demand response, it must be taken into account that for demand to compensate large DG levels, demand would have to be increased during certain periods instead of decreased. This is also viable but opposed to the typical approach with demand response programs and energy efficiency devices, which are typically used to decrease demand and facilitate the integration of higher demand levels.

8.3.3.2 Electric Vehicles

Meanwhile, electric vehicles are becoming more popular and are called to replace combustion engines in the long term. In the case of electric vehicles, very low penetrations of them might already be withstood by the existing networks. However, the demand of an electric vehicle can be equal or even higher than that of a household. This implies that for medium penetrations of electric vehicles, the required reinforcements can be very significant because demand (and thus power flows) can increase drastically [22]. This stresses the need to integrate efficiently the new resources in the networks to reduce their impact on the networks. To achieve this objective, several opportunities are appearing. For example, in the case of electric vehicles, smart charging approaches can be implemented reducing drastically the need for reinforcements [23]. Smart charging implies that instead of everyone charging the electric vehicle in the evening when they arrive home, the charging can be deferred to make use of demand valleys while ensuring that the vehicles are fully charged when the users need them. This idea can be implemented in different ways, from complex coordinated charging approaches to more distributed solutions, in which the users can decide to postpone the charging if appropriate incentives are put in place and automation devices are provided to ease differing the charging of the EVs.

Figure 8.6 compares three main types of charging strategies and their corresponding daily profiles, in a subset of the network. With peak charging the electric vehicles are charged in the evening right when the users come back home. With

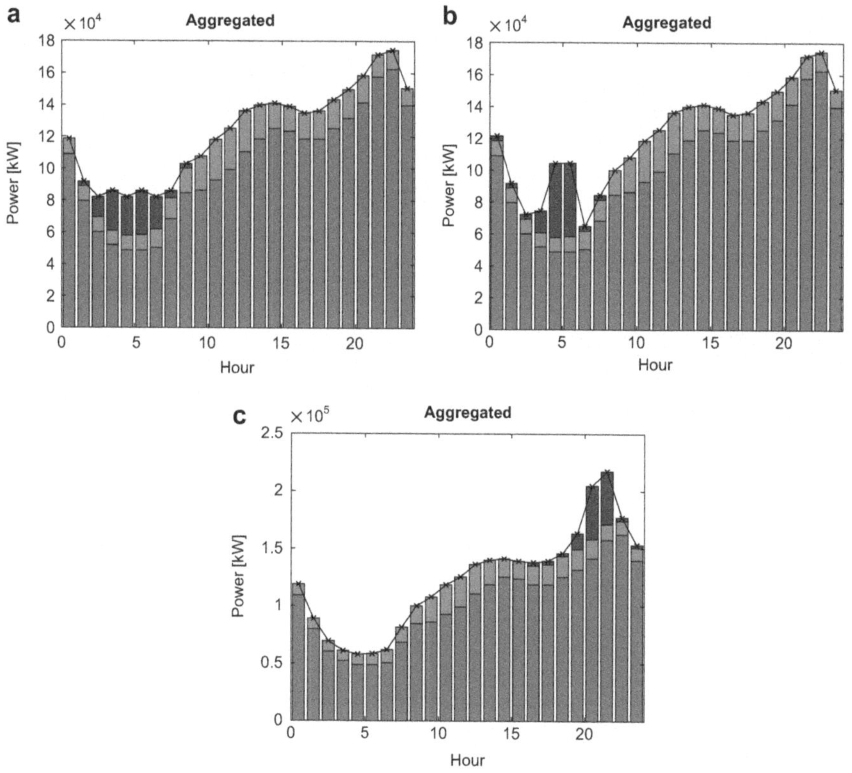

Fig. 8.6 Expected demand with electric vehicles: (**a**) smart charging, (**b**) valley charging, and (**c**) peak charging

valley charging, time-of-use tariffs can be defined, incentivizing consumers to delay the charging in order to use the valley of the electricity prices. With smart charging, the vehicles are coordinated in order to optimally fill in the valley.

Figure 8.7 compares the impact that these different strategies have in the need of distribution network reinforcements, assessed using a brownfield RNM. As it can be observed in this figure, peak charging can require a lot of reinforcements, which van be avoided under valley charging or smart charging paradigms. This is aligned with the fact that the distribution networks are sized mainly for peak demand. This requirement implies that it is critical to avoid electric vehicles charging during the peak. However, an optimally filled in valley is not necessary. This facilitates the implementation, being possible to obtain this solution with time-of-use tariffs, and technically requiring only a timer in the plug-in to let the user set the hour in which it is necessary to have the car fully charged. Then the car can decide when to start charging, considering only the constraint of having to be charged at the hour set by the user. This solution would not require a coordinated charging of all the electric vehicles, which would otherwise be more costly to implement. This analysis implies that valley charging strategies can achieve similar savings in the distribution network than ("smart") coordinated controls while reducing significantly the cost of implementation.

As shown in Fig. 8.7, there can be a high necessity of reinforcements with peak charging, which can be effectively deferred or avoided with the application of smart charging or valley charging strategies.

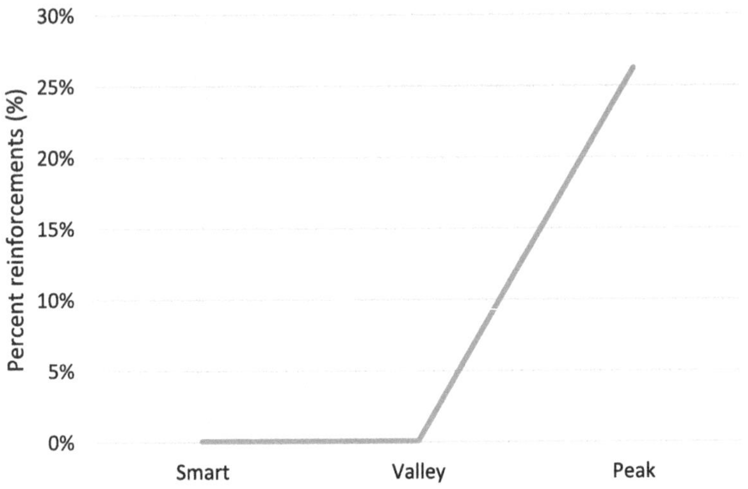

Fig. 8.7 Impact of different strategies of electric vehicle charging in the need for reinforcements in the distribution networks

8.4 Conclusions

Distributed energy resources are disrupting the power system, bringing opportunities, and posing new technical challenges. Renewable distributed generation can be of great value to comply with environmental targets, but it poses new challenges related to the reverse power flows and the uncontrollability of its generation. Storage and demand response can increase the flexibility of the system to accommodate more DG or allow for demand growth. Electric vehicles are also key to electrify the transport sector and make it more sustainable. However large penetrations of electric vehicles should be integrated efficiently into the distribution networks to prevent a drastic increase of the system costs. DERs are mainly located in the distribution network, requiring a migration from the "top-down" approach to a "bottom-up" approach. This makes more complex the operation and planning of the networks, among other reasons, due to the reverse power flows and the additional complexity to control voltage and operate storage or, in general, manage flexibility. In this chapter, the implications have been first discussed and then illustrated carrying out simulations using applying Reference Network Models in an example case study.

In the traditional approach, simulations show that distribution transformers are the first components requiring major reinforcements to accommodate demand growth. HV/MV substations and the HV network have step increments, implying a significant increase of the costs when demand growth exceeds certain thresholds (15% and 25% in the case study) requiring specific components to be reinforced. Energy losses trend is to increase, except when reinforcements take place. However, with the new paradigm of DG, losses have a U-shape decreasing for low DG penetration levels and increasing again for higher DG penetration. In the case of electric vehicles, it turns critical to avoid charging the EVs during the peak of the system, which would require huge reinforcements (up to 25% in the case study). This makes necessary to shift the charging of EVs to valley hours. However, simulations show that optimally filling in the valley is not critical as long as the system peak is not overpassed. This facilitates the implementation, being sufficient to incentivize valley charging with time-of-use tariffs and avoiding the need for sophisticated centralized control strategies.

Bibliography

1. I. Pérez Arriaga, C. Knittel et al, *Utility of the Future. An MIT Energy Initiative Response*, Massachusetts Institute of Technology, MA, USA. p. 382 (2016)
2. J.C. Do Prado, W. Qiao, L. Qu, J.R. Agüero, The next-generation retail electricity market in the context of distributed energy resources: Vision and integrating framework. Energies **12**, 24 (2019)
3. E. Ebrahimi, A.J. Ghanizadeh, M. Rahmatian, G.B. Gharehpetian, Impact of distributed generation on fault locating methods in distribution networks. Renew. Energy Power Qual. J. **1**(10), 1195–1199 (2012)

4. R. Li, W. Wang, Z. Chen, J. Jiang, W. Zhang, A review of optimal planning active distribution system: Models, methods, and future researches. Energies **10**(11), 1715 (2017)
5. UCTE, Final report of the Investigation Committee on the 28 September 2003 Blackout in Italy. Igarss 2014(1), 128 (2004)
6. S.V. Buldyrev, R. Parshani, G. Paul, H.E. Stanley, S. Havlin, Catastrophic cascade of failures in interdependent networks. Nature **464**(7291), 1025–1028 (2010)
7. J. Campisi, J. White, *Finally, 11 Months after Maria, Power Is Restored in Puerto Rico – Except for 25 Customers*, CNN (2018)
8. D. Harris, *Orlando Sentinel – Power Restored to More than 99% of Floridians after Hurricane Irma* (2017)
9. J.A. Daw, S.R. Stout, *Building Island Resilience through the Energy, Water, Food Nexus*, NREL/TP-7A40-74747, 1569216, NREL, USA. (2019). https://doi.org/10.2172/1569216
10. R. Schainker, J. Douglas, T. Kropp, Electric utility responses to grid security issues. IEEE Power Energ. Mag. **4**(2), 30–37 (2006)
11. E. I. Community, Cyber Resilience in the Electricity Ecosystem: Principles and Guidance for Boards In collaboration with Boston Consulting Group. *World Economic Forum*, no. January, 2019
12. A.O. Otuoze, M.W. Mustafa, R.M. Larik, Smart grids security challenges: Classification by sources of threats. J. Electr. Syst. Inf. Technol. **5**(3), 468–483 (2018)
13. Go-Gulf, *Cyber Crime Statistics and Trends*. NREL, USA (2013)
14. S. Livingston, S. Sanborn, A. Slaughter, P. Zonneveld, *Managing Cyber Risk in the Electric Power Sector*. Deloitte, USA (2018)
15. M. Mylrea, S.N.G. Gourisetti, Blockchain for smart grid resilience: Exchanging distributed energy at speed, scale and security, in *Proceedings of the 2017 Resilience Week, RWS 2017*, (2017), pp. 18–23
16. Y. Yan, Y. Qian, H. Sharif, D. Tipper, A survey on cyber security for smart grid communications. IEEE Commun. Surv. Tutorials **14**(4), 998–1010 (2012)
17. C. Mateo, T. Gómez, Á. Sánchez-Miralles, J.P. Peco, A. Candela, A reference network model for large-scale distribution planning with automatic street map generation. IEEE Trans. Power Syst. **26**(1), 190–197 (2011)
18. C. Mateo et al., European representative electricity distribution networks. Int. J. Electr. Power Energy Syst. **99**(January), 273–280 (2018)
19. Ministerio, Orden IET/2660/2015, de 11 de diciembre, por la que se aprueban las instalaciones tipo y los valores unitarios de referencia de inversión, de operación y mantenimiento por elemento de inmovilizado y los valores unitarios de retribución de otras tareas reg (2015)
20. L. González-Sotres, C. Mateo, T. Gómez, J. Reneses, M. Rivier, Á. Sánchez-Miralles, Assessing the impact of distributed generation on energy losses using reference network models, in *Cigré International Symposium. The Electric Power System of the Future. Integrating Microgrids and Supergrids*. Cigré, Italy (2011)
21. C. Mateo, P. Frías, R. Cossent, P. Sonvilla, B. Barth, Overcoming the barriers that hamper a large-scale integration of solar photovoltaic power generation in European distribution grids. Sol. Energy **153**, 574–583 (2017)
22. L. Pieltain Fernandez, T.G.S. Roman, R. Cossent, C.M. Domingo, P. Frias, Assessment of the impact of plug-in electric vehicles on distribution networks. Power Syst. IEEE Trans. **26**(1), 206–213 (2011)
23. C. Mateo, P. Frías, A. Sánchez-Miralles, Distribution planning with hourly profiles for analysing electric vehicle charging strategies. Int. J. Electr. Hybrid. Veh. **8**(1), 1 (2016)

Chapter 9
Stability Considerations for Transmission System Planning

Francisco M. Echavarren and Lukas Sigrist

9.1 Introduction

Transmission and distribution networks oversee transporting and delivering electrical energy from generation sites to final customers. Networks must be economic, adequate in terms of capacity, and reliable. Network expansion planning aims at developing the network as economically as possible while maintaining acceptable reliability levels in order to transport and deliver energy according to forecasted load and generation amounts and sites. Such developments might include reinforcement alternatives such as new AC lines, HVDC lines, and FACTS devices, for instance to accommodate power productions of large-scale PV plants, and the evaluation of their benefits. Currently, network expansion is typically bounded by deterministic criteria (N-1 and N-X) in order to guarantee adequacy and security. However, probabilistic planning might complement deterministic criteria.

Technical planning of network and generation expansion is correlated as seen from the previous example on accommodation of PV generation. Security assessment is a power system-wide assessment since it involves generation, loads, and networks. This assessment includes steady-state operation under normal and abnormal operating conditions as well as dynamic responses of the power system. The dynamic response of the power system depends on the dynamics of load, network and generation devices, and their interaction through the network. The dynamic behavior of the power system can limit the network's ability to transmit power.

The chapter is structured as follows. Section 9.2 contains a general description of power system stability phenomena including definitions, classification, and

F. M. Echavarren (✉) · L. Sigrist
Comillas Pontifical University, Madrid, Spain
e-mail: Francisco.Echavarren@iit.comillas.edu; Lukas.Sigrist@iit.comillas.edu

S. Lumbreras et al. (eds.), *Transmission Expansion Planning: The Network Challenges of the Energy Transition*, https://doi.org/10.1007/978-3-030-49428-5_9

illustrations of fundamental stability concepts using elementary power system configurations. Section 9.3 provides guidelines for stability studies during planning stage. Finally, Sect. 9.4 presents some study cases where the influence of transmission network configuration and topology among system stability is illustrated.

9.2 Power System Stability

A power system needs to be reliable and secure. Reliability is defined as the probability of the satisfactory operation of a power system, whereas security denotes the degree of risk in the ability of a power system to survive imminent disturbances [17]. Security relates to the robustness to imminent disturbances. Reliability requires evaluations over very long-time horizons, whereas time horizons related to security are relatively short.

Stability refers to the continuance of intact operation following a disturbance. Congruently, stability is related to both reliability and security. To be reliable, the power system must be secure most of the time. To be secure, the system must be stable but must also be secure against other contingencies. Formally, power system stability denotes the ability of an electric power system, for a given initial operating condition, to remain in a state of operating equilibrium under normal operating conditions and to regain a state of operating equilibrium after being subjected to a disturbance. Stability is a condition of equilibrium between opposing forces, and instability results when a disturbance leads to a sustained imbalance between the opposing forces.

A power system is a complex, highly non-linear system that operates in a constantly changing environment, where all the parameters and variables are constantly changing. Most of them undergo a continuous evolution through time, such as demand, renewable production, voltages, or frequency. Occasionally, some of them may register discrete changes, such as opening/closing branches and connecting/disconnecting devices or network faults. When subjected to a disturbance, the system must be able to operate satisfactorily under these conditions, whether the disturbance is small or large. Following a transient disturbance, the power system is stable if it reaches a new equilibrium state with all variables within reasonable limits. The actions of automatic controls and possibly human operators will eventually restore the system to normal state. On the other hand, if the system is unstable, it will result in a run-away or run-down situation with all the system gradually collapsing and leading to a blackout. In fact, a fault on a critical element followed by its isolation by protective relays will cause perturbations in power flows, bus voltages, and machine rotor speeds, among other magnitudes. Consequently, voltage variations will trigger voltage regulators, the generator speed variations will activate prime mover governors, and voltage and frequency perturbations will vary the system loads. Moreover, devices used to protect individual equipment may respond to variations in system variables and thereby affect the power system performance. Even though stability problem is a single problem, analysis in that way becomes an unattainable task. "Divide and rule" strategy is the most effective way to address the stability problem. Selection of the key variables in the problem, the best modeling for the devices implied, and adequate methodology require a proper classification into categories. Figure 9.1 shows a classification of the power system stability problem.

Classification of Fig. 9.1 is based on the physical nature of the instability, the size of the disturbance, and the devices, processes, and time span that must be taken into consideration. A disturbance is said to be large if the differential equations governing the system dynamics cannot be linearized. Relevant dynamics in short-term stability are those of synchronous generators and their primary controllers (voltage and frequency). Relevant dynamics in long-term stability are those of energy sources (boilers, reactors, etc.) and the secondary voltage and frequency controllers.

The different stability phenomena limit transfer capacity of networks. Steady-state loadability of a transmission line is bounded by its thermal limit. The thermal limit is usually the limiting factor with respect to the transfer capacity of short transmission lines. For medium and long lines, the power transfer is usually limited by voltage and angle stability considerations. Figure 9.2 qualitatively depicts a set of maximum power transfers according to the length of the line and the kind of limit considered.

Figure 9.2 shows how the thermal limit is independent from the line length, whereas the stability limits become shorter as the line length grows. In the case of short line, thermal limit is longer than stability limits, but for medium and long lines, stability limits are more restrictive than the thermal limit. Next subsections describe the main stability clusters identified in Fig. 9.1, i.e., angle stability, frequency stability, and voltage stability.

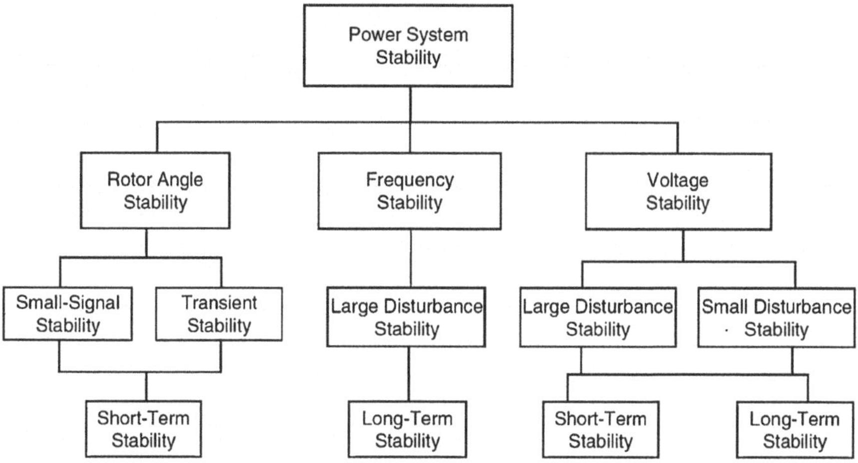

Fig. 9.1 Classification of power system stability [17]

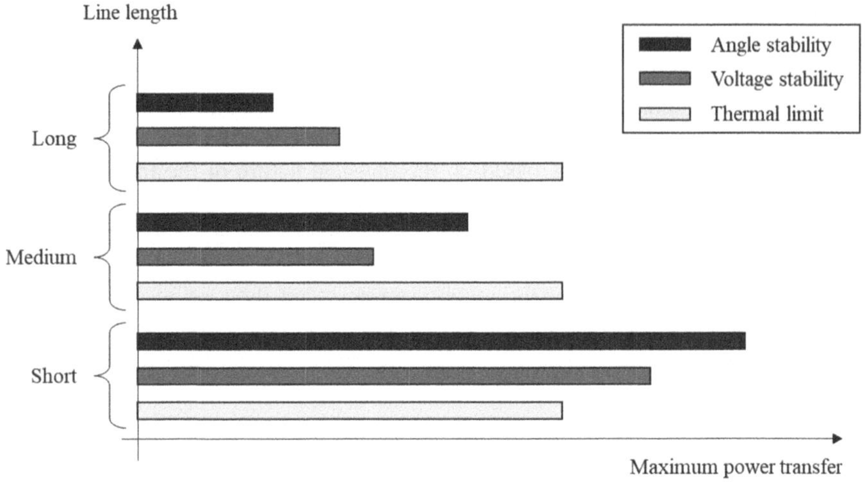

Fig. 9.2 Maximum power transfer for different criteria and line lengths

9.2.1 *Angle Stability*

In AC power systems, one of the key variables is the frequency. It is set by the angular speed of the synchronous generating units installed among the network. Operation of an AC power system requires that every AC device connected to the network rotates at the same speed, i.e., the frequency at every bus in the network must be equal to a nominal frequency. This nominal frequency is 50 Hz in most parts of the world and 60 Hz in the North, Center, and parts of the South of the Americas, and in few other countries in Western Asia.

When all the systems work at the nominal frequency, it is said that the system works in synchronism. Rotor angle stability is concerned with the ability of interconnected synchronous machines of a power system to remain in synchronism under normal operating conditions and after being subjected to a disturbance. It therefore depends on the ability of each synchronous machine in the system to sustain or to restore the equilibrium between electromagnetic torque and mechanical torque. Instability that may result occurs in the form of increasing angular swings of some generators, leading to their loss of synchronism with other generators.

The rotor angle stability problem involves the study of electromechanical oscillations inherent in power systems. To tackle the problem, it is important to understand the different forces acting and how they affect frequency. Each synchronous generator in the network consists of an inertial mass spinning at a nominal rotatory speed (frequency) under steady-state conditions, i.e., when generation and load are balanced. When this balance is broken, the rotatory inertial mass gains or loses rotatory speed depending on the sense of the unbalance. An excess of generation or a lack of demand will accelerate the system; thus, the frequency will rise. On the other hand, a lack of generation or an excess of demand will decelerate the system; thus, the frequency will fall. The immediate effect of having different frequencies among the network is that angular differences between buses become uncontrollable, so do the power flows through branches. The power-angle relationship is highly non-linear and, beyond a certain limit, an increase in angular separation is followed by a decrease in power transfer. Consequently, the angular separations get increased and lead to instability. For any given situation, the stability of the system depends on whether the deviations in angular positions of the rotors result in enough restoring torque or not. It should be noted that loss of synchronism can occur between one machine and the rest of the system, or between groups of machines, possibly with synchronism maintained within each group after separating from each other.

Synchronous operation of interconnected synchronous machines is analogous to a set of cars, one for each generating unit, speeding around a circular track [16]. All these cars are also joined to each other by rubber bands that represent the transmission lines. If all the cars run at the same speed, the rubber bands remain intact. But as soon as forces are applied to one or more of the cars, they will run at different speeds temporally. Different speeds imply that the rubber bands will stretch, decelerating the faster cars and accelerating the slower ones. Depending on some many factors (initial condictions, network topology, control and relying systems...), this unbalance can be temporally and eventually ending up in a new equilibrium point with all the cars running at the same speed again, or it can lead to a total blackout. In worst case scenarios, if one or more rubber bands reach their tensile strength and break, some of the cars will pull away, increasing the chances of instability.

9.2.1.1 Small-Signal Stability

Small-signal stability is concerned with the ability of the power system to maintain synchronism under small disturbances [17]. The disturbances are considered sufficiently small that linearization of system equations is permissible for purposes of analysis. Such disturbances are continually encountered in normal system operation, such as small changes in load.

Small-signal stability depends on the initial operating state of the system. Instability that may result can be of two forms: (i) increase in rotor angle through a non-oscillatory or aperiodic mode due to lack of synchronizing torque, or (ii) rotor oscillations of increasing amplitude due to lack of enough damping torque. Small-signal stability is assessed through modal analysis.

In today's practical power systems, small-signal stability is largely a problem of insufficient damping of oscillations [16]. The time frame of interest in small-signal stability studies is in the order of 10–20 s following a disturbance. The stability of the following types of oscillations is of concern:

- Local modes or machine-system modes, associated with the swinging of units at a generating station with respect to the rest of the power system. The term "local" is used because the oscillations are localized at one station or a small part of the power system.

- Inter-area modes, associated with the swinging of many machines in one part of the system against machines in other parts. They are caused by two or more groups of closely coupled machines that are interconnected by weak ties.
- Control modes, associated with generating units and other controls. Poorly tuned exciters, speed governors, HVDC converters, and static VAR compensators are the usual causes of instability of these modes.
- Torsional modes, associated with the turbine-generator shaft system rotational components. Instability of torsional modes may be caused by interaction with excitation controls, speed governors, HVDC controls, and series-capacitor-compensated lines.

To illustrate small-signal stability, the case of a generator connected to an infinite bus through a step-up transformer and a transmission line will be considered. Figure 9.3 shows the single line diagram and the equivalent circuit. This example illustrates a local oscillation mode.

In its simplest representation, the dynamics of the synchronous generator can be approximated by the dynamics of the rotor, which are described by the motion equation of a rotating rigid body:

$$2H\frac{d\omega}{dt} = p_m - p_e - D\left(\omega - 1\right) \tag{9.1}$$

where p_m, p_e, and ω are the mechanical power, the electrical power, and the angular speed, respectively. H and D represent the inertia and the damping coefficient of the synchronous generator, respectively. The electrical power delivered by the generator is

$$p_e = \frac{e' u_\infty}{x_e} \text{sen}\delta \tag{9.2}$$

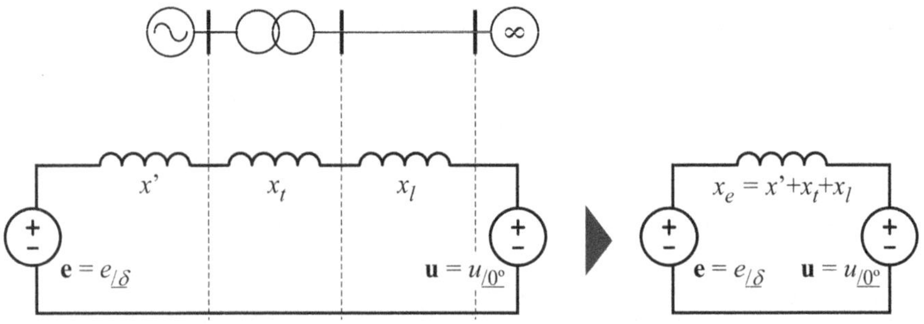

Fig. 9.3 Single line diagram and equivalent circuit of the synchronous generator connected to an infinite bus through a step-up transformer and a transmission line

where x_e is the equivalent reactance seen by the synchronous generator. This equivalent reactance is the sum of the transient reactance, the transformer reactance, and the line reactance. In this simple representation, the generator is represented by a voltage source with constant voltage magnitude. The connection of the mechanical and the electrical models of the synchronous generator is through the rotor angle a:

$$\omega = 1 + \frac{1}{\omega_0}\frac{d\delta}{dt} \tag{9.3}$$

Combining the three previous equations yields the following representation in matrix form:

$$\begin{bmatrix} \dot{\omega} \\ \dot{\delta} \end{bmatrix} = \begin{bmatrix} \frac{1}{2H} p_m - \frac{1}{2H}\frac{e'u}{x_e}\text{sen}\delta - \frac{D}{2H}(\omega - 1) \\ \omega_0 (\omega - 1) \end{bmatrix} \tag{9.4}$$

In case of small disturbances, the non-linear differential equations of the synchronous generator can be linearized around the operating point and characterize satisfactorily the dynamic behavior of the generator. Linearizing yields the following:

$$2H\frac{d\Delta\omega}{dt} = \Delta p_m - \Delta p_e - D\Delta\omega \tag{9.5}$$

$$\Delta p_e = \frac{e'_0 u_0}{x_e}\cos\delta_0 \Delta\delta = K\Delta\delta \tag{9.6}$$

$$\Delta\omega = \frac{1}{\omega_0}\frac{d\Delta\delta}{dt} \tag{9.7}$$

These equations can be written either as system of first order linear differential equations in matrix form:

$$\begin{bmatrix} \frac{d\Delta\delta}{dt} \\ \frac{d\Delta\omega}{dt} \end{bmatrix} = \begin{bmatrix} 0 & \omega_0 \\ \frac{K}{2H} & \frac{-D}{2H} \end{bmatrix} \begin{bmatrix} \Delta\delta \\ \Delta\omega \end{bmatrix} + \begin{bmatrix} 0 \\ \frac{1}{2H} \end{bmatrix} \Delta p_m \tag{9.8}$$

or as a second order differential equation:

$$\frac{2H}{\omega_0}\frac{d^2\Delta\delta}{dt^2} + \frac{D}{\omega_0}\frac{d\Delta\delta}{dt} + K\Delta\delta = \Delta p_m \tag{9.9}$$

Depending on the roots, the time response of a second order differential equation may or may not be oscillatory, with oscillations increasing (unstable) or decreasing (stable) in magnitude. The characteristic equation is as follows:

$$s^2 + \frac{D}{2H}s + \frac{K\omega_0}{2H} = 0 \tag{9.10}$$

The roots (eigenvalues) of the characteristic equation are

$$s_{12} = \frac{\frac{-D}{2H} \pm j\sqrt{4\frac{K\omega_0}{2H} - \left(\frac{D}{2H}\right)^2}}{2} \tag{9.11}$$

If $D = 0$ and $K > 0$ (the rotor angle is between 0 and 90°) the roots of the characteristic equation are pure complex numbers:

$$s_{12} = \pm j\sqrt{\frac{K\omega_0}{2H}} \tag{9.12}$$

It means that the generator response is oscillatory. The natural frequency of the oscillation is proportional to the inverse of both the equivalent reactance since $K = \left(e_0' u_0/x_e\right)\cos\delta_0$ and the rotor inertia. Higher equivalent reactance and higher inertia mean lower natural frequency. Whereas the simplified model is appropriate to illustrate the problem and even gives adequate values for natural frequencies, it is too oversimplified to reproduce the damping behavior since it neglects field circuits, the excitation system, and the damping windings. Figure 9.4 shows the impact of increasing line length through increasing xe on the electromechanical mode and its damping when considering a detailed generating unit model without power system stabilizer. Longer lines lead to less damped oscillations.

Now, the impact of a FACTS device is shown, again for the simplified model. The FACTS device considered is a thyristor-controlled switched capacitor (TCSC). Control of a TCSC for oscillation damping consists of modulating the series reactance using the speed deviation as input signal, i.e., the equivalent reactance x_e is now variable and controllable [25]. Figure 9.5 shows how series reactance modulation can be achieved with TCSC.

If damping coefficient of the synchronous generator D is assumed to be zero (no damping), linearizing the equations around the operating point by assuming series reactance variation results in the following:

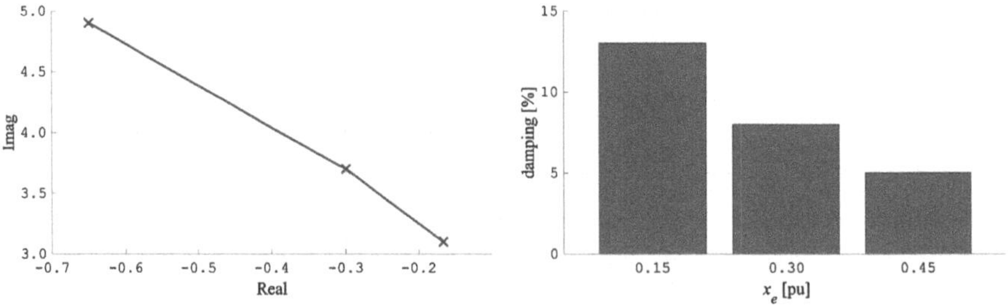

Fig. 9.4 Impact of increasing line length on the damping of the electromechanical mode

Fig. 9.5 Single line diagram of a synchronous generator connected to an infinite bus through a step-up transformer and a transmission line with a TCSC

$$2H\frac{d\Delta\omega}{dt} = \Delta p_m - \Delta p_e \tag{9.13}$$

$$\Delta p_e = \frac{e_0' u_0}{x_{e0}} \cos\delta_0 \Delta\delta - \frac{e_0' u_0}{x_{e0e}^2} \sin\delta_0 \Delta x_e = K\Delta\delta - K_x \Delta x_e \tag{9.14}$$

$$\Delta\omega = \frac{1}{\omega_0}\frac{d\Delta\delta}{dt} \tag{9.15}$$

If the series reactance is modulated using the speed deviation as input:

$$\Delta x_e = -K_{x,\omega}\Delta\omega \tag{9.16}$$

the second order differential equations that describe the dynamic behavior results are:

$$\frac{2H}{\omega_0}\frac{d^2\Delta\delta}{dt^2} + \frac{K_x K_{x,\omega}}{\omega_0}\frac{d\Delta\delta}{dt} + K\Delta\delta = \Delta p_m \tag{9.17}$$

This equation shows that TCSC control provides an equivalent damping $D_x = K_x \cdot K_{x,\infty}$ to the electromechanical oscillation. In other words, an adequate tuning of the controller's static gain, i.e., $K_{x,\infty}$, guarantees always the imposed damping. The impact of other FACTS devices is shown for example in references [24, 27].

9.2.1.2 Large Disturbance Stability

Large disturbance rotor angle stability, or transient stability as it is commonly referred to, is concerned with the ability of the power system to maintain synchronism when subjected to a severe transient disturbance [17]. The resulting system response involves large excursions of generator rotor angles and is influenced by the non-linear power-angle relationship. Solid three-phase short circuits put transient stability to the test.

Transient stability depends on both the initial operating state of the system and the severity of the disturbance. Transient stability is very much affected by the loading of the system, the distribution of the power flows, the network topology, the inertia, and the excitation system performances. Usually, the disturbance as well

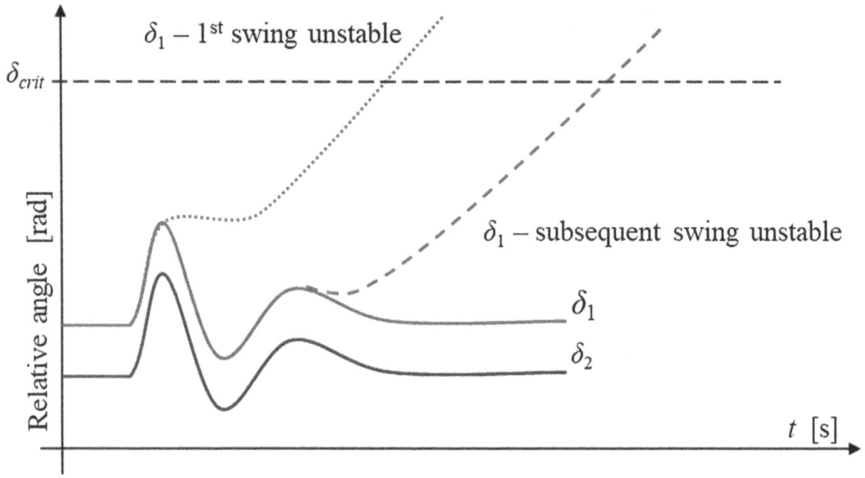

Fig. 9.6 Transient instability due to loss of synchronism

as the actuation of protection devices alter the system such that post-disturbance steady-state operation will be different from that prior to the disturbance. For example, a solid three-phase short circuit of a line is cleared by opening the breakers at its terminals, changing the topology of the grid. During faults, generating units accelerate, but this acceleration is not uniform and depends on the electrical distance of the generators to the fault, their inertia, controls, etc. Instability appears in the form of aperiodically diverging angles due to insufficient synchronizing torque and is referred to as first swing stability. In large power systems, transient instability may not always occur as first swing instability associated with a single mode; it could be as a result of increased peak deviation caused by superposition of several modes of oscillation causing large excursions of rotor angle beyond the first swing. Figure 9.6 illustrates transient instability due to the loss of synchronism.

The time frame of interest in transient stability studies is usually limited to 3–5 s following the disturbance. It may extend to 10 s for very large systems with dominant inter-area swings.

Transient stability is typically assessed through time-domain simulations. One way to quantify transient stability is by computing the critical clearing times [16]. The critical clearing time corresponds to the maximum duration of the fault before a generating unit loses synchronism. If the fault clearing time extends the associated critical clearing time, the acceleration of the rotor cannot be absorbed and the generating unit loses synchronism (i.e., it will not rotate at a constant, synchronous speed anymore). The critical clearing time further has a very practical use, since it determines the actuation of the protection devices, i.e., the time window within which they must act. Instantaneous line protections and the breaker act within about 100 ms. Second zone protections due to breaker failure clear the fault within 200 and 400 ms. A critical clearing time below the protection settings might affect the

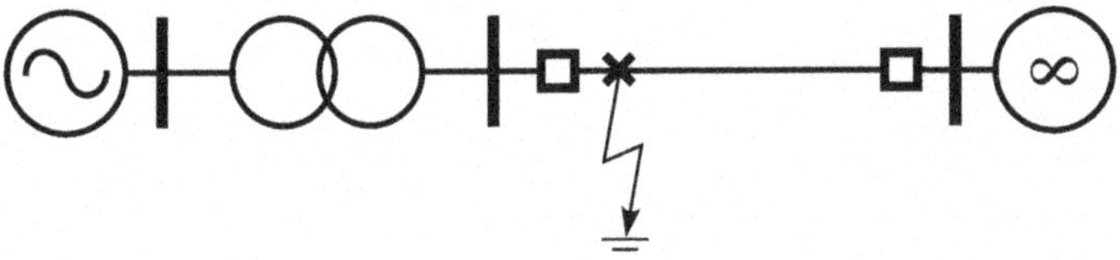

Fig. 9.7 Single line diagram of the synchronous generator connected to an infinite bus through a step-up transformer and a transmission line

design of the grid and require special protection schemes or operation measures (e.g., modify the dispatch).

To illustrate transient stability, the case of a generator connected to an infinite bus through a step-up transformer and a transmission line will be considered. The critical clearing time of a solid three-phase short circuit close to the HV bus of the transformer will be computed analytically. Figure 9.7 reproduces Fig. 9.3 by highlighting the location of the fault and the breakers. The case of multi-machine systems could be analyzed through Lyapunov stability theory, but time-domain simulations are still more practical.

The generator is again represented by a voltage source with constant voltage magnitude behind its transient reactance. Let us also assume that the mechanical power is constant. From Eq. (9.1), it becomes clear that during the fault, the generating unit accelerates since the electrical power becomes zero if the voltage becomes zero due to the short circuit. If the generating unit accelerates, the rotor speed increases and congruently, the rotor angle increases. Once the fault is cleared, the electrical power at that instance is larger than the mechanical power and the rotor starts to deaccelerate, and the rotor angle goes on increasing but at a decreasing pace and reaches eventually a maximum in the first-swing stable case. At this maximum, speed deviation is again zero. Computing the critical clearing times is then tantamount to finding the critical clearing angle. Equation (9.1) can be rewritten by omitting the damping factor as follows:

$$2H\frac{d\omega}{dt} = p_m - p_e = p_{acc} \tag{9.18}$$

where p_{acc} is the accelerating power. Multiplying both sides of Eq. (9.18) by the angle variation yields

$$2H\frac{d\omega}{dt}d\delta = p_{acc}d\delta \tag{9.19}$$

Or by using Eq. (9.3):

$$2H\omega_0\left(\omega - 1\right)d\omega = p_{acc}d\delta \tag{9.20}$$

Integrating both terms between the initial time and the final time results in

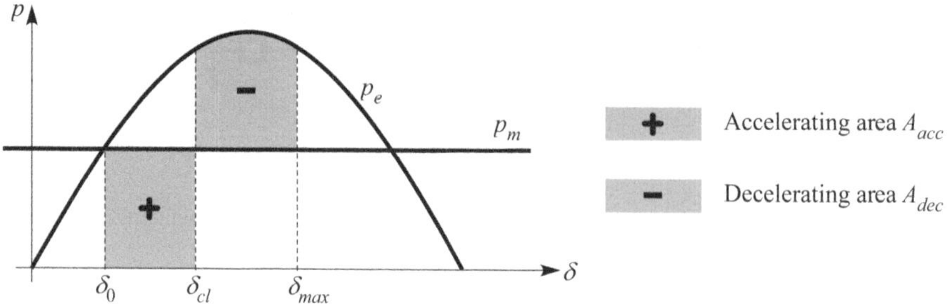

Fig. 9.8 Power-angle curve illustrating the EAC

$$\begin{aligned} 2H\omega_0 \int_{\Delta\omega_0}^{\Delta\omega} \Delta\omega d\omega &= \int_{\delta_0}^{\delta} p_{acc} d\delta \\ 2H\omega_0 \tfrac{1}{2}\left[\Delta\omega^2 - \Delta\omega_0^2\right] &= \int_{\delta_0}^{\delta} p_{acc} d\delta \end{aligned} \tag{9.21}$$

As seen before, the generator is stable if there exists a time when the speed deviation is zero. Since the initial speed deviation is zero (pre-fault condition), one obtains:

$$\int_{\delta_0}^{\delta} p_{acc} d\delta = 0 \Rightarrow \int_{\delta_0}^{\delta_{crit}} p_{acc} d\delta + \int_{\delta_{crit}}^{\delta_{max}} p_{acc} d\delta = 0 \Rightarrow \int_{\delta_0}^{\delta_{crit}} p_{acc} d\delta = \int_{\delta_{crit}}^{\delta_{max}} p_{dec} d\delta \tag{9.22}$$

where p_{dec} is equal to -p_{acc}. This equation is also known as the equal area criterion (EAC), which can also be derived by means of an energy function and by using Lyapunov stability theory. Figure 9.8 shows the power-angle curve. According to Fig. 9.8, the final equality in Eq. (9.22) actually expresses the equality of two areas, A_{acc} and A_{dec}. Note that δ_{max} is at most π -δ_0.

If now p_{acc} is again expressed in terms of p_e and p_m, Eq. (9.22) becomes:

$$\int_{\delta_0}^{\delta_{crit}} p_m d\delta = \int_{\delta_{crit}}^{\delta_{max}} (p_e - p_m) d\delta = \int_{\delta_{crit}}^{\pi-\delta_0} \left(\frac{e' u_\infty}{x_e} \text{sen}\delta - p_m\right) d\delta \tag{9.23}$$

The equation can be analytically solved for the critical angle:

$$\delta_{crit} = \arccos\left[\sin\delta_0 \left(2\pi - \delta_0\right) - \cos\delta_0\right] \tag{9.24}$$

Finally, the critical clearing time can be obtained by computing the time the angle needs to reach the critical angle.

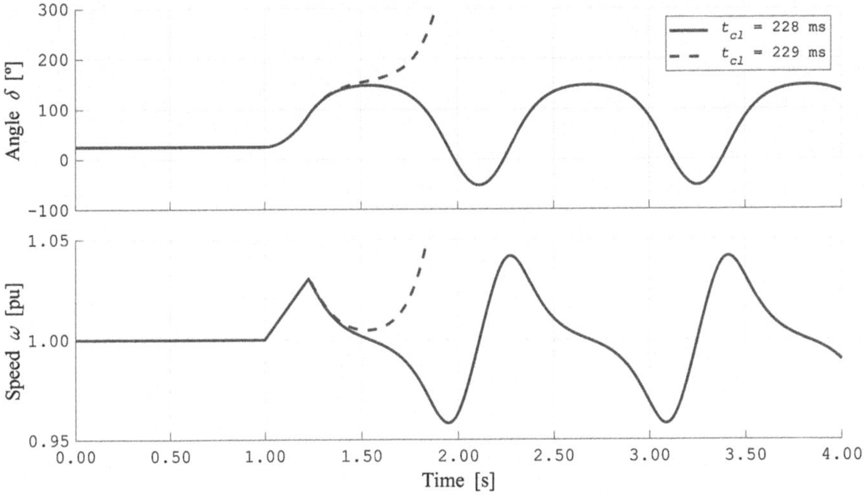

Fig. 9.9 Time-domain simulation of the critically stable case ($t_{cl} = 228$ ms) and an unstable case ($t_{cl} = 229$ ms) to a short circuit at the HV bus of the step-up transformer of the system in Fig. 9.7

$$t_{crit} = \sqrt{\frac{4H}{\omega_0 p_m}\left(\delta_{crit} - \delta_0\right)} \tag{9.25}$$

The critical clearing time in that simple case depends on the inertia, the loading, and the reactance (affecting the initial angle). Figure 9.9 shows time-domain simulation to a short circuit at the HV bus of the step-up transformer of the system in Fig. 9.7. The critically stable case is shown, where the clearing time is equal to the critical clearing time, as well as an unstable case, where the angle continuously increases after the first swing. Finally, Fig. 9.10 shows the impact of the line length by incrementing x_l on the critical clearing time. A larger line length leads to a reduction in the critical clearing time.

9.2.2 *Frequency Stability*

Frequency stability is concerned with the ability of a power system to maintain steady frequency within a nominal range following a severe system upset resulting in a significant imbalance between generation and load [29]. It depends on the ability to restore balance between system generation and load, with minimum loss of load. Severe system upsets generally result in large excursions of frequency, power flows, voltage, and other system variables, thereby invoking the actions of processes, controls, and protections that are not modeled in conventional transient stability or voltage stability studies. These processes may be very slow, such as

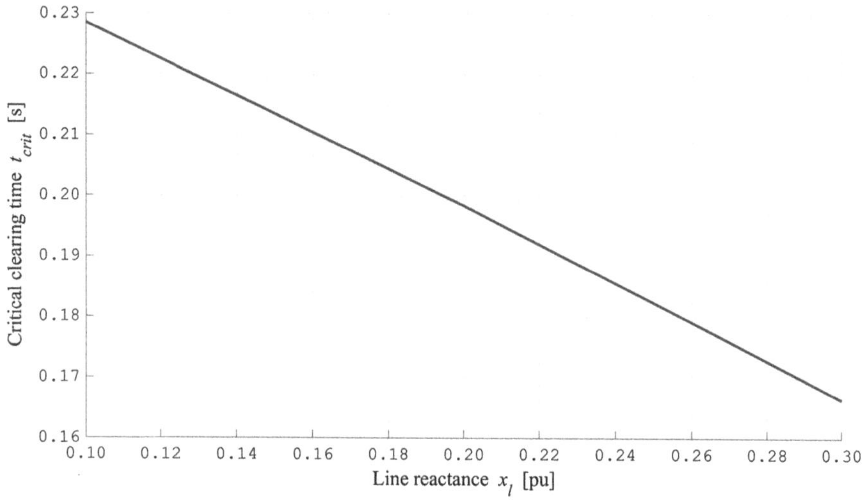

Fig. 9.10 Impact of the line length on the critical clearing time

boiler dynamics, or only triggered for extreme system conditions, such as volts-hertz protection tripping generators. In large interconnected power systems, this type of situation is most commonly associated with islanding. Stability in this case is a question of whether each island will reach an acceptable state of operating equilibrium with minimal loss of load or not. It is determined by the overall response of the island as evidenced by its mean frequency, rather than relative motion of machines. Generally, frequency stability problems are associated with inadequacies in equipment responses, poor coordination of control and protection equipment, and insufficient generation reserve. Examples of such problems are reported by [16]. Over the course of a frequency instability, the characteristic times of the processes and devices that are activated by the large shifts in frequency and other system variables will range from a matter of seconds, corresponding to the responses of devices such as generator controls and protections, to several minutes, corresponding to the responses of devices such as prime mover energy supply systems and load voltage regulators. Although frequency stability is impacted by fast as well as slow dynamics, the overall time frame of interest extends to several minutes.

The response of the generating units to active power imbalances depends on the electrical distance to the disturbance, the inertial response, and the performance of the turbine-governor systems [33]. Just after the disturbance, the share of each generator in meeting the power imbalance depends on its electrical distance to the disturbance. Closer generating units will face a higher electrical power request than far away units in case of a generation outage. According to Eq. (9.1), an increased electrical power results in about constant deceleration of the rotor speed. Figure 9.11 shows individual frequencies and the average frequency represented by the

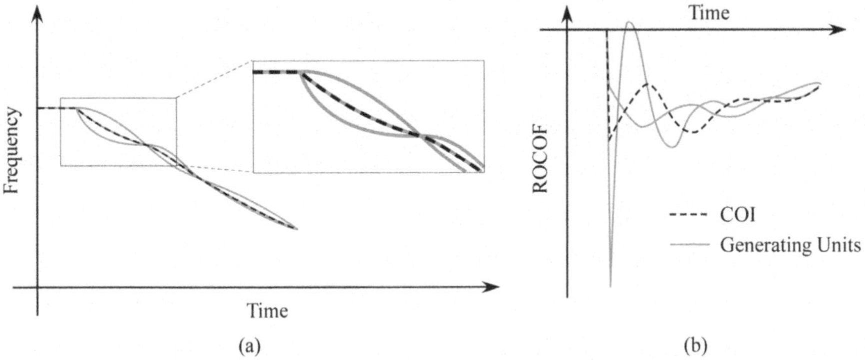

Fig. 9.11 Illustration of (**a**) individual frequencies and frequency at COI and (**b**) individual ROCOFs and ROCOF at COI

frequency of the center of inertia (COI) as well as the rate of change of frequencies (ROCOFs) and the ROCOF at COI. Although frequencies (and generator speeds) at different locations are different, the overall response in terms of the average frequency is of interest for frequency stability. This becomes truer if long-term dynamics are considered such as in AGC studies, where frequency differences throughout the system are not significant.

Since turbine-governor systems exhibit slower dynamics (time constants in the order of several seconds) and control dead bands, they intervene with some delay. Turbine-governor systems provide a means of controlling power and frequency. The contribution of each generator is basically a function of its governor speed drop, the speediness of the turbine-governor system response, and the amount of available spinning reserve. Short-term frequency dynamics are thus mainly affected by the inertia and turbine-governor systems of conventional generating units, and they depend much less on the grid.

Frequency stability is typically assessed through time-domain simulations. In case of islands or an islanded part of an interconnected system, so-called system-frequency response models can be used to simulate frequency dynamics [32]. These models are reduced-order models and in the simplest case, they assume uniform frequency and reduce generator representation to their rotor and turbine-governor system. Such models are complementary to more detailed models and they allow estimating the impact of disturbances on frequency stability under different generation scenarios easily. Figure 9.12 shows a system-frequency response model, where each generating unit is represented by a generic non-linear second order model, where k_i is the inverse of the droop. H is the system equivalent inertia, arising from the assumption of uniform frequency, and D is the load damping factor. The models also represent the response of the underfrequency load shedding (UFLS) scheme. The response of the UFLS scheme might affect the load damping factor [28].

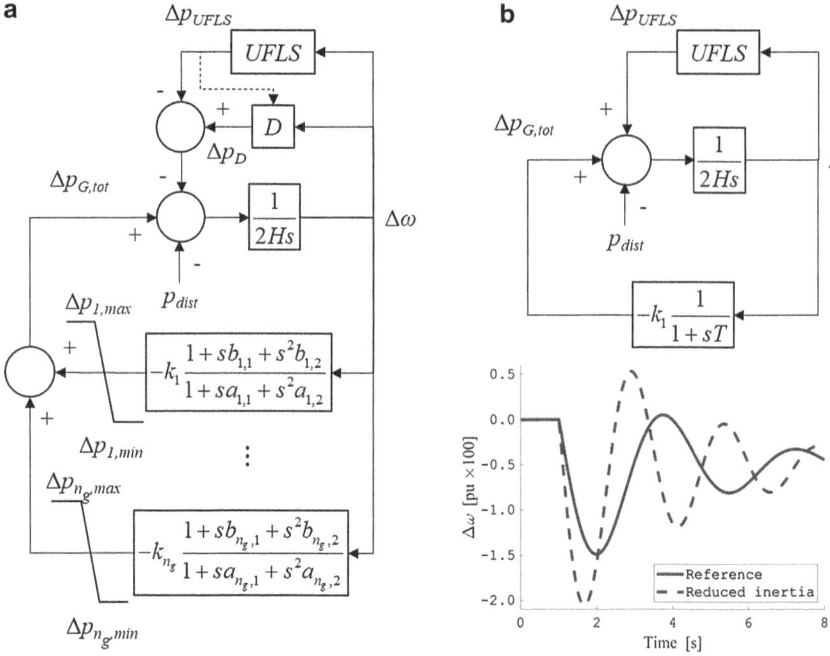

Fig. 9.12 Block diagram of a generic system-frequency response model (**a**) and a system represented by an equivalent unit (**b**)

The response in terms of frequency deviation to an active power imbalance can be easily illustrated in case of a single equivalent generating unit as shown in Fig. 9.12, represented by a first order system (i.e., $b_{i,1}$, $b_{i,2}$, and $a_{i,2}$ are equal to zero). Under this assumption and by neglecting the load damping factor, the transfer-function between p_{dist} and $\Delta\omega$ is

$$\Delta\omega(s) = \frac{1}{2HT} \frac{1 + sT}{s^2 + \frac{1}{T}s + \frac{k}{2HT}} p_{dist}(s) \tag{9.26}$$

Figure 9.11 shows a response in terms of frequency to an active power imbalance of 10%. Minimum frequency deviation reaches −0.01 pu (49.5 Hz in 50 Hz system). In steady state, frequency deviation is equal to

$$\Delta\omega = \frac{p_{dist}}{k} \tag{9.27}$$

During the first few instants, the response of the generating unit can be further approximated by

$$\Delta p_{G,tot}(s) = -\frac{k}{sT}\Delta\omega(s) \tag{9.28}$$

With this approximation, the minimum frequency deviation can be estimated as follows [30]:

$$\Delta\omega_{\min} = -\frac{T}{k}\sqrt{\frac{k}{2HT}}p_{dist} \tag{9.29}$$

Equation (9.29) illustrates how the minimum frequency deviation depends on the disturbance, the turbine-governor system performance through parameters T and k, and the rotor by means of the equivalent inertia H. A reduction in inertia while maintaining the remaining parameters constant is also shown in Fig. 9.12.

9.2.3 Voltage Stability

Voltage stability refers to the ability of a power system to maintain acceptable bus voltages under normal conditions and after being subjected to a disturbance [16]. Low voltages imply high currents to maintain the power demand. Since power systems are non-linear, not only do voltage profiles decrease with demand growth, but also their slope; thus, the voltage fall will be as fast as the increase in demand. This situation usually leads to a cascade failure of the different system devices, followed by a system blackout.

Voltage collapse can occur over a wide variety of time frames, as Fig. 9.13 depicts [14]. In the transient region, voltages and angular stabilities are closely related in some cases, since low voltages may accelerate angular instability; likewise, the loss of synchronism may result in voltage depression. However, transient voltage collapse can also emerge without loss of angular stability. In the long-term region, angles of voltages are also affected, but only in the very previous moments before the complete blackout. In most of the cases, analysis and prevention of long-term voltage collapse during planning of the operation also reduces the possibility of transient voltage collapse situations.

From an electrical point of view, there are different reasons why voltage collapse can occur in a power system [35, 37]. Key factor is how the branches in the network, i.e., lines and transformers, are being operated. In highly loaded scenarios, one or more branches in the network may be supporting high power flows and large voltage drops that undergo voltage collapse. In addition, the higher the branch current, the lower the network voltage profile, which implies a reduction in the maximum load that the branch can transfer without collapsing. In the light of the foregoing, principal causes of voltage collapse can be grouped as follows:

- **Active power**: In highly loaded case scenarios, large amounts of active power need to be transferred from generation resources to the costumers. However not

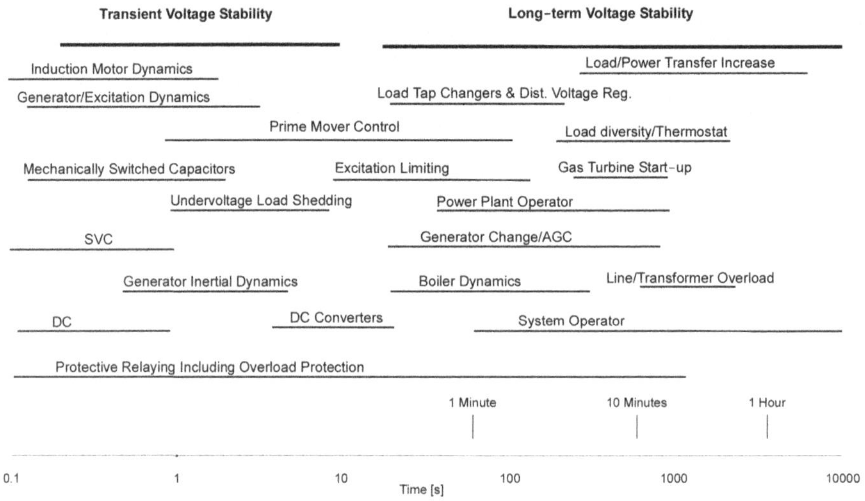

Fig. 9.13 Time frames for voltage stability phenomena

only is a matter of quantity, but also of how the active power flows are distributed among the network and the presence of bottlenecks. Likewise, interconnected networks can establish active power interchanges larger than the limit before collapse, which is why the computation of the Available Transfer Capability often includes voltage stability constraints.

- **Bus voltages**: In all those buses where there are generating units, the bus voltage is maintained if reactive power generated or absorbed is in limits. For the rest of the network buses, their voltages are uncontrolled, which means that they are subject to the state of the system. For a constant power, the lower the voltage the higher the current; thus, it is important to maintain load buses within reasonable voltage limits. For that purpose, different voltage control devices, such as generating units, shunt devices, and SVC devices, need to be optimally adjusted to obtain an adequate voltage profile that prevents the collapse of the system.
- **Reactive power**: In transmission networks, bus voltages are closely related to the reactive power balance of the network. In few words, injecting reactive power into the network causes an increase in the voltages of nearby buses, and therefore, extracting reactive power from the network implies a decrease. Most of the voltage control devices work by injecting or absorbing reactive power as bus voltage varies. In addition, transmission lines generate reactive power with low active power flow and consume it with high active power flow. Thus, in highly active power loaded scenarios, there is also a great reactive power consumption among loads, lines, and transformers. These scenarios require great reserves of reactive power. Severe lack of reactive power margin is by far the shortest way

to a voltage collapse situation. Voltage control devices improperly adjusted or reactive power limits saturation in generation units put voltage stability at risk.

- **The network**: In the case of transmission networks, they are designed and operated as mesh networks, i.e., the energy has different paths to complete the generation-demand transfer. To maximize the robustness of the network, an optimal design is strongly required. The design must also consider N-1 or even N-2 situations, i.e., when one or more network devices get disconnected from the network under a contingency.

Voltage stability is mainly a dynamic phenomenon that implies detailed models of each device of the power system: generators, transformers, and loads, etc. However, voltage instability phenomenon can be analyzed using the steady-state model of the network, and thus, preventive control actions can be optimized to maintain the system far enough from the voltage collapse point. From a mathematical point of view, voltage collapse phenomenon is due to the proximity of the system operating point to a saddle-node bifurcation in the power system equations [8]. Under secure operating conditions, voltages barely descend as the demand grows, but as systems get close to a saddle-node bifurcation, this descend becomes relevant and voltages decrease dramatically until the collapse point. Far from this point, the steady-state equations of the power systems have no solution. Therefore, the analysis of how far the system is from the voltage collapse point represents a key tool to keep the system safe from voltage instability situations.

9.2.3.1 Concept and Definition of Voltage Collapse Margin

From a steady-state point of view, voltage collapse is illustrated by the trajectories of bus voltages as system demand grows, usually called "nose curves." When active power demand grows and so does active power generation, voltages tend to decrease. This is because increasing the power flow implies more current through lines and transformers, and hence higher voltage drop. Nevertheless, this behavior becomes highly non-linear as the load of the systems grows, so the voltage slope tends to be vertical and the voltage fall emerges as critical. This boundary is known as voltage collapse point, and the distance from the starting base case to the maximum load is considered as the margin to voltage collapse.

Figure 9.14 depicts the evolution of the active power demand of an area of the Spanish electric power system, and the voltage trajectory of the area pilot bus, in a long-term close-to-collapse scenario occurred in 2001. At 16:40 the area demand reaches a local minimum of 4000 MW, and then starts to grow. As a consequence of an ill-conditioned system, pilot bus voltage falls dramatically for several minutes from 400 to 350 kV. At 18:30, since the voltages are falling out of control, an emergency load shedding of almost 300 MW is applied to stabilize the system. Once the load has been shed, the system voltages change the tendency and recover reasonable values and hence their stability.

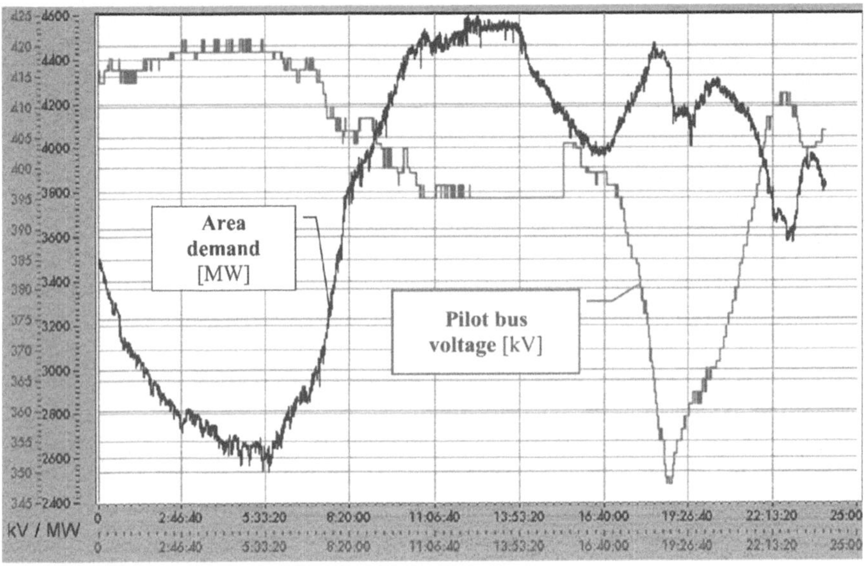

Fig. 9.14 Time trajectories of area demand and pilot bus voltage in an actual scenario of the Spanish power system

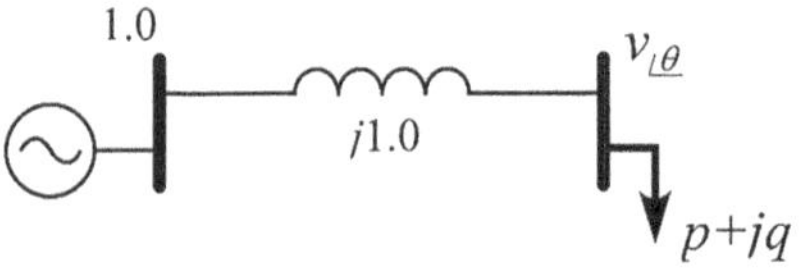

$$p = -v \cdot \sin\theta$$
$$q = -v^2 + v \cdot \cos\theta$$

Fig. 9.15 Two-buses network diagram and steady-state equations

To illustrate the voltage collapse concept, consider a two-buses network, where a generation unit and a load are interconnected by a 1.0-pu reactance. In addition, the generation unit voltage is set to 1.0 [37]. The diagram of the network and corresponding power flow equations are represented in Fig. 9.15.

Depending on the active (p) and reactive (q) power demands, the system described in Fig. 9.15 can present two solutions (feasible), one solution (critical), or no solution (unfeasible). Figure 9.16 describes this phenomenon by depicting different full voltage v manifolds as active power demand p increases for some values of the load power factor. These curves are also known as "nose curves."

In Fig. 9.16, each load power factor corresponds with a single manifold, i.e. a nose curve. Each of these manifolds contains two different trajectories for the voltages considering the same demand evolution. The difference between them is that the upper trajectory (continuous line) is stable and the lower (dashed line) is unstable. The two trajectories collide into a singularity, a saddle-node bifurcation where the two roots of the system become one and after that they vanish. The upper curves give the picture of how an actual electric power system works. When the

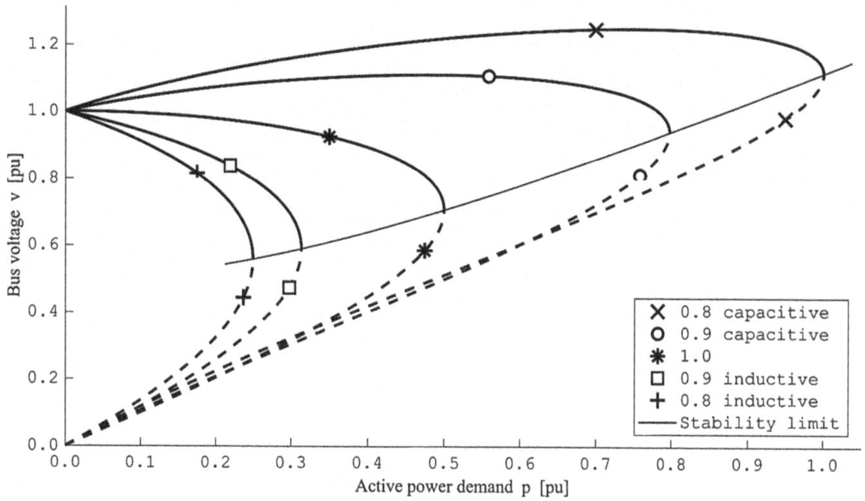

Fig. 9.16 Voltage trajectories for demand grow, considering different load power factor

network is on an off-peak scenario, bus voltages tend to slowly vary when demand changes. More concretely, when the load is inductive, which happens in most of the cases, bus voltages fall when the demand grows, and the more inductive the load more negative the slope. In case the load is capacitive, bus voltages rise when the demand grows, and the more capacitive the load the more positive is the slope. However, if the demand grows further than normal operation conditions, then the slopes of all the curves tend to be dramatically negative, even the capacitive ones. At this point, not only bus voltages but also the system variables begin to sustain large perturbations that may become uncontrollable. Indeed, if the demand exceeds the limit, i.e. the tip of the nose curve, the system will eventually lead to a blackout.

To better illustrate the coupling between reactive power and voltage, Fig. 9.17 shows the reactive power demand required as bus voltage grows for different active power demand values. Active power demand values are expressed in % of its maximum value for pure resistive power factor.

In Fig. 9.17, for each of the active power demand values considered correspond to a manifold that contains two different trajectories, a stable one (continuous line) and an unstable one (dashed line). The system remains stable if the point of operation is located to the right of the stability limit, i.e., where the sensitivity dq/dv is negative. As reactive power demand grows, both roots tend to collide into the saddle-node bifurcation. If the reactive power demand is larger than the maximum, the voltages collapse, and system goes to blackout.

The abov e example has been limited to a radial system because it presents a simple picture of the voltage stability problem. It has provided the basics to understand the voltage stability phenomenon among transmission networks. They are designed to work considering voltages close to 1.0 pu and negative gradients with respect to demand. However, in complex meshed transmission networks,

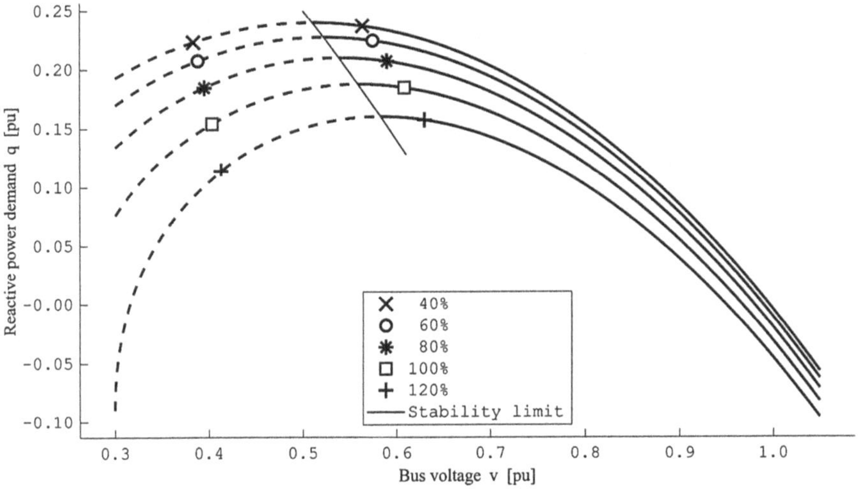

Fig. 9.17 Reactive power demand required as bus voltage grows for different active power demand values

many factors contribute to the process of system collapse because of voltage instability, as it has been pointed out at the beginning of this section: load and power transfer levels, characteristics of reactive compensating and voltage control devices, topology, and parameters of the network, etc. and even in the case of having all devices optimally set and the network far from collapse, some dramatic events may take the system from the stable trajectory to an unstable one, undergoing an immediate voltage collapse. Those events are the outage of one or more network devices, such as generation units, lines, and transformers, or the reactive power saturation in one or more generation units, among others.

Power flow steady-state equations represent the power balance between the net power produced $\mathbf{S}^0$ and the sum $\mathbf{S}$ of power injected into the network at each bus.

$$\mathbf{S}^0 = \mathbf{S}(\mathbf{V}, \boldsymbol{\theta}) \tag{9.30}$$

In equation (30) power $\mathbf{S}$ represents either active or reactive power. In the LHS of equation (30) there is net power $\mathbf{S}^0$ produced at the bus, i.e. the difference between the generated and demanded power. In the left-hand-side of equation (30) there is net power $\mathbf{S}^0$ produced at the bus, i.e. the difference between the generated and demanded power. In the right-hand-side there is the power injected into the network, as function of the state variables of the system, i.e. magnitude and angle of bus voltages. In the case of actual networks, voltage collapse concept is the same, i.e., there exists a feasibility boundary in the bus power space, which represents a saddle-node bifurcation. Therefore, any power dispatch inside the boundary represents a feasible power flow problem, and any power dispatch outside the boundary corresponds with an unfeasible power flow problem.

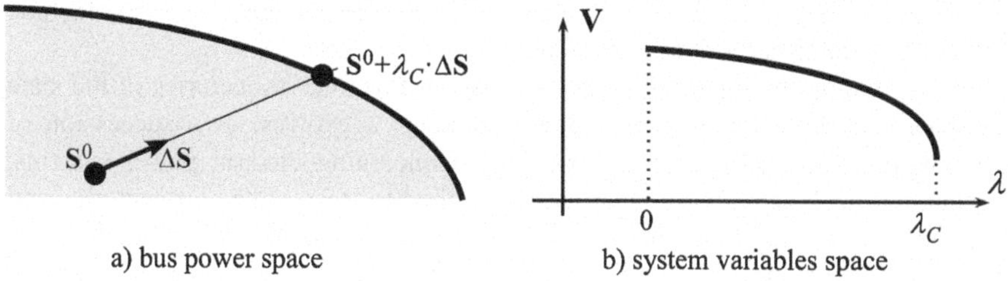

Fig. 9.18 Bus power dispatch parameterization and concept of critical load margin

To evaluate the risk of a voltage collapse, it is not necessary to compute the whole feasibility boundary among the bus power space. The most widely used procedure to evaluate voltage stability is to assume that the power dispatch of the system is going to evolve in a concrete direction. This evolution vector **ΔS** is parameterized using a factor λ; thus, the bus power dispatch becomes a linear function of λ. Rearranging (30) and introducing the power dispatch variation vector:

$$\mathbf{S}^0 + \lambda \cdot \Delta\mathbf{S} - \mathbf{S}(\mathbf{V}, \boldsymbol{\theta}) = 0 \tag{9.31}$$

System (31) will be feasible or unfeasible depending on the parameter λ, as Fig. 9.18 depicts.

In Fig. 9.18 (a), the bus power dispatch evolves from an initial point of work $\mathbf{S}^0$ following the direction defined by **ΔS**. For each value of the load margin λ, a new feasible power dispatch is defined, until the critical load margin λ_C is reached. At this value of the load margin, the corresponding bus power dispatch belongs to the feasibility boundary, i.e., the saddle-node bifurcation. Therefore, the resulting distance $\lambda_C \cdot \Delta\mathbf{S}$ from the base case to the feasibility boundary is considered as the margin to voltage collapse.

In Fig. 9.18 (b), the state variables, i.e., the magnitude and angle of bus voltages, evolve from the initial point of work following a concrete path defined by the system of equations in (31). The end point of state variables trajectories is the voltage collapse point, located at the critical load margin λ_C. At this point, the slope of the trajectories becomes infinite, since the voltage collapse point corresponds with a saddle-node bifurcation, and hence, the Jacobian matrix of the system is singular.

9.2.3.2 Margin to Voltage Collapse Computation Techniques

Most of the critical load factor computation techniques are based on the same idea: how large can the load factor be without crossing the feasibility boundary. Depending on how to get close to the feasibility boundary, there are two main sets of techniques: continuation methods and direct methods. The former follows the trajectories of state variables until the voltage collapse point, the latter computes the voltage collapse point by solving a set of non-linear equations that include singularity condition for the Jacobian matrix.

On one hand, continuation techniques are used to trace trajectories of the state variables **x** as long as an independent parameter λ evolves, as a succession of different roots of the state equations [26]. Concerning electric power systems, continuation techniques have been widely applied to trace voltage trajectories and detect the voltage collapse point, using independent parameter λ to control the evolution of both the demand and generation [3]. The aim is to find a new root starting from a previous equilibrium point. This objective gets more difficult as far as the system gets close to the voltage collapse point. As far as the system evolves toward its saddle-node bifurcation, the Jacobian matrix conditioning gets worse. This fact makes it more difficult to converge non-linear equation system solvers. To overcome this difficulty, continuation techniques split the new root finding process into two steps: predictor and corrector. Figure 9.19 shows complete continuation iteration, including predictor and corrector steps.

First approximation of the new root is computed using the predictor step. This approximation represents a good starting point for the corrector step, where the system of equations is solved using a standard non-linear equation system solver and adding additional equations that control the conditioning of the problem; thus, convergence to the new solution is improved.

On the other hand, direct methods to compute load margin are closely related with the mathematical characteristics of the voltage collapse point, i.e., the singularity of the Jacobian matrix. From a mathematical point of view, the voltage collapse point represents a saddle-node bifurcation in the power flow equations. In a saddle-node bifurcation, two points of work collide and annihilate each other; thus, the system jumps from being feasible with two roots to being unfeasible. At the saddle-node bifurcation, the system will have only one root, and the Jacobian matrix of the state equation system will be singular. To compute the voltage collapse point using direct methods, a set of additional equations is required to force the Jacobian matrix to be singular. To achieve this, the state equation system is augmented by forcing the Jacobian matrix to have an eigenvector associated to a singular eigenvalue [1]. This could be either the left or the right eigenvector.

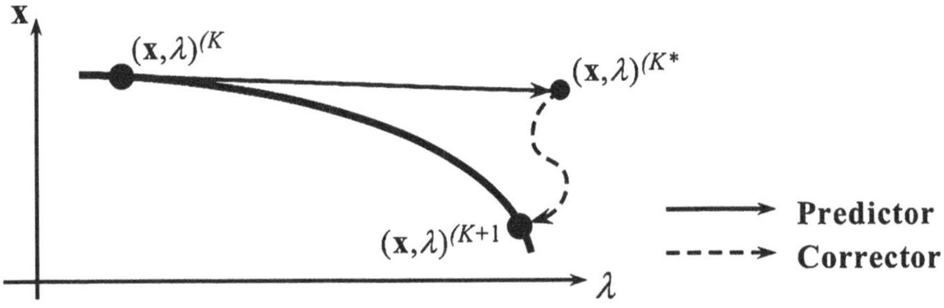

Fig. 9.19 Complete continuation iteration

$$\begin{aligned} &\mathbf{g}(\mathbf{x}, \lambda) = \mathbf{0} \\ &\mathbf{w}^{\mathbf{T}} \cdot \mathbf{g}_{\mathbf{x}} = 0 \\ &\|\mathbf{w}\| \neq 0 \end{aligned} \tag{9.32}$$

System (32) is formed by three conditions. Firstly, the parameterized power flow equations. Secondly, the singularity condition, which is formulated by doing zero the product of the Jacobian matrix and vector **w**, which represents the left eigenvector of the Jacobian. Finally, vector **w** has to be set as a non-zero vector, controlling either its module or one or more of its components. Solution of (32) will hence represent a saddle-node bifurcation of the system, i.e., the voltage collapse point, and **w** will be the corresponding left eigenvector associated to the singular value of the Jacobian matrix. A system equivalent to (32) may be formulated using the right eigenvector. The direct methods can also be formulated as a set of optimality conditions. The computation of the margin to the voltage collapse point can be formulated as an optimization problem [10, 11], where the objective is to maximize the load margin, subject to the system of parameterized steady-state equations.

$$\begin{aligned} &\max \lambda \\ &s.to\ \mathbf{g}(\mathbf{x}, \lambda) = \mathbf{0} \quad : \quad \mu \end{aligned} \tag{9.33}$$

9.2.3.3 Other Voltage Stability Indices

Despite the load margin, a lot of voltage stability indices have been defined in the literature [2, 6]. For the most of them, their formulation is based on the key characteristic of voltages collapse point. i.e. it corresponds with a saddle-node bifurcation where the Jacobian of the state equations is singular. Those indices are usually easy and fast to compute. However, the drawback is that they do not provide any information about the variable trajectories; thus, any perturbation in the system will not be considered, such as the saturation of the reactive power production in generation units. In addition, some of those indices provide useful mathematical information about the distance to the singularity, but with no physical interpretation on the power system operation.

One set of alternative indices are those based on the algebraic decomposition of the Jacobian matrix of the state equations, such as the Eigen Decomposition (*ED*) or the Singular Value Decomposition (*SVD*) [7, 19]. The *ED* consists of factorizing the Jacobian matrix of the state equations in a set of associated values called eigenvalues. If the matrix decomposed is singular, at least one of their eigenvalues will be zero. Therefore, if a power system gets close to voltage collapse, the Jacobian matrix will present an eigenvalue tending to zero. In addition, eigenvectors corresponding with the singular value provide useful information about critical buses:

- Largest elements of the left eigenvector correspond with the buses where power injection will undergo the greatest perturbation on state variables.

- Largest elements of the right eigenvector correspond with the buses that present the state variables with the highest sensitive to power system injections.

Concerning the singularity of the Jacobian matrix in the voltage collapse point, some authors propose to work with non-singular reduced versions [4]. By selecting the adequate bus, corresponding rows and columns in the Jacobian matrix are vanished and the new Jacobian recalculated. From that point, the determinant is computed and compared against the determinant of the original Jacobian matrix. Also here exist methods to reduce the bus impedance matrix of a network into its two-bus equivalent model at a referred bus ([18]); thus, equations of the equivalent model could then be derived to facilitate an equivalent nodal analysis at the referred bus to determine its voltage stability limit. Other approaches, after convenient Jacobian matrix manipulation on critical bus, use this reformulated Jacobian matrix to build test functions. These test functions give an approximation of the maximum load margin but without state variable trajectories [13].

Other methods compute the distance to voltage collapse by means of Lyapunov functions, also known as energy functions [12, 15]. Lyapunov functions are scalar functions used to prove the stability of an equilibrium point of an ordinary differential equation system. Lyapunov functions have been widely used for small-signal stability analysis, as well as for voltage stability problem. The idea is to associate the power system state equations to a dissipation function that represents the gradient of the energy function; thus, the equilibrium point represents a minimum energy point for the Lyapunov function. In addition, energy functions allow to trace state variables trajectories, but both the computational effort and the mathematical difficulties on energy function construction render those energy function-based methods less suitable than others, such as the continuation techniques.

Finally, most recent works on the Holomorphic Embedding Load-flow Method (*HELM*) present it as a very useful tool to trace trajectories of system variables [36]. The *HELM* consists of solving the complex power flow equations by defining complex voltages as holomorphic functions of a complex parameter. This complex parameter is used to expand the voltages as polynomial functions. Therefore, the computing of each coefficient of the polynomial will be carried out by solving a linear equation system involving all the coefficients with lower degrees. To date, no works on using *HELM* with voltage collapse point detection have been carried out, but the results on the power flow problem solution present *HELM* with great potential.

9.3 Guidelines for Stability Studies During Planning Stage

Integration of stability studies as one more of the habitual planning tasks requires a deep understanding of how stability analysis works and what are their inherent difficulties. Modeling and analysis of stability phenomena are common to operation and planning studies. The tools to carry out these tasks are then the same. Operation studies focus on modeling and parametrizing existing components such that they adequately reflect the behavior of the real system in order to tune settings of the existing controllers to guarantee stability. Planning studies must additionally model and parameterize future equipment with the purpose of establishing best control strategy and settings for the new installations and overall system. Planning studies thus aim at identifying the technical requirements to cope with the so-called credible contingencies and outlining the emergency actions to be triggered at the occurrence of extreme contingencies. The final goal is to minimize the limitations in the power transfers related to dynamic constraints. This section provides an overview of the procedure for stability studies and a classification of the different dynamic analysis that may need to be carried out during planning studies [5].

9.3.1 Procedure

Stability studies may involve large sets of data, models, and scenarios. Stability studies require a structured procedure to tackle all the objectives purchased. The decisions taken during any of the stages in the study can be critical for the final outcomes. Figure 9.20 shows an overview of the procedure for stability studies.

The first step is to clearly define the objectives of the study. It is important to identify what stability mechanism is in play, what is the time horizon needed, and what are the variables affected (see Sect. 9.3.2). It is also relevant to review the different policies and guidelines from Transmission System Operators (TSO) and other Authorities to observe the different constraints. For instance, decisions of interconnecting different systems are surely driven by economic analyses on the profitability of energy exchanges. Dynamic constraints can however lead to the adoption of alternative technical solutions (e.g., HVDC or HVAC interconnections) with, in some cases, an additional burden in the investments, which could even lead to the non-profitability of the whole project. Interconnecting systems can create new dynamic interactions and very complex transients can be triggered in large systems, causing the need for reviewing the setting or even the philosophy of control loops. Long distance transmission of bulk power in AC requires adequate solutions such as series compensation. Series compensation can however lead to subsychronous resonance. This phenomenon can be also found in weakly connected wind farms using DFIG. Finally, adequate tools must be identified to run the simulations.

Selection of models is the most relevant part of the process. It is necessary to identify what and how the devices are affected and select adequate models of all of them: generating units, loads, the network, control, and protection devices, etc. In this case, adequate models must be distinguished from full-detailed models. Adequate models must reliably represent how the devices behave under the conditions

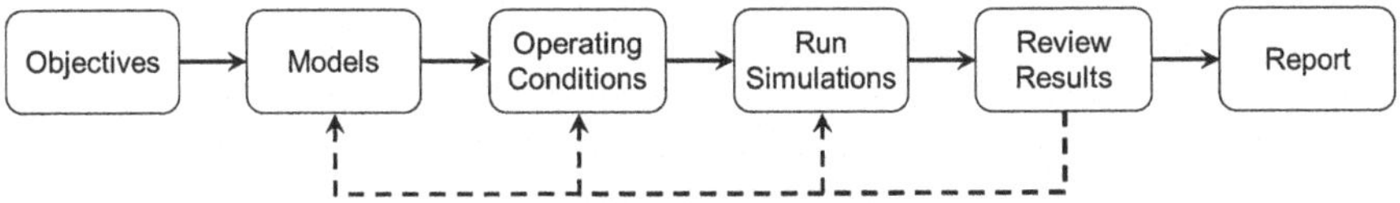

Fig. 9.20 Overview of the procedure for stability studies

Table 9.1 Equipment modeling requirements

		Time frame				
		SSR < 1 s	Transient < 10 s	Small signal < 20 s	Long term mins	Voltage stability 1 s - mins
Equip.	Generator	y	y	y	y	y
	Shaft system	y	x	?	x	x
Excitation system	AVR	?	y	y	y	y
	PSS	?	y	y	?	y
	Limiter	x	?	x	y	y
	PV/Var	x	x	x	y	y
Turbine control	Speed governor	x	x	?	y	x
	Overspeed control	x	x	x	y	x
	AGC	x	x	x	?	x
	Prime mover	x	x	x	y	x
	Loads	x	?	?	y	y

described in the study. This decision also relies on the time horizon. Modeling stage is often a tricky task, since dynamic characteristics of electric equipment are the less used and therefore the most difficult to collect in proper conditions. Most of the commercial platforms implement standard library models. IEC 61400-27 [34] presents for instance generic wind turbine models. A survey has shown that mostly dynamic models suitable only for short-term dynamics (transient stability) are used. Slow control loops, like over-underexcitation current limiting circuits, AGC or SVR, which are essential to simulate mid-long-term dynamics spanning over tens of seconds are modelled only in a limited number of cases. Further, the lack of data of future devices such as generating units or loads and its controls as well as the absence of validated models complicate this task. For instance, distributed generation is fully visible to the system and establishing dynamic equivalents is not straightforward.

Table 9.1 provides guidelines for equipment modeling for different stability study types, where "**y**" indicates that a dynamic model is needed, "**x**" indicates that it is not needed, and "**?**" indicates that a dynamic model may be included.

Before running the models, it is also important to define the operating conditions. Therefore, it is recommended to consider different ranges for the operating conditions, i.e., the demand and generation profiles, scheduled energy interchanges, point of work of control devices, and topology, etc. Another aspect is to establish the stability limits and define the minimum acceptable conditions to consider the system stable or not. Finally, some additional considerations may rise, such as if contingency situations are contemplated.

During the running model stage, one of the key aspects of stability analysis is the initial point. The outcome of a stability analysis, as being the solution of a system of differential equations, depends on the initial point of operation. Furthermore, in some cases, slight perturbations in the initial point of operation may lead to radically different outcomes. Another issue in running the model is the acceptance criteria, i.e., to validate if the analysis techniques used are adequate and identify if the outcome represents reality properly. Finally, results must be analyzed looking for other different stability mechanisms than may be distorting the outcomes.

Final stage is to review the results and adjust as necessary. In most of the cases, a dynamic analysis is a cycling process where the obtained results lead to different adjustments at some steps of the process: parameters of the models, time horizon and time step of the simulation, initial operation conditions, etc.

9.3.2 Dynamic Analyses Considered

Nowadays, the tendency is for electric power systems to grow and expand, especially for the liberalized ones. This expansion comes with a geographical expansion that brings formerly isolated networks closer, facilitating new interconnections. This tendency of building new interconnections and reinforcing the already existing corridors increases the complexity of power systems. Consequently, the understanding of the dynamic behavior of the system after a disturbance becomes more and more complex every day. A single perturbation may trigger many different dynamic phenomena, often overlapped with each other. The key role of the dynamic analyses is to discriminate the different phenomena and detect which is the most critical variable for the system stability. Through appropriated model reductions, it is for example possible to differentiate in case of slow dynamics between transient angle instability and voltage instabilities, leading to a voltage collapse. An exact understanding of the nature of this critical phenomenon is also very important in order to adopt the most suitable countermeasures both in planning and in operation stage. These measures consist usually of the adoption of operation rules in conjunction with possible automatic control actions to guarantee system security and alleviate emergency situations.

Even more challenging, when executing dynamic studies in planning stage, the planner must also correctly anticipate the system behavior under the uncertainty in the future development of generation, demand, and network. An incomplete or inaccurate dynamic diagnosis of the system can jeopardize the expected revenue of the investment: generating units working with strong limits with respect to its capability curve, interconnecting lines forced to operate at very poor power flow levels to avoid inter-area instability, or even new lines kept switched off until adequate stabilizing measures are undertaken.

To facilitate dynamic studies, the different underlying phenomena of stability are separated, and each one is analyzed by means of different methods (see Sect. 9.2). Table 9.2 summarizes the key steady-state and dynamic analyses to be carried out during planning and for which purposes [21].

Table 9.2 Main analyses performed in power system planning

		Analysis	Condition	Purpose
Static analyses		Load flows	Peak load	Assessment of power flows, voltage profile, reactive power margins
			Off-peak	Identification in a deterministic way of security constraints; bottlenecks on power flows, violations on voltage limits or limits on generators' capability curves
			Other "binding" conditions, such as unbalanced generation dispatching	Assessment of steady-state security: evaluation of security margins considering the pre-established planning criteria (e.g., N-1 security criterion)
				Definition of preliminary measures (e.g., new shunt var compensation, network reinforcements)
		Short circuit	Maximum Pcc	Sizing of circuit breakers or network configuration
			Minimum Pcc	Assessment of voltage variations
		Optimum Var compensation	Normally, same conditions examined in the load flow computations	Definition of the optimal location, size, and preliminary control design of shunt Var compensation
				Preliminary indication on the kind of devices to be installed
		Reliability analyses	Time span of 1 year	Assessment of reliability indexes
				Identification of possible network bottlenecks and related costs
Dynamic analyses	**Rotor angle**	Transient stability	Three-phase/unbalanced faults with reclosure at a predefined distance from the generator	Assessment of critical fault clearing time (CFCT) and compliance of CFCT with the adopted stability criterion
				Control and protection design
		Steady-state stability	Conditions of inter-area power transfers and active power dispatch	Assessment of system stability against small perturbations. Stability enhancement through appropriate location and tuning of PSS or other control loops associated to HVDC links or FACTS devices
	Frequency & voltage	Frequency stability	Tripping of large generators	Assessment of frequency deviations
			Tripping of lines causing islanding of part of the system	Checking of the compliance of load shedding scheme with the expected frequency deviations
				Checking of the generator stability facing maximum over/under-speed and tuning of relevant protection systems

(continued)

Table 9.2 (continued)

		Analysis	Condition	Purpose
		Voltage stability	Tripping of large generators	Assessment of reactive power margins of the generators or other Var compensation devices to cope with the new reactive power demand
			Tripping of highly loaded lines or lines causing islanding of part of the system	Redistribution of reactive power margins among generators and compensation devices and/or enhancement of margins
	Others	Subsynchronous resonance (SSR)	Analysis required in the presence of series compensation nearby thermal generation	Identification of measures to decouple identified resonance between network oscillation frequencies and the mechanical oscillation frequencies of the turbine-alternator system
		Electromagnetic transients (switching overvoltages, temporary overvoltages, lighting)	Energization of lines and cables (switching maneuvers) and transformers (in-rush currents) Lightning	Definition of the basic insulation level (BIL) of network components

9.4 Case Studies

This section presents some study cases where the influence of transmission network configuration and topology among system stability is illustrated.

9.4.1 Impact of HVDC on Transient Stability

From the transient stability point of view, the robustness of a system is measured by looking at the clearing time of a fault. The clearing time is the maximum duration of the fault for which the power system remains stable. Evaluation of the relative angles is observed to figure out by time-domain simulations whether the system has turned unstable. Since instability is equivalent to a continuous and non-oscillatory increase of relative angles, a critical angle is defined; thus, when one relative angle exceeds it, the system is considered unstable. This critical angle understandably needs to be over the maximum of the first swing, but it should be neither too large to avoid unnecessarily increasing the simulation time. Firstly, the impact of different current limiting strategies and other control modes of VSC-HVDC links on a small-scale power system will be illustrated. Afterwards, simulation results of a reduced model of the French-Iberian power system will be presented.

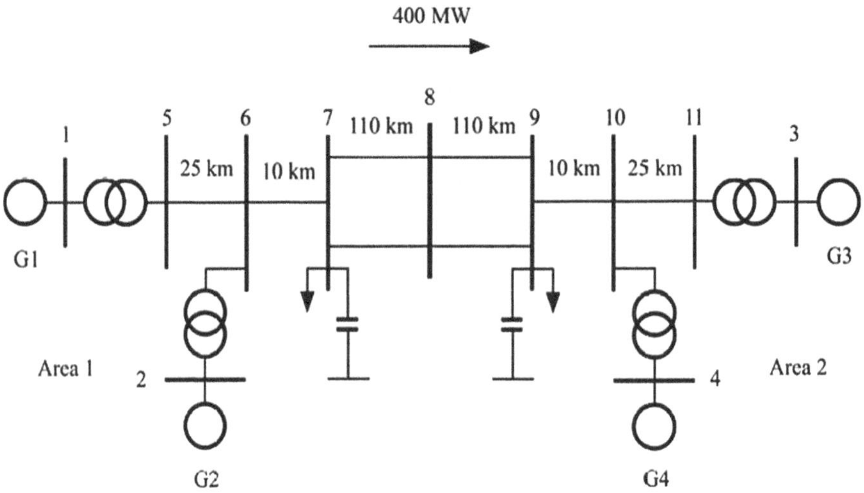

Fig. 9.21 Two-area test system

9.4.1.1 Small-Scale Power System

A small-scale test system proposed for fundamental studies of power system stability [16] has been considered (see Fig. 9.21). This small-case test system includes four generators and eleven buses, distributed among two areas. There are two generators in each area, and a weak link between them used to transmit power from area 1 to area 2. This test system has been completed by interconnecting the two areas using a VSC-HVDC link (see Fig. 9.22) to obtain conclusions about the impact of their control schemes in the system performance.

For both converters, the control has been set as current-controlled. Current control is carried out by using a Park transform in rotating reference system. The set point in current control depends on the external control loops. One converter controls the dc-link voltage, whereas the other controls the active power flow. Reactive power control loop of VSC-HVDC links can be set to control either reactive power or AC voltage. Figure 9.23 shows the set point modulation strategies for the active power and reactive power to improve transient stability. The active power depends on the frequency difference of the two terminals, such as in [22]. The reactive power depends on the difference of each terminal's frequency and the average frequency of both terminals, as for example suggested in [22, 23, 31]. Average frequency is employed to forecast the frequency of the center of inertia.

In steady-state operation, frequency deviations and differences are zero and consequently active and reactive power set points remain constant. However, during the transient period, active and reactive power set points are modulated as shown in Fig. 9.23. To illustrate the performance of the VSC-HVDC link, Figs. 9.24 and 9.25 display the active power flow in the AC line for a fault cleared after the critical clearing time of the original two-area system and of the two-area system with the

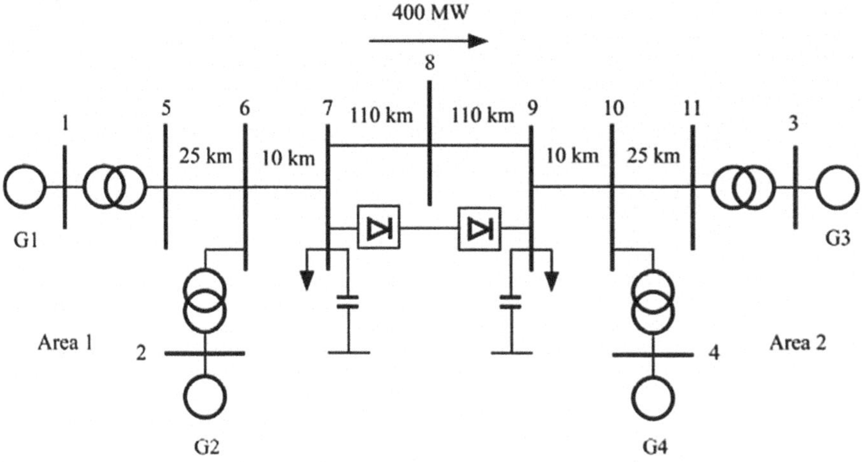

Fig. 9.22 Two-area test system with HVDC link

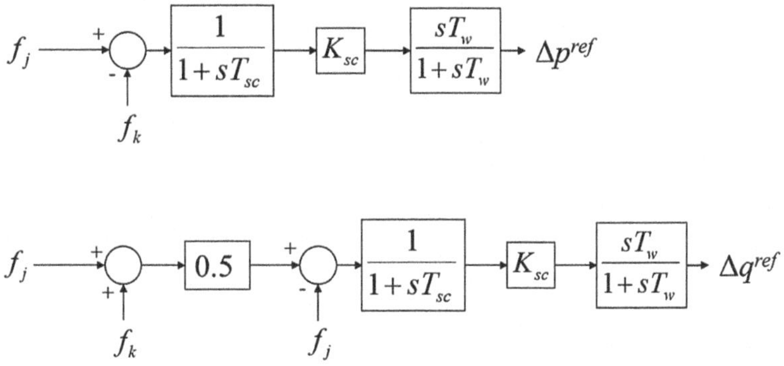

Fig. 9.23 Active and reactive power set point modulation

VSC-HVDC link, respectively. The VSC-HVDC link helps to sustain the power flow during the fault.

Table 9.3 provides the critical clearing times of the test case with parallel VSC-HVDC link under different control schemes. Compared with the original case (0.99 s), all control schemes except one present higher critical clearing times. The results show how reactive power control provides better results than voltage control. Further, when current limits are activated, it is prejudicial to reduce the active power component of the current, reducing inter-area power flow. Finally, voltage control at the terminal within the exporting area (bus 7) and reactive power control at the terminal within the importing area (bus 9) further improves the critical clearing time. On one hand, voltage control at the exporting area keeps high voltages and slows down the acceleration with respect to the importing area. On the other hand, reactive power control at the importing area keeps reactive power constant and maintains voltages low at the importing area during the fault.

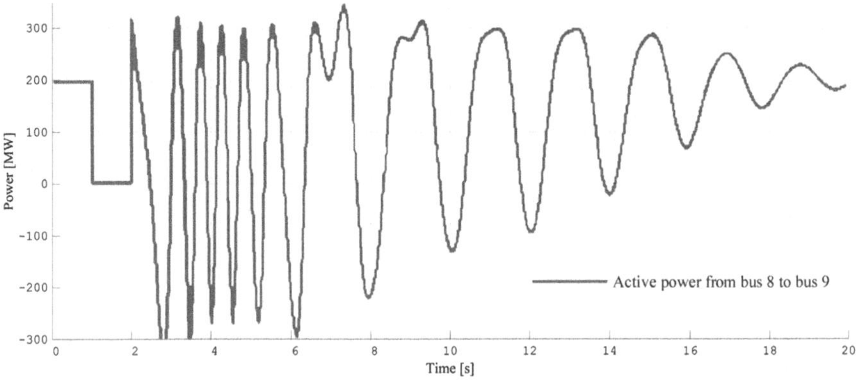

Fig. 9.24 Response of the original two-area test system for a fault cleared after the critical clearing time

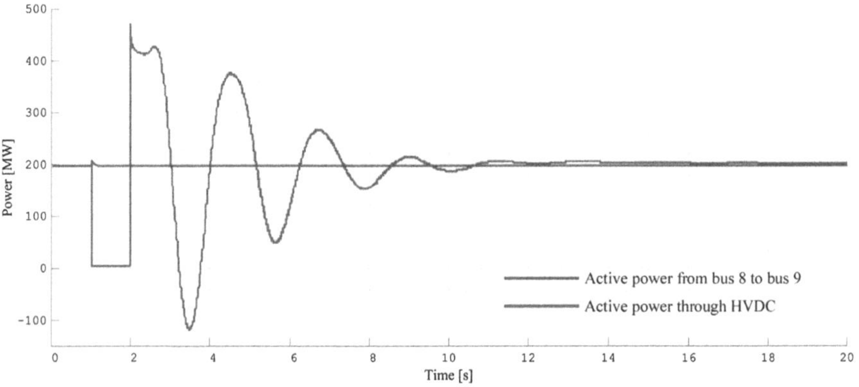

Fig. 9.25 Response of the two-area test system with VSC-HVDC link for a fault cleared after the critical clearing time

Table 9.3 Impact of control schemes of VSC-HVDC links on the critical clearing time

Original	V-control		Q-control		V-control at 7 Q-control at 9
	decrease P	decrease Q	decrease P	decrease Q	
0.990	0.585	1.130	1.380	1.380	2.165

9.4.1.2 French-Iberian Power System

Reduced sized models are obtained from a large model of the French-Spanish-Portuguese power system. Clusters have been obtained using geographical distance. Physical tie-lines among clusters are kept. Figure 9.26 shows the 24-cluster model.

Figure 9.27 contains the critical clearing times of the 24-cluster model. The lowest critical clearing time is reached in case of a fault at cluster 20 (0.24 s).

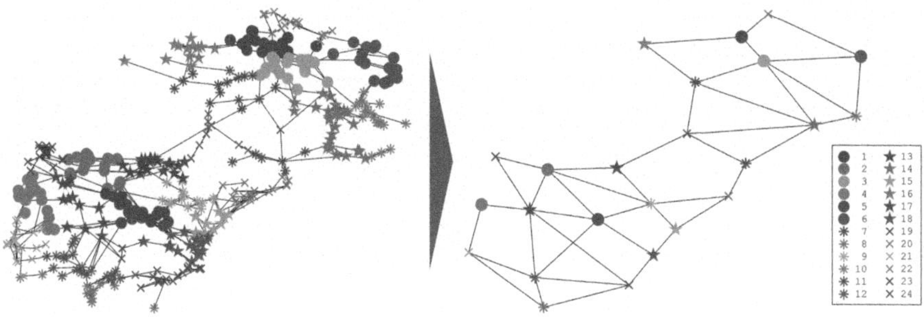

Fig. 9.26 24-cluster model

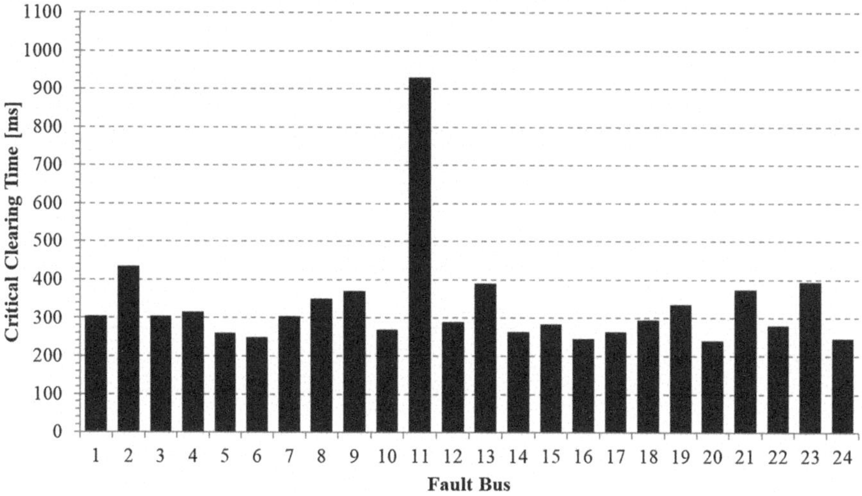

Fig. 9.27 Critical clearing times of the 24-cluster model

The critical clearing times are in accordance with the most demanding ENTSO-e fault ride through requirement. Figure 9.28 shows the responses in terms of relative angle of one generator per cluster (including the critically stable one) in case of the critically stable fault at cluster 20. Finally, Fig. 9.29 displays the generator responses in case of the critically unstable fault at cluster 20. Clearly, a generator loses synchronism during the first power swing.

The reduced-order model of the French-Iberian power system has been modified to accommodate an additional 4000 MW of wind power generation. The export from Spain to France increases from 1000 MW to 4000 MW in this case. Figure 9.30 shows and compares the critical clearing times of the 24-cluster model with current and a largely increased export. In this case, the lowest critical clearing time is reached in case of a fault at cluster 15 (35 ms). Critical clearing times are clearly below the most demanding ENTSO-e fault ride through requirement.

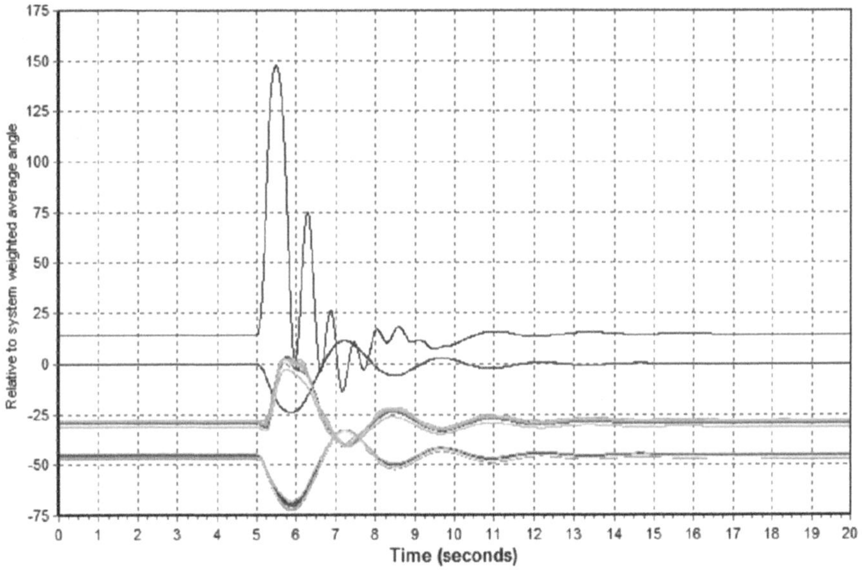

Fig. 9.28 24-cluster model – Generator responses in case of the critically stable fault at cluster 20

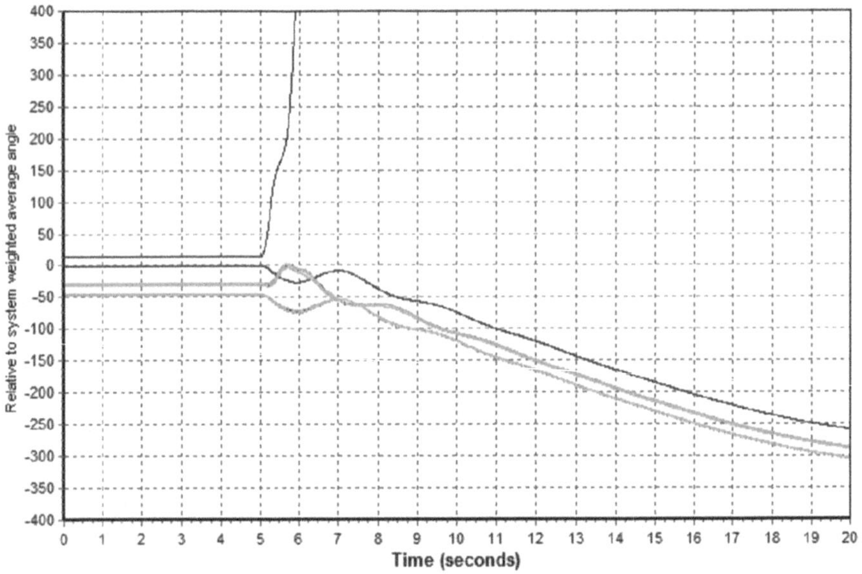

Fig. 9.29 24-cluster model – Generator responses in case of the critically unstable fault at cluster 20

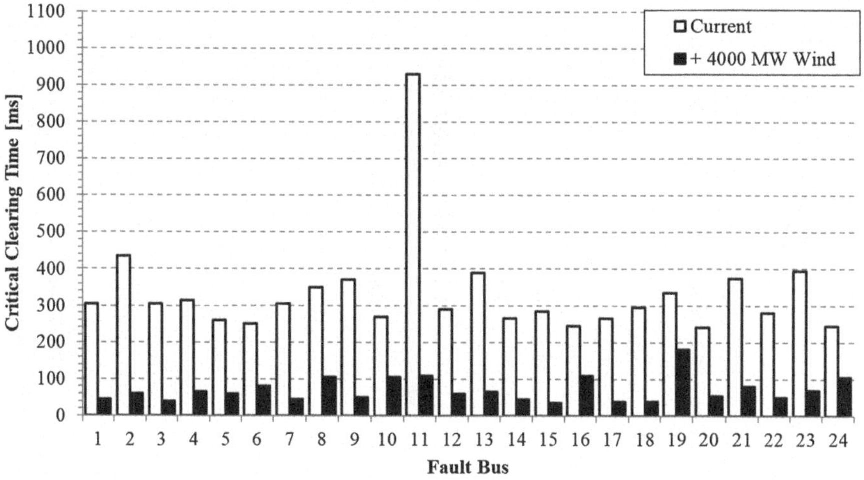

Fig. 9.30 Comparison of the critical clearing times of the 24-cluster model with current and a moderately increased export

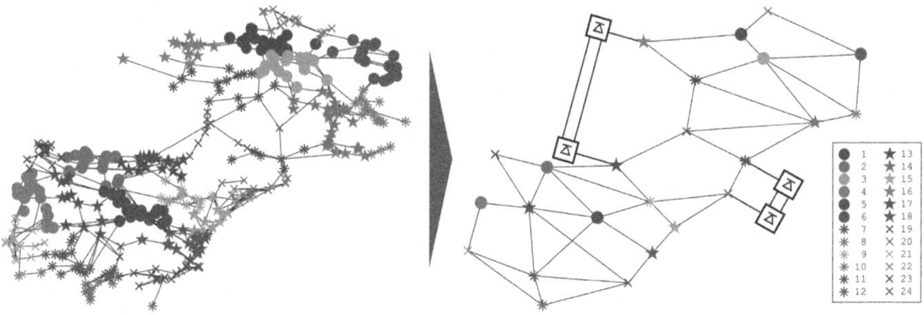

Fig. 9.31 24 cluster model with two VSC-HVDC links

Finally, the reduced-order model has been modified to accommodate an additional 4000 MW of wind power generation and by including two VSC-HVDC links. The export from Spain to France increases from 1000 MW to 4000 MW, where 2000 MW flows through the HVDC links. Figure 9.31 shows the modified 24-cluster model.

Figure 9.32 shows and compares the critical clearing times with the current and a largely increased power export to France with two HVDC links in the latter case. By comparing Figs. 9.32 and 9.30, it can be seen that the HVDC links increase the critical clearing times and bring them back to values prior to the 2increase of power export.

Figure 9.33 shows and compares the critical clearing times for the case of a largely increased export with two HVDC links but considering the limited fault ride through capability of the converters. In other words, for faults at or close to the converter terminals, the converter trips after 250 ms. The limited fault ride through

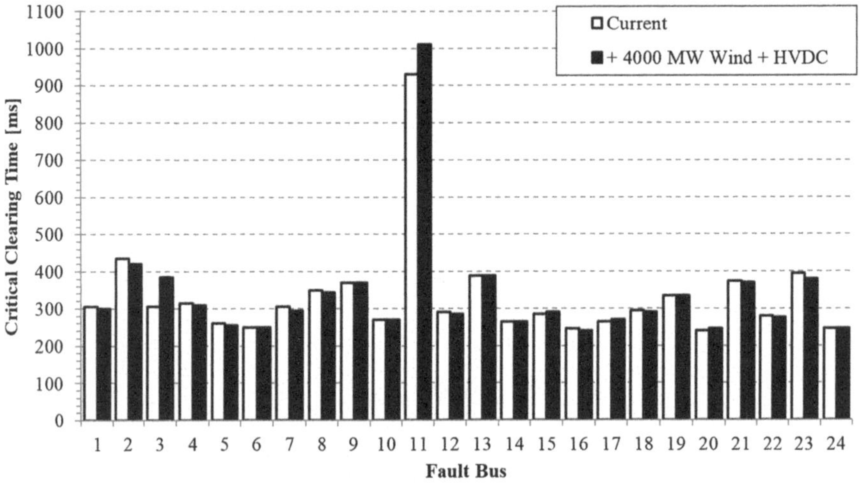

Fig. 9.32 Comparison of the critical clearing times of the 24-cluster model with current and a largely increased export with two HVDC links

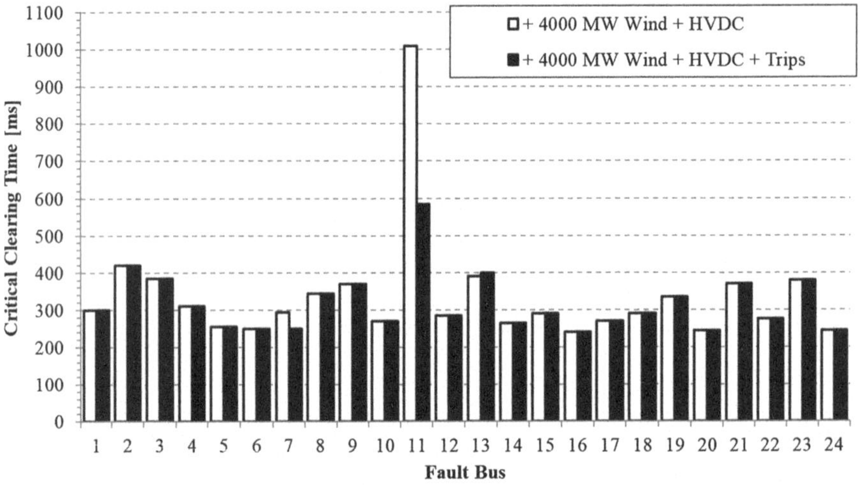

Fig. 9.33 Comparison of the critical clearing times of the 24-cluster model with two HVDC links with and without contemplating HVDC tripping

capability only affects critical clearing times for faults at buses 7, 11, and 13, where the HVDC links are connected to (the impact of fault at bus 9 is negligible).

A possible way to mitigate the impact of low voltage HVDC tripping is to redistribute the lost power flow on the remaining HVDC link. A centralized control scheme would be necessary, detecting the HVDC tripping and sending the new power set point to the intact HVDC link. Figure 9.33 shows the critical clearing times considering HVDC tripping and redistribution of active power flow. In fact,

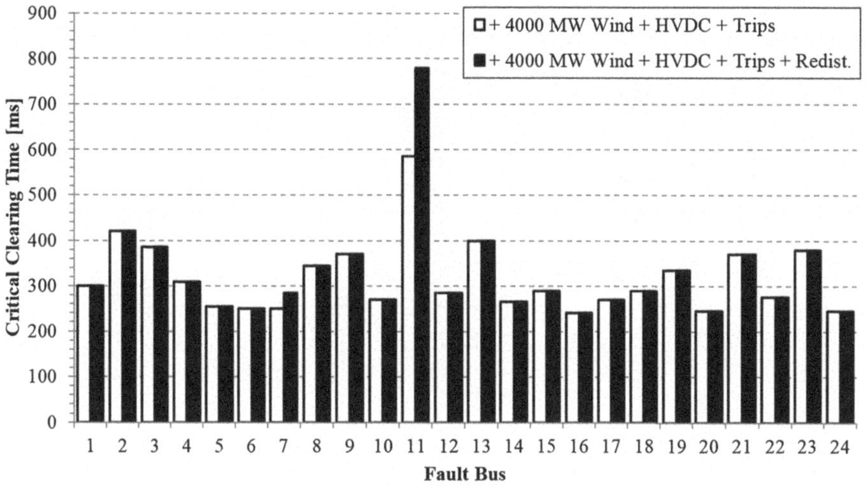

Fig. 9.34 Comparison of the critical clearing times of the 24-cluster model with two HVDC links contemplating HVDC tripping with and without controlling active power of the remaining HVDC link

the power flow through the remaining link has been increased by about 1000 MW (each pole has been modeled as a proper link) (Fig. 9.34).

9.4.2 *Impact of HVDC on Small-Signal Stability*

Modal analysis is used to evaluate small-signal stability. Instability is equivalent to a mode with a negative damping, i.e., with a positive real part. Small-signal stability of the Cigré Nordic 32A test system will be analyzed. The Nordic 32A system represents a 32-bus equivalent system of the Swedish grid. A description and the data of the Nordic32A system (including full generator + control models) can be found in [20]. The system has been modified by including a three-terminal DC grid as shown in Fig. 9.35 [22]. The impact of the multi-terminal DC (MTDC) grid with and without supplementary controls will be studied. In either case, the MTDC system applies a distributed droop control scheme for the DC voltage control.

The Nordic 32A system exhibits three inter-area modes. Two modes are dominated by oscillations between the North area and Central and South areas, and one mode is dominated by the oscillations between North and South areas and the Central area. Figure 9.36 shows the mode shape of the one of the inter-area modes between the North area and Central and South areas. The mode shape graphically shows the right eigenvectors of the electromechanical state variables, i.e., the relative activities of these state variables when the inter-area mode is excited. Figure 9.36 illustrates how the activity of the rotor speeds of the machines in the North area opposes in phase the activity of the rotor speeds of the machines in the South and Central areas.

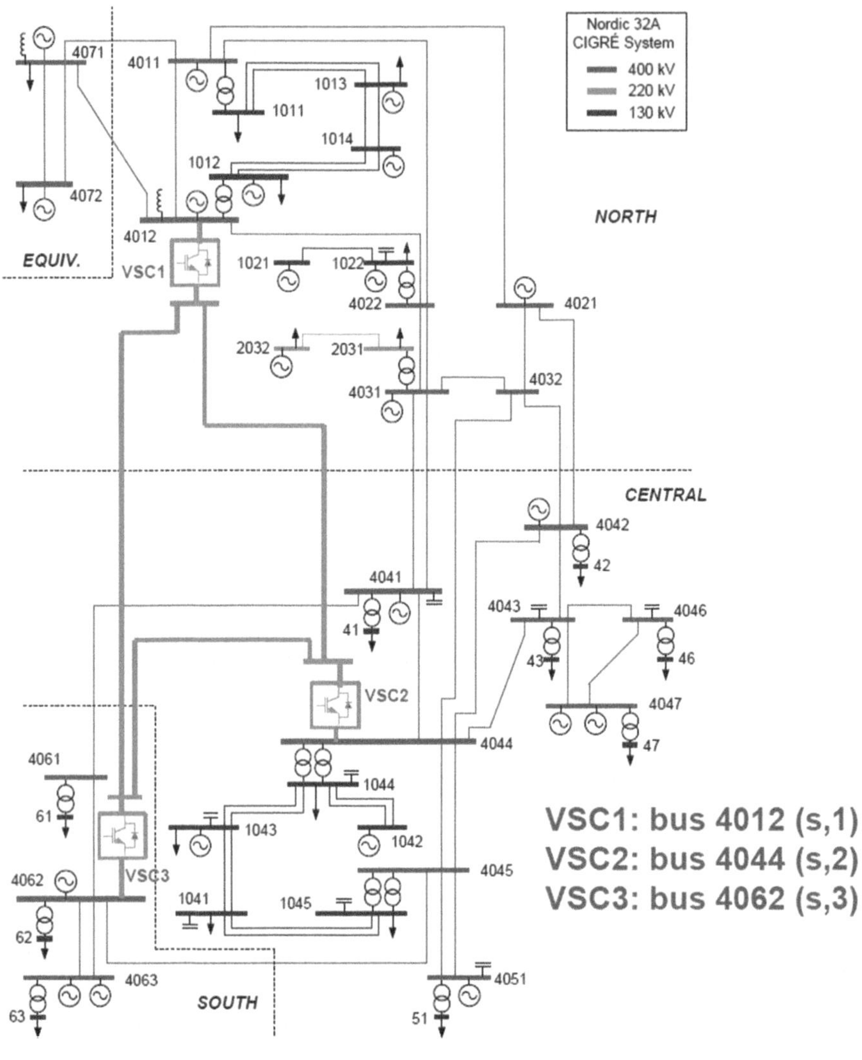

Fig. 9.35 Nordic 32 A power system [22]

Supplementary control of the MTDC system consists of varying the active power reference according to the difference of the frequency of an AC terminal of the MTDC system with respect to a reference frequency. The reference frequency could be either a fixed reference value or the weighted average of the terminal frequencies of the MTDC system AC terminals. Figure 9.37 shows the supplementary active power set point modulation of the MTDC system [22]. In case of the Nordic 32A system, a gain K_{sc} of 200, a time constant T_{sc} of 0.1 s, and a wash-out T_w of 15 s have been used.

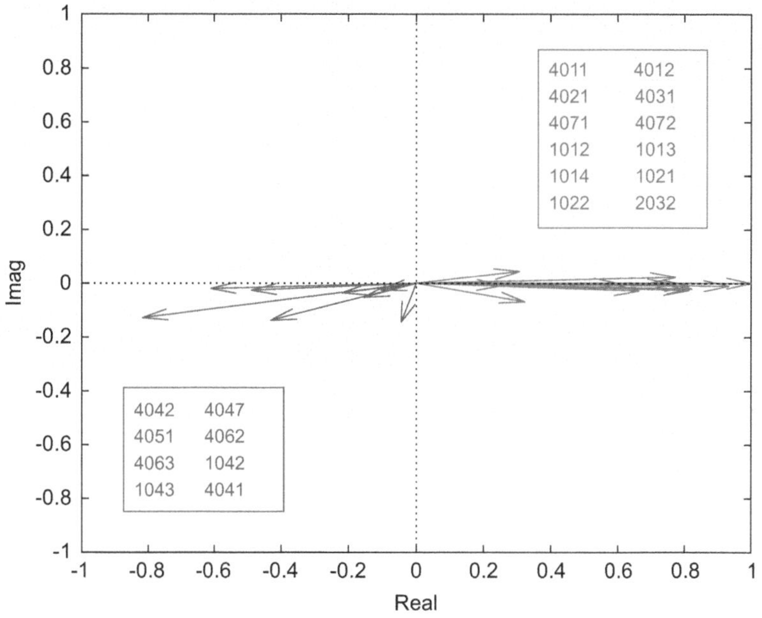

Fig. 9.36 Mode shape of the inter-area mode North against Central-South without multi-terminal DC grid

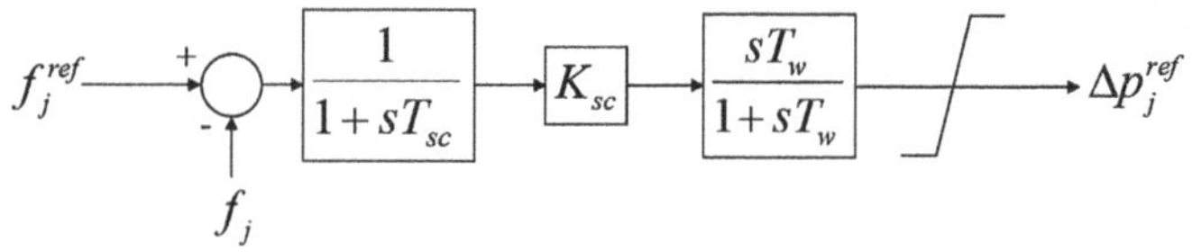

Fig. 9.37 Supplementary active power set point modulation of the MTDC system

Table 9.4 Inter-area mode North against Central-South with and without multi-terminal DC grid and supplementary active power control

	Real	Imag	Damp [%]	Freq [Hz]
No MTDC	−0.156	3.726	4.18	0.59
MTDC – without supplementary control	−0.159	4.010	3.96	0.64
MTDC – with supplementary control	−0.529	4.919	10.69	0.78

Table 9.4 shows the eigenvalue, damping, and frequency of the inter-area mode North against Central-South with and without multi-terminal DC grid and supplementary active power control. Without MTDC, the inter-area mode is slightly damped with a damping of 4.18%. The inclusion of a MTDC system does not significantly affect the damping of the inter-area mode. By contrast, the supplementary control increases the damping to 10.69%.

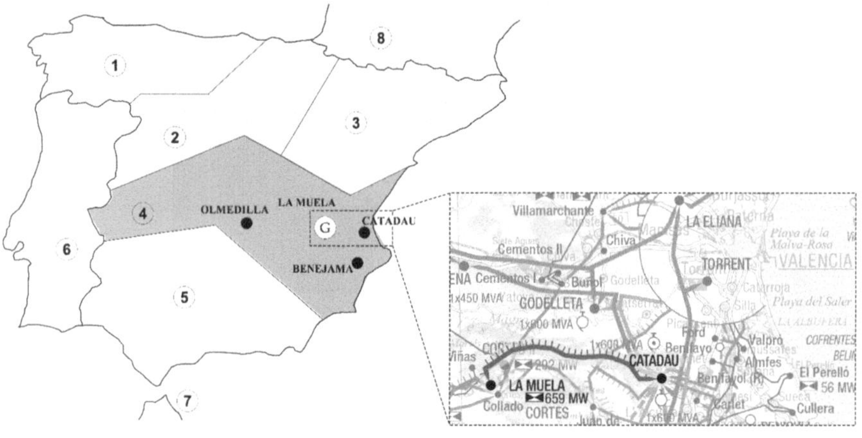

Fig. 9.38 Spanish power system areas, and geographical location of La Muela, Catadau, Olmedilla, and Benejama substations

9.4.3 *Impact of Transmission Lines on Voltage Stability*

A meshed transmission network provides robustness to the voltages in the system [9]. A meshed transmission network will have higher voltages, and lines and transformers will be less loaded. This effect causes an increase in the voltage's stability indices, such as the load margin. To illustrate this, this section considers a highly loaded scenario of the operation of the Spanish power system and compares critical load margins for different network configurations. The Spanish power system model contains 1097 buses, 1683 branches, and 428 generators and includes representations of the French, Portuguese, and Moroccan systems.

The Spanish power system and the external systems are divided into eight areas. Northwest and north areas (area 1 and 2) are usually exporting areas, and east, center, and south areas (area 3, 4 and 5) are importing areas. Areas 6, 7, and 8 correspond to the areas representing the Portuguese, Moroccan, and French systems, respectively. The analysis will be focused in the 400 kV double circuit La Muela-Catadau, located in area 4 (Center area), near to Valencia. This branch is used to evacuate part of the energy produced by the hydraulic generating units from La Muela (659 MW) to the Catadau substation, where the energy is spread among different 400 kV, 220 kV, and 132 kV circuits. To watch the evolution of system voltages, 400 kV buses of Catadau, Olmedilla, and Benejama have been selected as pilot buses. Figure 9.38 depicts the geographical location of the eight areas in the Spanish power system model and of the La Muela, Catadau, Olmedilla, and Benejama substations. In Fig. 9.38, the Center area has been highlighted. Then, technical data for each of the two 400 kV circuits La Muela-Catadau are presented in Table 9.5, including length in km, parameters (series resistance, series reactance and shunt susceptance in pu), and rate in MVA.

Table 9.5 Parameters of each of the two 400 kV circuits placed between Catadau and La Muela substations

From	To	Length [km]	Rs [pu]	Xs [pu]	Bp [pu]	Rate [MVA]
La Muela	Catadau	43	0.00054	0.00749	0.25723	1560

Table 9.6 Active power balance [MW] for the Base Case scenario

Area	GEN	DEM	LOSS	EXP
1-NORTHWEST	9464.5	3328.6	181.0	5954.9
2-NORTH	3878.4	3712.5	278.5	−112.6
3-EAST	5978.2	7380.2	155.8	−1557.7
4-CENTER	7638.5	11230.0	332.9	−3924.4
5-SOUTH	3249.6	4548.8	139.3	−1438.5
TOTAL	30209.1	30200.1	1087.5	−1078.4
6-PORTUGAL	6761.2	6227.0	83.3	450.9
7-MOROCCO	3248.0	3023.1	9.8	215.2
8-FRANCE	9859.6	9272.8	174.5	412.3

Table 9.7 Load margin to voltage collapse variation due to changes in the network topology

	Initial Demand [MW]	Maximum Demand [MW]	Load Margin [MW]
Original	30200.1	35318.5	5118.5
1 circuit opened	30200.1	34943.7	4743.7
2 circuits opened	30200.1	33664.4	3464.4

To study the effects of transmission lines over voltage stability, a highly loaded hourly scenario of the operation of the Spanish power system has been selected. Table 9.6 contains the active power balance broken down by areas.

It can be seen in Table 9.6 that the Spanish network imports energy from all external systems, i.e., the Portuguese (451 MW), the French (412 MW), and the Moroccan (215 MW). Within the Spanish system, only area 1 (northwest) exports energy, whereas the rest of areas import energy. The center area appears to be the most demanding area (3924 MW). The total load of the Spanish system sums up to 30,209 MW.

To study how the double circuit selected influences voltage stability, nose curves have been computed considering three different scenarios: the original base case scenario with both circuits working; with one of the circuits opened; and with both circuits opened. Table 9.7 presents the initial (base case) and maximum (point of collapse) Spanish demands, and corresponding load margin, for each of the three scenarios considered.

Table 9.7 shows that the load margin gets reduced as long as the circuits of the La Muela-Catadau 400 kV line are opened one at a time. From an initial demand of 30,200 MW, the system is capable of increasing its demand up to 35,319 MW before collapse, i.e., the load margin is 5119 MW (16.9%). When one of the circuits in the La Muela-Catadau 400 kV line is opened, the maximum demand drops down to 34,944 MW; thus, the load margin gets reduced by 375 MW. Finally, in case

Table 9.8 Voltages [kV] at pilot buses variation due to changes in the network topology

Bus	Original	1 circuit opened	Variation	2 circuit opened	Variation
CATADAU	374.8	370.4	−4.4	355.0	−19.8
BENEJAMA	370.7	368.5	−2.2	360.8	−9.9
OLMEDILLA	384.7	383.2	−1.6	377.6	−7.1

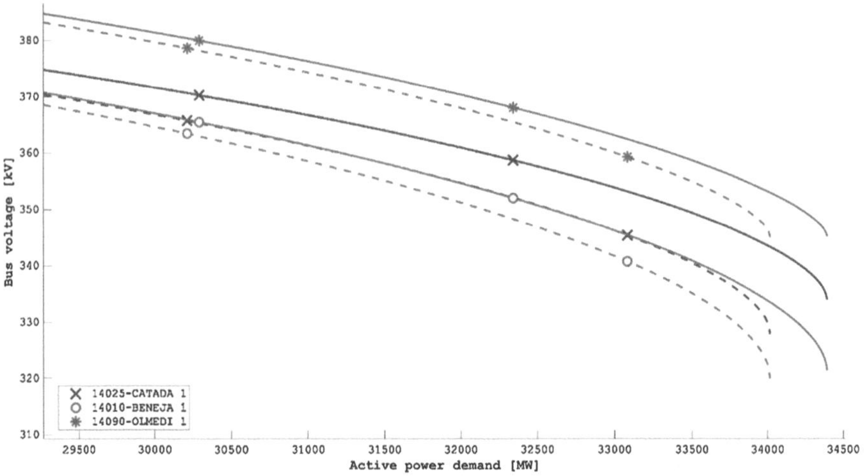

Fig. 9.39 Nose curves for pilot buses when one of the Catadau-La Muela circuits is opened

both circuits are opened, then the system can increase its demand to up to only 33,664 MW, which implies a new load margin of only 3464 MW (11.5%), 1655 MW lower than the base case load margin.

To illustrate the behavior of bus voltages, Table 9.8 presents the voltages at the pilot buses in kV for the different three scenarios considered and corresponding differences. Then, Figs. 9.39 and 9.40 compare the nose curves at the pilot buses obtained for base case and those obtained for the one and two circuits opened scenarios.

Table 9.8 shows how pilot bus voltages fall as long as the circuits are opened. Focusing on the most affected bus (Catadau), its initial voltage of 375 kV drops to 370 kV when one of the circuits is opened, and to 355 kV if both of the circuits are out. It can be seen in the non-linear behavior of removing/adding lines, since one circuit lost implies a drop of only 5 kV, but on removing both of the circuits, the voltage drop increases to 20 kV.

A useful index to test the robustness of the system voltages is the slope of the nose curves, i.e., the sensitivity of bus voltages with the active power demand. Figures 9.39 and 9.40 illustrate the previous results obtained, i.e., how removing two circuits implies much more load margin reduction (see Table 9.7) and larger voltages drops (see Table 9.8). However, Figures 9.39 and 9.40 also show how the slope of each of the nose curves becomes more negative as long as the circuits are removed. In Figure

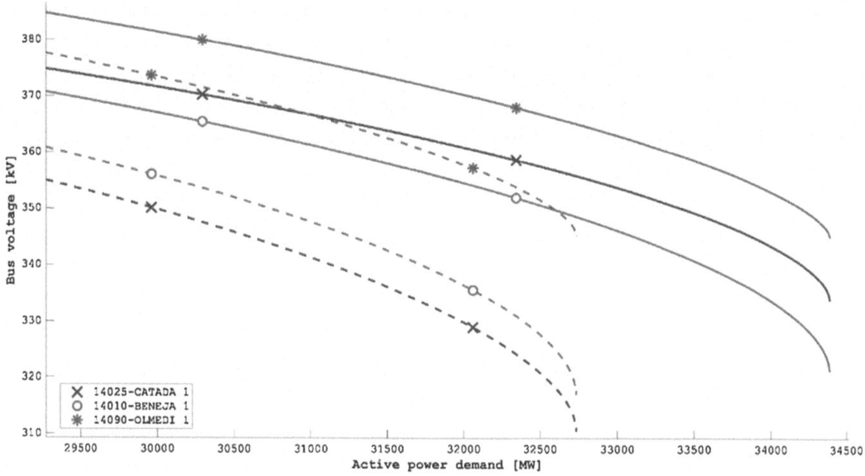

Fig. 9.40 Nose curves for pilot buses when both of the Catadau-La Muela circuits are opened

9.39, the nose curves without one of the circuits (dashed line) are almost equal to the nose curves of the base case scenario (continuous line). However, in Fig. 9.40, the difference between the nose curves without both of the circuits (dashed line) and the nose curves of the base case scenario (continuous line) becomes remarkable. It can be noticed that the sensitivities of bus voltages with the active power demand considering both circuits opened are clearly more negative than the sensitivities in the base case. All these results highlight how relevant a line can be in terms of the robustness of the stability of system bus voltages.

9.5 Conclusions

In the present chapter, the relevance of stability studies to elaborate the planning schedule for the network in various time horizons, from short-term to long-term, has been exposed. Stability of a power system is concerned with its ability to maintain acceptable network magnitudes (bus voltages, branch power flows…) under normal conditions and after being subjected to a disturbance. Stability is related with both reliability (mid-term and long-term) and security (short-term).

Traditionally, stability studies have been considered as a short-term issue; thus, TSOs elaborated the long-term planning schedule for the network without considering stability studies. However, this perspective is gradually changing due to the sustained growth of the demand, the expansion of transmission networks, the increasing number of power electronics associated with renewable energy resources and advance control devices, and higher inter-area energy interchanges, among other diverse factors. In addition, network expansion planning goal is to develop the network as economically as possible while maintaining acceptable reliability levels. Nowadays, planning schedules include not only traditional reinforcements (synchronous plants, AC lines, transformers . . .) but also reinforcement alternatives such as HVDC lines, FACTS devices, large-scale PV plants, and wind farms, among others. All the arguments listed in this chapter allow us to conclude that stability studies have become key tasks in the long-term planning schedule for transmission and distribution networks.

To support the conclusions of this chapter, some study cases have been presented where the different types of stability are covered. All these cases illustrate the influence of transmission network configuration and topology among system stability. Firstly, the impact of an HVDC link on the transient stability has been illustrated using a small-scale test system with two areas weakly connected and reinforced with a VSC-HVDC link between areas, and the French-Iberian power system reinforced with an additional 4000 MW of wind power generation and by including two VSC-HVDC links. For both networks, critical fault clearing times have been compared with and without the corresponding reinforcements. Secondly, the impact of an HVDC link on small-signal stability has been studied by carrying out a modal analysis in the Nordic 32A system, which represents an equivalent system of the Swedish grid, modified by including a three-terminal DC grid. Modal analysis shows the impact of the MTDC grid with and without supplementary controls. Finally, the impact of transmission lines on voltage stability is analyzed using an actual scenario of the Spanish electric power system. Distance to voltage collapse point was computed and compared with and without considering different expansion alternatives.

References

1. C.A. Cañizares, On bifurcations, voltage collapse and load modeling. Power Syst. IEEE Trans. **10**(1), 512–522 (1995)
2. Cañizares, C.A., 2002. Voltage Stability Assessment: Concepts, Practices and Tools
3. C.A. Cañizares, F.L. Alvarado, Point of collapse and continuation methods for large AC/DC systems. IEEE Trans. Power Syst. **8**(1), 1–8 (1993)
4. K.E. Chen, A. Hussein, M.E. Bradley, H. Wan, A performance-index guided continuation method for fast computation of saddle-node bifurcation in power systems. IEEE Trans. Power Syst. **18**(2), 753–760 (2003)
5. B. Cova, O. Fosso, D. D'hoop, E. Ãhlund, A. Braun, K. Lakmeeharan, E. Lerch, J. Lrivacoba, T. Nguyen, Y. Noro, V. Sermanson, Y. Sankar, W. Sattinger, F. Silvestro, Application and required developments of dynamic models to support practical planning. ELECTRA, 230 (2007)
6. A.C.Z. De Souza, C.A. Canizares, V.H. Quintana, New techniques to speed up voltage collapse computations using tangent vectors. Power Syst. IEEE Trans. **12**(3), 1380–1387 (1997)
7. I. Dobson, Observations on the geometry of saddle node bifurcation and voltage collapse in electrical power systems. Circ. Syst. I Fundam. Theory Appl. IEEE Trans. **39**(3), 240–243 (1992)
8. I. Dobson, H. Glavitsch, L.I.U. Chen-Ching, Y. Tamura, K. Vu, Voltage collapse in power systems. Circ. Devices Mag. IEEE **8**(3), 40–45 (1992)

9. F.M. Echavarren, E. Lobato, R. Rouco, Contingency analysis: Feasibility identification and calculation algorithm. IEE Proc. Gener. Transm. Distrib. **152**(5), 645–652 (2005)
10. F.M. Echavarren, E. Lobato, L. Rouco, T. Gomez, Formulation, computation and improvement of steady state security margins in power systems. Part I: Theoretical framework. Int. J. Electr. Power Energy Syst. **33**(2), 340–346 (2011a)
11. F.M. Echavarren, E. Lobato, L. Rouco, T. Gómez, Formulation, computation and improvement of steady state security margins in power systems. Part II: Results. Int. J. Electr. Power Energy Syst. **33**(2), 347–358 (2011b)
12. I.A. Hiskens, D.J. Hill, Energy functions, transient stability and voltage behaviour in power systems with nonlinear loads. IEEE Trans. Power Syst. **4**(4), 1525–1533 (1989)
13. C. Hsiao-Dong, R. Jean-Jumeau, Toward a practical performance index for predicting voltage collapse in electric power systems. IEEE Trans. Power Syst. **10**(2), 584–592 (1995)
14. IEEE Power System Relaying Committee, *Voltage Collapse Mitigation* (1996)
15. G.D. Irisarri, X. Wang, J. Tong, S. Mokhtari, Maximum loadability of power systems using interior point nonlinear optimization method. IEEE Trans. Power Syst. **12**(1), 162–172 (1997)
16. P. Kundur, N.J. Balu, M.G. Lauby, *Power System Stability and Control* (McGraw-Hill, 1994)
17. P. Kundur, J. Paserba, V. Ajjarapu, G. Andersson, A. Bose, C. Canizares, N. Hatziargyriou, D. Hill, A. Stankovic, C. Taylor, T. Van Cutsem and V. Vittal, 2004. Definition and Classification of Power System Stability IEEE/CIGRE Joint Task Force on Stability Terms and Definitions
18. D.H.A. Lee, Voltage stability assessment using equivalent nodal analysis. IEEE Trans. Power Syst. **31**(1), 454–463 (2016)
19. P.A. Lof, T. Smed, G. Andersson, D.J. Hill, Fast calculation of a voltage stability index. IEEE Trans. Power Syst. **7**(1), 54–64 (1992)
20. M. Stubbe (Convener). *Long Term Dynamics, Phase II, Cigré Task Force 38.02.08 – TB 102.* CIGRE (1995)
21. J. Machowski, P. Kacejko, S. Robak, P. Miller, M. Wancerz, Power system stability screening for long-term planning of transmission network development, in *2013 12th International Conference on Environment and Electrical Engineering 2013*, (2013), pp. 100–105
22. Renedo, J., García-Cerrada, A. and Rouco, L., 2016a. Active Power Control Strategies for Transient Stability Enhancement of AC/DC Grids with VSC-HVDC Multi-Terminal Systems
23. J. Renedo, A. García-Cerrada, L. Rouco, *Reactive-Power Coordination in VSC-HVDC Multi-Terminal Systems for Transient Stability Improvement* (IEEE Transactions on Power Systems, 2016b), pp. 1–1
24. L. Rouco, Coordinated design of multiple controllers for damping power system oscillations. Int. J. Electr. Power Energy Syst. **23**, 517–530 (2001)
25. L. Rouco, F. Pagola, An eigenvalue sensitivity approach to location and controller design of controllable series capacitor for damping power system oscillations. Power Syst. IEEE Trans. **12**, 1660 (1997)
26. R. Seydel, A Continuation Algorithm with Step Control, in *Numerical Methods for Bifurcation Problems.*, Aug 22–26, 1983, (1984), pp. 480–494
27. L. Sigrist, L. Rouco, *Damping Electromechanical Oscillations with Both Ratio and Phase Modulation of Static Synchronous Series Sources: Controller Design* (2007), p. 316
28. L. Sigrist, L. Rouco, An induction motor model for system frequency response models. Int. Trans. Electr Energy Syst. **27** (2017)
29. L. Sigrist, I. Egido, L. Rouco, A method for the design of UFLS schemes of small isolated power systems. IEEE Trans. Power Syst. **27**, 951–958 (2012)
30. L. Sigrist, I. Egido, L. Rouco, Principles of a centralized UFLS scheme for small isolated power systems. IEEE Trans. Power Syst. **28**(2), 1779–1786 (2013)
31. L. Sigrist, F. Echavarren, L. Rouco, P. Panciatici, A fundamental study on the impact of HVDC lines on transient stability of power systems, in *2015 IEEE Eindhoven PowerTech*, (2015), pp. 1–6
32. L. Sigrist, E. Lobato, F.M. Echavarren, I. Egido, L. Rouco, *Island Power Systems* (CRC Press, 2016)

33. L. Sigrist, L. Rouco, F. Echavarren, A review of the state of the art of UFLS schemes for isolated power systems. Int. J. Electr. Power Energy Syst. **99**, 525–539 (2018)
34. Sorensen, P., Fortmann, J., Jiménez Búendía, F., Bech, J., Morales, A. and Ivanov, C., 2014. Final Draft International Standard IEC 61400-27-1
35. C.W. Taylor, N.J. Balu, D. Maratukulam, *Power System Voltage Stability* (McGraw-Hill Ryerson, Limited, 1994)
36. A. Trias, 2012. The Holomorphic Embedding Load Flow Method
37. T. Van Cutsem, *Voltage Stability of Electric Power Systems* (Springer, 2007)

Chapter 10
Energy Storage Systems in Transmission Expansion Planning

Reza Hemmati

10.1 Introduction

The transmission expansion planning (TEP) is an optimization programming that determines optimal location, size, capacity, time, type, and number of new installed lines in the network. This program is run to deal with load growth in the network or connecting a new generation unit to the network. The TEP is a long-term plan, and the planning horizon covers about 5–20 years [10, 12]. Over the recent years, the fast development of industry has created a big demand of energy, and such energy needs new transmission networks. As a result, TEP has attracted much more attention recently. In the USA, the investment cost on transmission lines has been increased from about 3.5 billion dollar at 1996 to about 22 billion dollar at 2016 [1].

The TEP is a mathematical programming that can be expressed as a mixed integer nonlinear programming or a mixed integer linear programming. In the linear model, the nonlinear equations are linearized to achieve a linear model. Each of the nonlinear and linear models has its own advantages and disadvantages. In the nonlinear model, it is not necessary to linearize the equations and making the final model is easier. However, solving the nonlinear model is difficult, and there is not any guarantee to find the global optimal solution. In such problems, the output may be a local optimal solution. On the other hand, the linear model needs a complicate mathematical process to convert the nonlinear equation to a linear model. As well, the linearization always comes with some errors and approximations. However, solving the linear model is easy, and the global optimal solution is guaranteed [10, 12].

The TEP can be solved by available mathematical methods for solving the linear and nonlinear constrained optimization programming. The mathematical methods

R. Hemmati (✉)
Kermanshah University of Technology, Kermanshah, Iran

S. Lumbreras et al. (eds.), *Transmission Expansion Planning: The Network Challenges of the Energy Transition*, https://doi.org/10.1007/978-3-030-49428-5_10

are often available in the available optimization software such as GAMS [2]. The common solvers for mixed integer linear model are CPLEX, and it is BARON for mixed integer nonlinear model [4]. The heuristic algorithms can also be properly applied to solve TEP. Various heuristic algorithms have been tested to solve TEP such as genetic algorithm [25], particle swarm optimization [15], ant colony [17], harmony search algorithm [21], artificial bee colony algorithm [23], and shuffled frog-leaping algorithm [8]. The heuristic optimization algorithms often are easier to model and implementation. However, there is not the mathematical proof to confirm the global optimality of their solution. On the other hand, the mathematical models are harder to implementation. But their optimal solution can be verified by mathematical proofs [10, 12].

The TEP deals with transmission network lines; it is therefore associated with power flow in the network. One of the main constraints on the TEP is to satisfy the network power flow equations under all scenarios and conditions. Two kinds of power flows have been modeled in TEP including AC power flow and DC power low. The AC power flow is the complete model including both active and reactive powers [11], and it provides much more realistic results. However, the equations of AC power flow are completely nonlinear, and it is somehow difficult to include AC power flow in the studies that aims to present a linear model. In the nonlinear programming, application of AC power flow is not a challenge; however in the linear programming, it would be a major challenge, and it needs mathematical process to convert all nonlinear equations of power flow to the linear ones. In order to deal with such trouble, DC power flow is applied in some studies. The DC power flow only includes active power, and its equations are completely linear [10, 12].

The uncertainty and its sources are the other important issues in the TEP [27]. The main sources of uncertainty may be the renewable energies such as large-scale wind farms and solar power plants [22]. The output powers produced by such renewable energy resources are not deterministic, and they follow the stochastic models. As a result, it is necessary to include such uncertainty in the TEP [20]. TEP has been addressed including both solar resources and wind farms [19]. The stochastic programming is the main method to deal with such uncertainty [14]. However, some advanced methods have also been offered to handle uncertainty such as robust programming [26] and adaptive robust programming [3].

Reliability is the other important parameter in the long-term problems such as TEP [16]. Various reliability indexes have been modeled in TEP such as LOLE, LOLP, and EENS. The security criterion such N-1 contingency may also be included in the problem [10, 12].

The TEP has been broadly addressed in electricity market [7]. In such environment, the problem aims to maximize profit of the stockholders. The nodal process is often considered in such environment, and the planning would be carried out including one or more private agents [18]. All the process needs to be confirmed by independent system operator (ISO). The ISO often checks and confirms security constraints of the network such as congestion and voltage limits, and it does not interfere in the operation of private agents [10, 12].

The TEP can be classified as static, semi-dynamic, and dynamic programming. In the static programming, all the transmission lines are installed in the first stage of the planning horizon. In the semi-dynamic planning, the years of planning horizon are divvied into several time sections each time section more than 1 year. For instance, 20-year planning horizon is classified into 5 time sections each one including 4-year. Eventually, the dynamic programming considers each time section equal to 1 year [10, 12].

The TEP can consider congestion of transmission lines and deal with this problem [24]. In such programming, the problem may relieve the congestion or the operational cost related to congestion. The congestion can be considered as a term of objective function, or it could be incorporated as a constraint to TEP [10, 12].

Application of FACTS devices in TEP has also been addressed. The FACTS devices can alter network parameters such as voltage magnitude, impedance, and voltage angle. As a result, they can make positive impacts on TEP and reduce the planning cost or postpone investment on new lines [10, 12].

The energy storage systems (ESS) are one of the new developed technologies in electric power systems. These devices store different types of energy such as mechanical, electrical, chemical, and thermal and then restore the energy when necessary [13]. The ESSs have operational time in terms of second to days. The capacitors, flywheels, and some batteries have fast and short-term operation, and they are often applied to improve power quality. On the other hand, the pumped-storage hydroelectricity (PSH), compressed air energy storage (CAES), and some batteries have slow and long-term operation, and they are usually utilized to deal with energy problems such as energy shifting over hours or days [28].

In this chapter, the purpose is to study the impacts of ESSs on TEP. A typical test network is considered to study the model. The TEP is modeled as a mixed integer linear programming. The objective function is to minimize the investment cost on new lines and new ESSs. The security constraints of the network and ESSs are also taken into account. The proposed mixed integer linear programming is solved by GAMS-CPLEX. The nonlinear equations in the model are converted to the linear equations by BIGM and binary variables. The dynamic planning with 6-year planning horizon is adopted. The results demonstrate that installing ESSs on TEP significantly reduces the planning cost as well as improves the line congestion. The ESSs operate properly to shift energy over the day hours. Such energy arbitraging defers the network reinforcement and reduces the planning cost.

10.2 Modelling and Formulation

This chapter presents TEP including ESSs. The problem finds optimal location, capacity, number, and time of new installed lines. As well, it finds optimal power, capacity, location, and charging-discharging pattern for new installed ESSs. This problem is expressed as a mixed integer linear programming and implemented in GAMS software. The objective function of the problem is the investment cost on

ESSs as shown by (10.1) and the investment cost on lines as shown by (10.2). The final objective function is achieved as (10.3). The Eqs. (10.1), (10.2), and (10.3) show the investment costs, and they are the annualized costs given as "$/year." In (10.1), the investment cost is considered on both rated power (kW) and rated capacity (KWh) of the ESSs. The rated capacity indicates the capability of ESS to store energy, and the rated power presents the capability of ESS to exchange energy at each time section usually 1 h. In (10.2), the investment cost on new lines is calculated considering length of line as well as capacity of line.

$$TIc_{ess} = \sum_{y \in Y} \sum_{n \in N} \begin{pmatrix} Er_{ess}^{y,n} \times Ic_{ess}^{c} + \\ \Pr_{ess}^{y,n} \times Ic_{ess}^{p} \end{pmatrix} \times K_{ess}^{an} \tag{10.1}$$

$$TIc_{line} = \sum_{y \in Y} \sum_{n \in N} \sum_{np \in N} \begin{pmatrix} K_{\exp}^{t,n,np} \times L_{line}^{n,np} \times \\ Ic_{line}^{n,np} \end{pmatrix} \times K_{line}^{an} \tag{10.2}$$

$$TIc_{tep} = TIc_{line} + TIc_{ess} \tag{10.3}$$

The parameters of the above equations are defined as follows:

n: Index of buses in the network
N: Set of buses in the network
y: Index of years in the planning horizon
Y: Set of years in the planning horizon
np: Index of buses in the network
TIc_{ess}: Total investment cost on new installed esss ($/year)
$Er_{ess}^{y,n}$: Rated capacity of ESS (mwh)
Ic_{ess}^{c} Investment cost on capacity of esss ($/mwh)
$\Pr_{ess}^{y,n}$ Rated power of ESS (MW)
Ic_{ess}^{p}: Investment cost on power of esss ($/MW)
K_{ess}^{an}: Coefficient for converting total cost to annual cost for esss
TIc_{line}: Total investment cost on new installed lines ($/year)
$K_{\exp}^{t,n,np}$: Expansion level of lines
$L_{line}^{n,np}$: Length of line between bus "n" and bus "np" (km)
$Ic_{line}^{n,np}$: Investment cost on line between bus "n" and bus "np" ($/km)
K_{line}^{an}: Coefficient for converting total cost to annual cost for lines
TIc_{tep}: Total investment cost of planning ($/year)

The power through the transmission lines is calculated by (10.4). It should be pointed out that the proposed planning comprises two timescales. One timescale for long-term planning that shows the planning horizon, it is 6 years. The other timescale for short-term operation of network that shows the daily hours, it is 24 h. The index "y" shows the planning horizon years, and the index "t" indicates the daily hours. As a result, the power flow is run under all days at all years.

$$P_{line}^{y,t,n,np} = \frac{\theta_v^{y,t,n} - \theta_v^{y,t,np}}{x_{line}^{n,np}} \forall n, np \in N, t \in T, y \in Y \tag{10.4}$$

The parameters of the above equation are defined as follows:

$P_{line}^{y,t,n,np}$: Power flow through line (p.u.)
$\theta_v^{y,t,n}$: Voltage angle (Radian)
$x_{line}^{n,np}$: Reactance of line (p.u.)
t: Index of time periods
T: Set of time periods

The capacity of transmission lines is limited by (10.5), and the generated power by generators is limited by (10.6). The magnitude of voltage on all buses is checked to lie in the permitted range as shown by (10.7). Eventually, the equilibrium of power on all buses of the network is confirmed by (10.8).

$$\left| P_{line}^{y,t,n,np} \right| \le MP_{line}^{n,np} \ \forall n \in N, np \in N, t \in T, y \in Y \tag{10.5}$$

$$P_{gen}^{y,t,n} \le MP_{gen}^{y,n} \ \forall n \in N, t \in T, y \in Y \tag{10.6}$$

$$\underline{V}_{bus} \le V_{bus}^{y,t,n} \le \overline{V}_{bus} \ \forall n \in N, t \in T, y \in Y \tag{10.7}$$

$$\begin{aligned} &P_{gen}^{y,t,n} + Pd_{ess}^{y,t,n} = P_{load}^{y,t,n} + \sum_{np \in N} \left(P_{line}^{y,t,n,np} \right) + Pc_{ess}^{y,t,n} \\ &\forall n \in N, t \in T, y \in Y \end{aligned} \tag{10.8}$$

The parameters of the above equations are defined as follows:

$MP_{line}^{n,np}$: Capacity of line (p.u.)
$MP_{gen}^{y,n}$: Capacity of generator (p.u.)
$V_{bus}^{y,t,n}$: Magnitude of voltage (p.u.)
$\underline{V}_{bus}, \overline{V}_{bus}$: Minimum and maximum voltage of bus (p.u.)
$Pc_{ess}^{y,t,n}$: Charging power of ESS (p.u.)
$Pd_{ess}^{y,t,n}$: Discharging power of ESS (p.u.)
$P_{gen}^{y,t,n}$: Generated power by generator (p.u.)
$P_{load}^{y,t,n}$: Power of load (p.u.)

The constraints (10.4), (10.5), (10.6), (10.7), and (10.8) show the conventional power flow formulation in the electrical networks. It means that the power flow is modeled as set of constraint on the problem and the power flow must be passed and satisfied under all scenarios and conditions.

The expansion of lines is determined by (10.9). It determines location, capacity, time, and number of new lines in the network. This integer variable is optimized by the planning to minimize the investment cost.

$$K_{\exp}^{t,n,np} = \sum_{k \le t} K_{\exp}^{k,n,np} \; \forall n \in N, np \in N, t \in T, k \in T \tag{10.9}$$

In the above equation, $k$is the index of time periods.

The TEP problem is a very flexible programming, and it can include different factors as constraint or as a part of objective function. For instance, the budget limitations can be included in TEP. The investment cost on new lines may be limited as (10.10), the investment cost on ESSs can be limited as (10.11), and total investment cost of the planning would also be limited as (10.12). As well, the other constraints, such as security, reliability, environmental pollutions, and etc., can be added to the problem.

$$TIc_{line} \le TB_{line} \tag{10.10}$$

$$TIc_{ess} \le TB_{ess} \tag{10.11}$$

$$TIc_{tep} \le TB_{tep} \tag{10.12}$$

The parameters of the above equations are defined as follows:

TB_{ess}: Total budget on ESSs ($/year)
TB_{line}: Total budget on lines ($/year)
TB_{tep}: Total budget on planning ($/year)

The ESS operation is modeled through (10.13), (10.14), (10.15), (10.16), (10.17), (10.18), and (10.19). In (10.13) and (10.14), the rated power of ESS is modeled. Constraints (10.13) and (10.14) show that the charging power and discharging power must be less than the rated power of ESS at all times and scenarios. The rated power of ESS is a design variable, and it is calculated by the planning as (10.15).

$$Pd_{ess}^{y,t,n} \le \mathrm{Pr}_{ess}^{y,n} \; \forall y \in Y, t \in T, n \in N \tag{10.13}$$

$$Pc_{ess}^{y,t,n} \le \mathrm{Pr}_{ess}^{y,n} \; \forall y \in Y, t \in T, n \in N \tag{10.14}$$

$$\mathrm{Pr}_{ess}^{y,n} = \sum_{k \le y} \mathrm{Pr}_{ess}^{k,n} \; \forall y \in Y, k \in Y, n \in N \tag{10.15}$$

The ESS efficiency is given by (10.16), and the ESS energy is calculated as (10.17). The constraint (10.18) indicates that the ESS energy must be less that the rated capacity of ESS at all times and scenarios. The rated capacity of ESS is calculated in (10.19). It is a design variable of the problem and optimized by the planning.

$$\eta_{ess}^{y,n} = \frac{\sum_{t \in T} Pd_{ess}^{y,t,n}}{\sum_{t \in T} Pc_{ess}^{y,t,n}} \quad \forall y \in Y, n \in N \tag{10.16}$$

$$E_{ess}^{y,t,n} = E_{ess}^{y,t-1,n} + \left(Pc_{ess}^{y,t,n} - Pd_{ess}^{y,t,n}\right) \times T_d^t \\ \forall y \in Y, t \in T, n \in N \tag{10.17}$$

$$E_{ess}^{y,t,n} \leq Er_{ess}^{y,n} \ \forall y \in Y, t \in T, n \in N \tag{10.18}$$

$$Er_{ess}^{y,n} = \sum_{k \leq y} Er_{ess}^{k,n} \ \forall y \in Y, k \in Y, n \in N \tag{10.19}$$

The parameters of the above equations are defined as follows:

$\eta_{ess}^{y,n}$: Efficiency of ESS (%)
$E_{ess}^{y,t,n}$: Energy of ESS (MWh)
T_d^t: Duration of time section (hour)

10.3 Test Network and Data

In this chapter, IEEE 24-bus test network is considered as test case. Figure 10.1 shows single line diagram of the network. Table 10.1 shows the bus data of test network, and Table 10.2 lists the line data. The data are taken from [6]. Figure 10.2 shows the load growth over the planning horizon, and it is clear that 6-year planning horizon is adopted. The generation growth is also assumed the same as load growth. The daily loading profile is shown by Fig. 10.3 [9]. The other data used in the planning are listed in Table 10.3 [5]. The base power for per unit system is 100 MVA.

10.4 Numerical Results and Discussions

Before providing the results of network expansion, it would be useful to study the impacts of line capacity on power flow. Because the network expansion is carried out to deal with line capacity limitation, Table 10.4 shows the generated power by each generator under different line capacities. It shows that the generated powers of generators are different under various line capacities. As a result, the line capacity

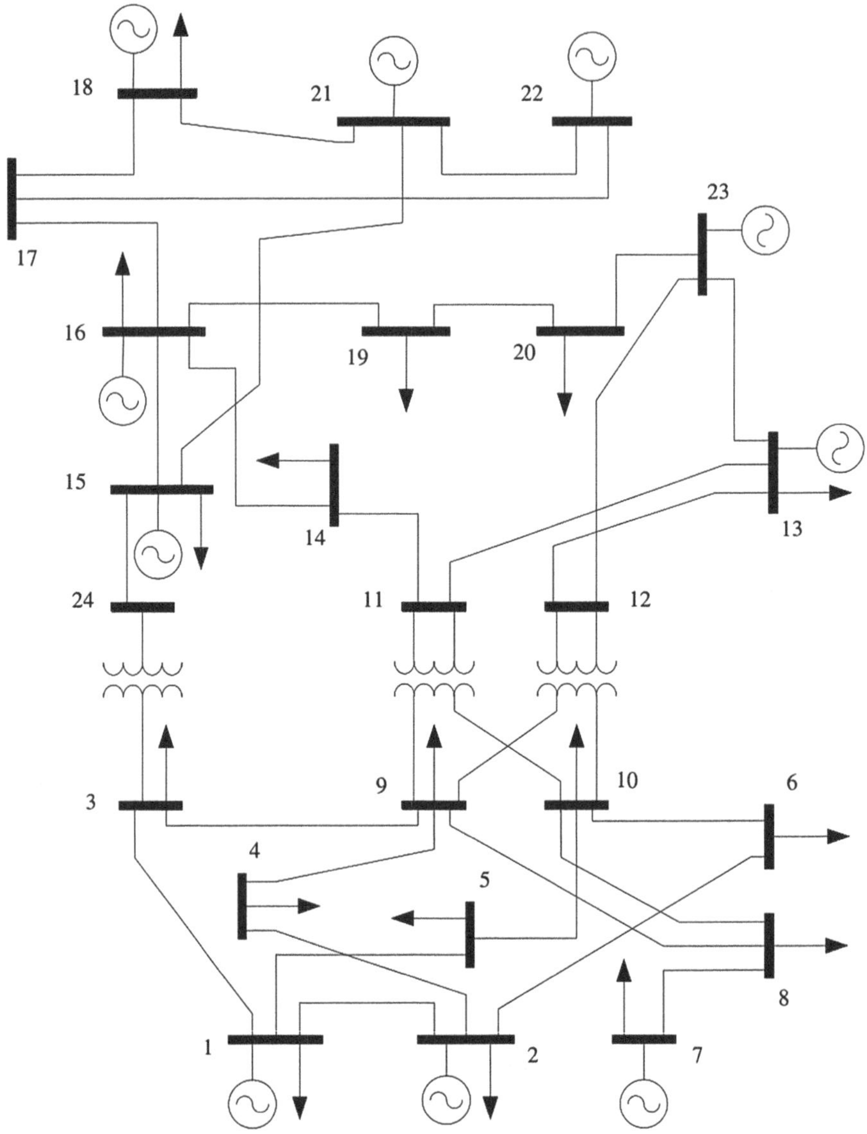

Fig. 10.1 IEEE 24-bus test network

makes significant impacts on the network operation, and it is necessary to have enough line capacity in the network in order to optimal utilization of the network. Table 10.5 also shows the generated power by each generator under different loading conditions. It is demonstrated that the generators change their production pattern to supply the demand following different scenarios.

Table 10.1 Bus data of test network

Bus number	Load (MW)	Maximum capacity of generator (MW)
1	108	192
2	97	192
3	180	0
4	74	0
5	71	0
6	136	0
7	125	300
8	171	0
9	175	0
10	195	0
11	0	0
12	0	0
13	265	591
14	194	0
15	317	215
16	100	155
17	0	0
18	333	400
19	181	0
20	128	0
21	0	400
22	0	300
23	0	660
24	0	0

10.4.1 TEP Without ESS

In order to better focus on the results, first the results of the planning without ESS are presented. The planning costs without ESS are listed in Tables 10.6 and 10.7 shows the installed lines without ESS. It is clear that six new lines are installed in the network to supply load growth over the planning horizon. Installing the lines at different years confirm that the proposed programming is a dynamic planning that considers time as a dimension of the planning. In the static planning, all new lines are installed at first year of the planning horizon, and it will increase the planning cost. However, the proposed dynamic programming is more flexible and optimal.

In order to show the impacts of line expansion on the network operation, the passed power through the line between bus 6 and bus 10 at different years of the planning horizon is listed in Table 10.8. It should be noted that this corridor (between bus 6 and bus 10) has been reinforced with a new line at second year as listed in Table 10.7. As well, the capacity of this corroder is 100 MW (1 p.u.) as shown in Table 10.2. The results show that the maximum power through this line

Table 10.2 Line data of test network

Line number	From bus	To bus	Admittance (p.u.)	Capacity of line (MW)
1	1	2	71.942	100
2	1	3	4.735	110
3	1	5	11.834	110
4	2	4	7.893	110
5	2	6	5.208	110
6	3	9	8.403	110
7	3	24	11.919	300
8	4	9	9.643	110
9	5	10	11.325	110
10	6	10	16.529	100
11	7	8	16.287	110
12	8	9	6.057	110
13	8	10	6.057	110
14	9	11	11.919	300
15	9	12	11.919	300
16	10	11	11.919	300
17	10	12	11.919	300
18	11	13	21.008	300
19	11	14	23.923	312.5
20	12	13	21.008	312.5
21	12	23	10.352	312.5
22	13	23	11.561	312.5
23	14	16	25.707	312.5
24	15	16	57.803	300
25	15	21	40.816	300
26	15	24	19.268	300
27	16	17	38.61	300
28	16	19	43.29	300
29	17	18	69.444	300
30	17	22	9.497	300
31	18	21	76.923	300
32	19	20	50	300
33	20	23	92.593	300
34	21	22	14.749	300

at first year is 1 p.u. at hours 19–20 (on-peak loading), and it means that the line is congested at these hours. However, the corridor is installed with one new line at second year, and its capacity is increased to 2 p.u. at second year of the planning horizon. As a result, in the next years, the maximum power of line is increased, and it is about 1.5 p.u. at last year of the planning horizon. These results mean that the proposed corridor needs expansion, and the network needs to pass more power

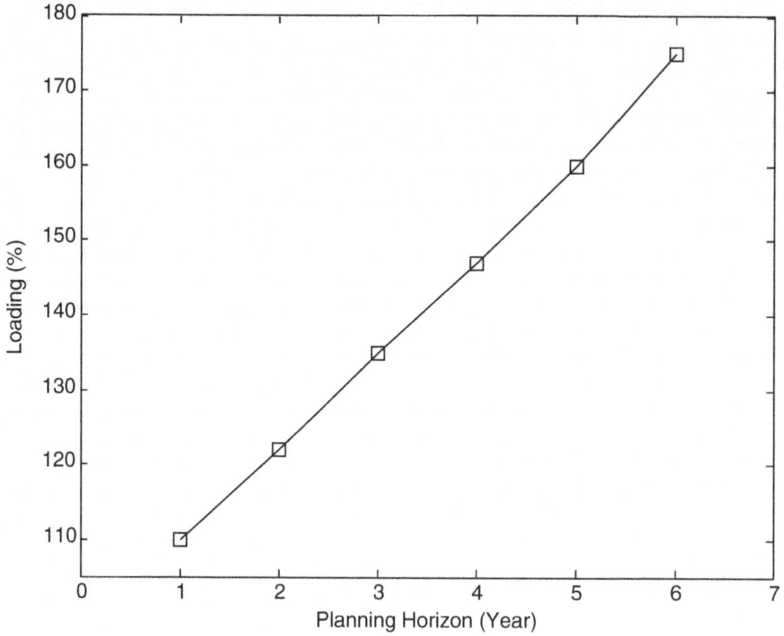

Fig. 10.2 Load growth over the planning horizon

Table 10.3 Data of the proposed TEP including ESS

Parameter	Level
ESS efficiency (%)	100
Initial energy of ESS (MWh)	0
ESS lifetime (year)	8
Transmission line lifetime (year)	15
Discount rate (%)	10
Length of every line (km)	150
Investment cost on line ($/km)	500,000 for line with 300 MW capacity (for the other lines the investment cost is decreased or increased linearly)
Investment cost on ESS power ($/MW)	120,000
Investment cost on ESS capacity ($/MWh)	120,000

through this corridor to achieve the optimal operation. As a result, the proposed planning has properly expanded this corridor at second year.

The generated power by generator 15 at different years of the planning horizon is shown by Fig. 10.4. The results show that the produced power follows different patterns at different years. As well, the generated power is increased along with the planning horizon to supply load growth. As a result, increasing generation needs more line capacity to transfer power to the load centers. Such requirement

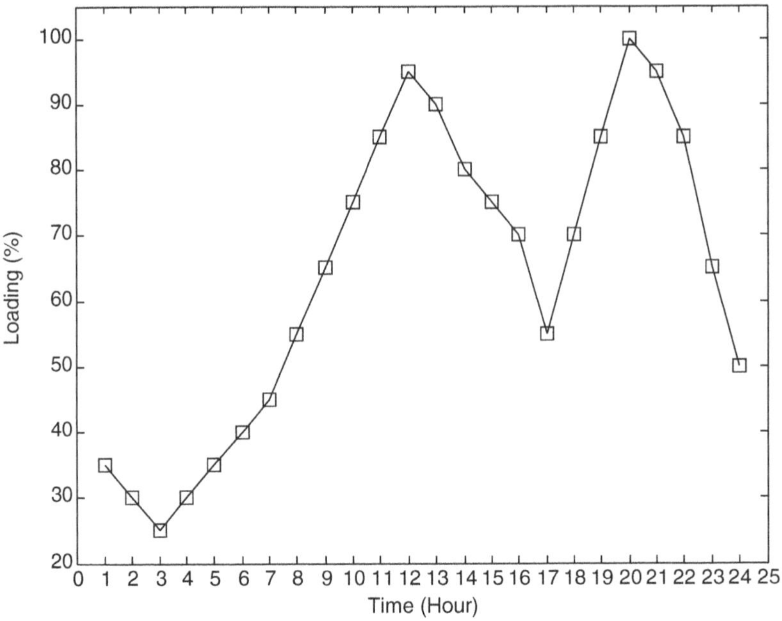

Fig. 10.3 Daily loading profile

Table 10.4 Generated power by each generator under different line capacities (p.u)

Generator number	200% capacity of lines	100% capacity of lines	90% capacity of lines
1	1.92	1.92	1.92
2	1.92	1.92	1.92
7	3	0.927	2.218
13	5.91	5.91	5.91
15	0.6	1.953	2.15
16	1.55	1.55	1.55
18	0	4	4
21	4	1.528	1.114
22	3	3	3
23	6.6	5.793	4.717

is successfully met by the proposed TEP by installing new lines in conjunction with the planning horizon.

10.4.2 TEP with ESS

The ESSs are added to the planning, and results are given in Table 10.9. In order to compare the results, the results of TEP without ESS are also presented. It is clear

Table 10.5 Generated power (p.u.) by each generator under different loading conditions (p.u)

Generator number	100% loading	110% loading
1	1.92	2.112
2	1.92	2.112
7	0.927	2.265
13	5.91	6.501
15	1.953	2.365
16	1.55	1.705
18	4	4.4
21	1.528	1.271
22	3	3.3
23	5.793	5.319

Table 10.6 Planning costs without ESSs

Cost of installed lines (million $/year)	Cost of installed ESSs (million $/year)	Total cost of plan (million $/year)
38.890	–	38.890

Table 10.7 Installed lines without ESS

Year	Installed lines
1	Line 15–21
2	Line 6–10
3	Line 16-14
4	Line 16–17
5	–
6	Line 7–8 (double line)

that installing ESSs reduces the cost by about 22%. Table 10.10 also shows the installed lines and ESSs over the planning horizon. The results show that only three lines are installed on the network, while many ESSs are installed at different years. The ESSs increase flexibility of the planning and reduce the planning cost.

The rated capacities of the ESSs are listed in Table 10.11. It is clear that the ESSs have different capacities and such capacities are optimized by the planning.

The ESSs follow different charging-discharging patterns to shift energy over the day hours and deal with line congestion and afterward defer the expansion planning. The ESS installed on bus 5 at year 3 only shows two operations during the day hours as shown by Fig. 10.5. This ESS charges energy at hour 1 (off-peak hour) and discharges it at hour 20 (on-peak loading). This minimum operation of ESS increases its lifetime, and it is achieved by the proposed planning.

Figure 10.6 shows the charging-discharging operation of ESS on bus 6 at year 3. It is demonstrated that the ESS operates at many hours to shift energy and deal with peak loading. It charges energy at off-peak hours and restores it to the grid during the on-peak time periods. Such procedure and operation help the network to deal with loading with minimum line capacity. The stored energy inside this ESS is shown by Fig. 10.7, and it is clear that the initial energy of ESS is zero and its final energy is zero too.

Table 10.8 Hourly power flow from bus 6 to bus 10 at different years (p.u)

Hour	Year 1	Year 2	Year 3	Year 4	Year 5	Year 6
1	−0.124	−0.633	−0.733	−0.799	−0.928	−0.936
2	−0.518	−0.575	−0.622	−0.679	−0.746	−0.813
3	−0.426	−0.472	−0.488	−0.139	−0.627	0.247
4	−0.531	−0.534	−0.597	−0.232	−0.278	−0.807
5	−0.565	−0.633	−0.733	−0.793	−0.39	−0.452
6	−0.531	−0.276	−0.457	−0.905	−0.615	−1.069
7	−0.256	−0.37	−0.928	−0.956	−1.101	−1.07
8	−0.951	−0.573	−0.734	−0.537	−0.605	−0.558
9	−0.254	−0.748	−0.752	−1.438	−0.94	−1.273
10	−0.856	−0.98	−0.546	−1.244	−0.936	−0.944
11	−0.682	−1.142	−1.021	−0.849	−0.983	−1.142
12	−0.807	−0.926	−0.993	−1.413	−1.378	−1.372
13	−0.715	−0.864	−0.849	−1.415	−1.089	−1.323
14	−0.983	−0.574	−0.694	−1.33	−1.099	−1.177
15	−0.867	−0.48	−1.472	−0.654	−0.82	−1.187
16	−0.749	−0.818	−0.878	−0.56	−0.975	−1.3
17	−0.09	−0.67	−0.611	−0.718	−0.848	−1.054
18	−0.398	−0.397	−0.467	−1.485	−0.958	−0.898
19	−1	−0.717	−0.739	−0.848	−0.983	−1.139
20	−1	−1	−1.109	−1.154	−1.303	−1.489
21	−0.808	−0.904	−1.013	−1.044	−1.196	−1.399
22	−0.643	−0.743	−0.739	−1.297	−0.983	−1.142
23	−0.64	−0.405	−0.395	−1.446	−0.558	−1.229
24	−0.4	−0.024	−0.59	−1.12	−1.053	−1.349

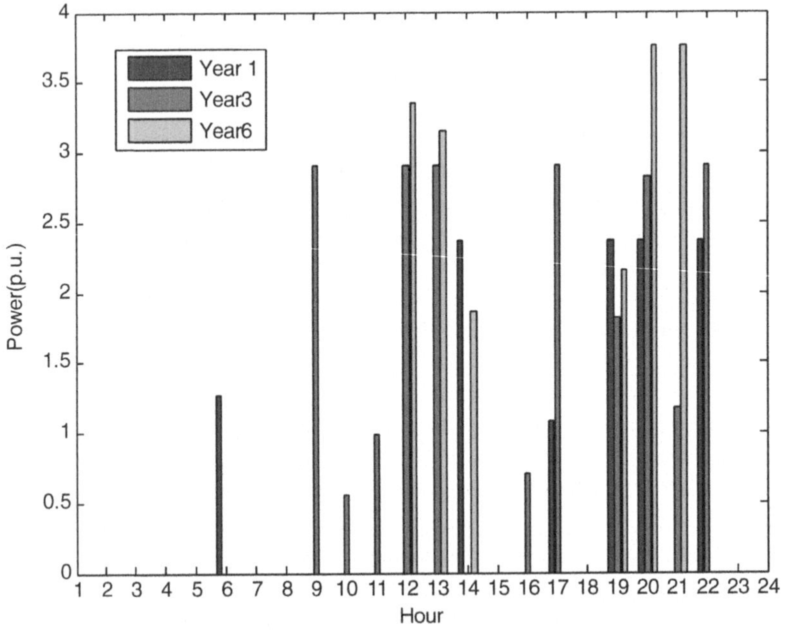

Fig. 10.4 Generated power by generation 15 at different years

Table 10.9 Planning costs with ESS

	Cost of installed lines (million \$/year)	Cost of installed ESSs (million \$/year)	Total cost of plan (million \$/year)
TEP with ESS	16.408	14.905	31.313
TEP without ESS	38.890	–	38.890

Table 10.10 Locations of Lines and ESSs

Year	Installed lines	Installed ESSs
1	–	One ESS on bus 6 One ESS on bus 20
2	–	One ESS on bus 14 One ESS on bus 20
3	Line 15–21	One ESS on bus 5
4	–	One ESS on bus 14
5	Line 7–8 (double line)	One ESS on bus 6
6	–	One ESS on bus 5 One ESS on bus 6 One ESS on bus 14 One ESS on bus 20

Table 10.11 Rated capacity of ESSs (MWh)

Year	Bus 5	Bus 6	Bus 14	Bus 20
1	0	108.8	0	8.6
2	0	0	65.4	24.2
3	7.7	0	0	0
4	0	0	29.7	0
5	0	10.7	0	0
6	7.7	3.8	74.1	48.7

The hourly power from bus 6 and bus 10 is listed in Table 10.12. This power was already shown for the case without ESS in Table 10.8. Now, comparing Tables 10.8 and 10.12 is beneficial. In this case, TEP with ESS, the corridor between bus 6 and bus 10 is not installed with new line. As a result, the maximum power of the corridor at all hours and years is less than 1 p.u., while in the previous case, TEP without ESS, the corridor is installed with one new line, and the maximum power through the corridor increases to more than 1 p.u.

10.4.3 Sensitivity Analysis

In order to study the impacts of the parameters on the planning, a sensitivity analysis is carried out on several important parameters, and results are summarized in Table 10.13. For better analysis, the results of TEP with ESS and TEP without ESS are

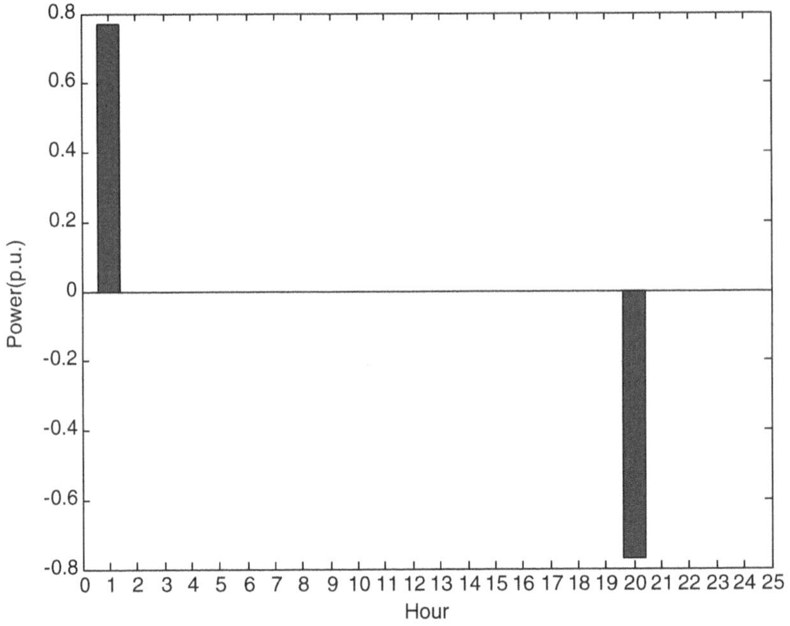

Fig. 10.5 Charging-discharging operation of ESS on bus 5 at year 3

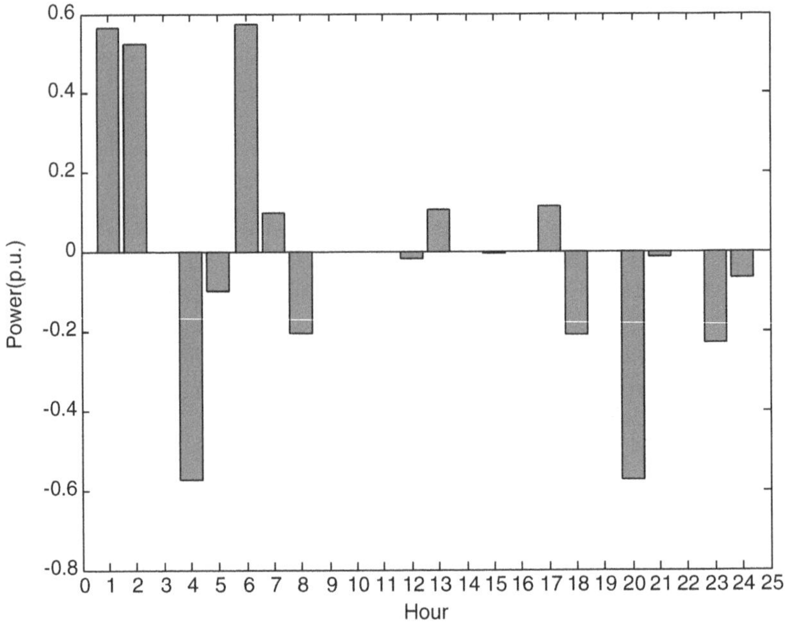

Fig. 10.6 Charging-discharging operation of ESS on bus 6 at year 3

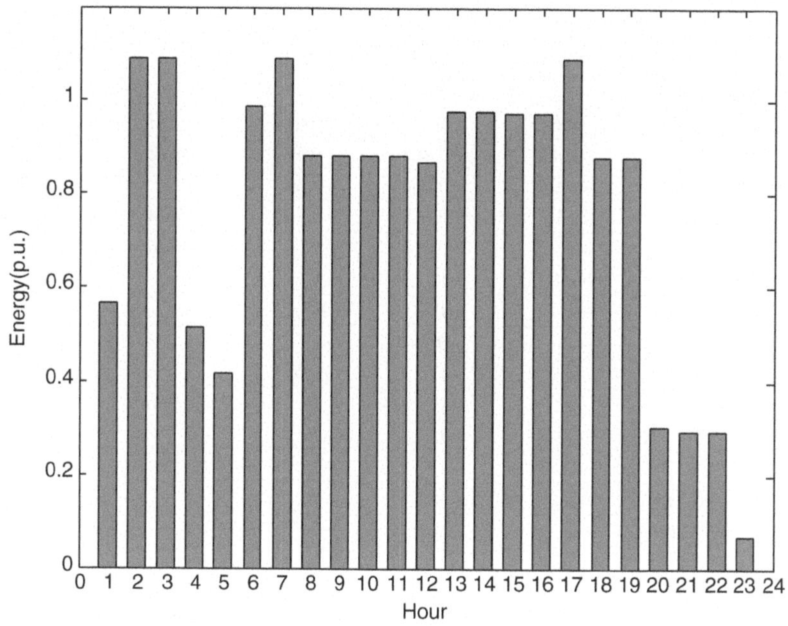

Fig. 10.7 Energy of ESS on bus 6 at year 3 (initial energy is zero)

also listed. The first analysis is carried out on ESS efficiency. The results confirm that the ESS efficiency is an important parameter and it makes significant impacts on the outputs. As a result, considering accurate and realistic efficiency is inevitable for ESS to achieve the realistic outputs. The ESS lifetime is also analyzed, and it is demonstrated that decreasing lifetime makes more impacts on the plan. As a result, it is necessary to consider accurate ageing of ESS in the planning. The results indicate that the aging and efficiency are important in the ESS planning. However, the ageing itself makes impact on the efficiency and together with increasing age of ESS, its efficiency is reduced. As a result, it would be more practical to consider this correlation in the planning and presenting adaptive ageing and efficiency for ESS. The lifetime of transmission line is also important in the planning, and the results indicate that decreasing lifetime makes significant impacts on the outputs.

10.5 Conclusions

This chapter presents a mixed integer linear programming for TEP including ESSs. The introduced problem finds optimal location, capacity, number, and time of new lines and ESSs, while it optimizes the short-term operation of ESSs. The simulation results are carried out on IEEE 24-bus test network. The results demonstrate that line capacity makes a great impact on the network power flow and generation pattern.

Table 10.12 Hourly power flow from bus 6 and bus 10 with ESS (p.u)

Hour	Year 1	Year 2	Year 3	Year 4	Year 5	Year 6
1	−0.627	−1	−1	−1	−1	−1
2	−0.504	−0.615	−1	−0.617	−0.737	−1
3	−0.445	−0.42	−0.468	−0.58	−0.602	−0.678
4	−0.479	−0.94	−0.254	−0.673	−1	−0.814
5	−0.638	0.131	−0.688	−0.792	−0.044	−0.991
6	−0.682	−1	−0.766	−1	−0.486	−0.895
7	−0.757	−0.875	−1	−1	−1	−1
8	−0.986	−1	−1	−1	−1	−1
9	−1	−1	−0.767	−1	−1	−1
10	−1	−0.978	−1	−0.654	−1	−1
11	−1	−1	−1	−1	−0.986	−1
12	−1	−1	−1	−1	−1	−1
13	−1	−0.831	−1	−1	−1	−1
14	−0.984	−1	−1	−1	−1	−1
15	−0.869	−1	−1	−1	−1	−1
16	−1	−1	−1	−1	−1	−1
17	−0.919	−1	−0.746	−1	−1	−1
18	−0.393	−0.408	−1	−0.929	−0.644	−1
19	−1	−1	−0.805	−0.877	−1	−1
20	−0.918	−0.763	−0.965	−0.895	−1	−1
21	−0.805	−0.922	−1	−1	−1	−1
22	−1	−1	−1	−0.967	−1	−1
23	−0.722	−1	−1	−1	−1	−1
24	−0.863	−0.936	−1	−1	−1	−1

Table 10.13 Sensitivity analysis on important parameters of the planning

Case	Total cost of plan (million $/year)
Nominal case 1 (TEP with ESS)	31.313
Nominal case 2 (TEP without ESS)	38.890
ESS efficiency is set on 50%	36.920
ESS lifetime is increased by 25%	29.233
ESS lifetime is decreased by 25%	34.665
Transmission line lifetime is increased by 30%	30.121
Transmission line lifetime is decreased by 30%	33.211

In order to optimize the network operation, it is mandatory to have enough capacity for the lines, and this point is dealt by expansion planning. The results show that the TEP with ESS installs three lines, while the TEP without ESSs needs six lines to deal with load growth. As a result, in the TEP with ESS, a number of installed lines are reduced by about 50%, and the investment cost on lines is properly deferred. As well, the planning cost of TEP with ESS is about 31 million $/year, while it is about 38 million $/year for TEP without ESSs. As a result, application of ESSs shows about 22% cost reduction. The generated power by generators at different years of the planning horizon follows different patterns, and the results reveal that the generated power is increased along with the planning horizon to supply load growth. Such generation needs more line capacity to transfer power to the load centers, and this need is met by the proposed TEP. The results show that some ESSs only operate at off-peak and on-peak hours (minimum charging-discharging cycles), and on the other hand, some ESSs operate at different hours to shift energy. Such procedure increases lifetime of ESSs as well as helps the network to shift energy from on-peak periods to the off-peak time sections. Such energy arbitraging helps the network to deal with peak loading with minimum line capacity, and the network expansion will be postponed by the ESSs. The sensitivity analysis on the major parameters of the planning demonstrate that the ESS efficiency and lifetime make significant impacts on the outputs, and it is essential to consider accurate ageing and efficiency for the ESSs in the problem.

References

1. U.S. Energy Information Administration (EIA). www.eia.gov
2. J. Aghaei, N. Amjady, A. Baharvandi, M.-A.J.I.T.O.P.S. Akbari, Generation and transmission expansion planning: MILP–based probabilistic model. IEEE Trans Power Syst **29**(4), 1592–1601 (2014)
3. N. Amjady, A. Attarha, S. Dehghan, A.J.J.I.T.O.P.S. Conejo, Adaptive robust expansion planning for a distribution network with DERs. IEEE Trans Power Syst **33**(2), 1698–1715 (2018)
4. M.R. Bussieck, A.S. Drud, A.J.I.J.O.C. Meeraus, MINLPLib—A collection of test models for mixed-integer nonlinear programming. ORSA J. Comput. **15**(1), 114–119 (2003)
5. H. Chen, T.N. Cong, W. Yang, C. Tan, Y. Li, Y. Ding, Progress in electrical energy storage system: A critical review. Prog. Nat. Sci. **19**(3), 291–312 (2009)
6. R. Christie, The UW Power System Test Case Archive provides World Wide Web access to power system data (test cases). The site is maintained by Richard D. Christie, an Associate Professor at the University of Washington located in Seattle, Washington, USA. (1999). https://labs.ece.uw.edu/pstca/
7. S. De La Torre, A.J. Conejo, J.J.I.T.O.P.S. Contreras, Transmission expansion planning in electricity markets. IEEE Trans Power Syst **23**(1), 238–248 (2008)
8. M. Eghbal, T.K. Saha, K.N. Hasan, Transmission expansion planning by meta-heuristic techniques: A comparison of shuffled frog leaping algorithm, PSO and GA, IEEE Power and Energy Society General Meeting (IEEE PES-GM), Detroit, MI, United States, San Diego, United States. IEEE, USA, pp. 24–28 (2011). https://doi.org/10.1109/PES.2011.6038998
9. R. Hemmati, Optimal design and operation of energy storage systems and generators in the network installed with wind turbines considering practical characteristics of storage units as design variable. J. Clean. Prod. **185**, 680–693 (2018)
10. R. Hemmati, R.-A. Hooshmand, A. Khodabakhshian, State-of-the-art of transmission expansion planning: Comprehensive review. Renew. Sust. Energ. Rev. **23**, 312–319 (2013a)
11. R. Hemmati, R.-A. Hooshmand, A. Khodabakhshian, Coordinated generation and transmission expansion planning in deregulated electricity market considering wind farms. Renew. Energy **85**, 620–630 (2016)
12. R. Hemmati, R.A. Hooshmand, A. Khodabakhshian, Comprehensive review of generation and transmission expansion planning. IET Gener. Transm. Distrib. **7**(9), 955–964 (2013b)

13. R. Hemmati, H. Saboori, Emergence of hybrid energy storage systems in renewable energy and transport applications–a review. Renew. Sust. Energ. Rev. **65**, 11–23 (2016)
14. J.M. Home-Ortiz, O.D. Melgar-Dominguez, M. Pourakbari-Kasmaei, J.R.S.J.I.J.O.E.P. Mantovani, E. Systems, A stochastic mixed-integer convex programming model for long-term distribution system expansion planning considering greenhouse gas emission mitigation. Int. J. Electr. Power Energy Syst. **108**, 86–95 (2019)
15. R.-A. Hooshmand, R. Hemmati, M. Parastegari, Combination of AC transmission expansion planning and reactive power planning in the restructured power system. Energy Convers. Manag. **55**, 26–35 (2012)
16. Y. Lei, P. Zhang, K. Hou, H. Jia, Y. Mu, B.J.I.T.O.P.S. Sui, An incremental reliability assessment approach for transmission expansion planning. IEEE Trans Power Syst **33**(3), 2597–2609 (2018)
17. A.M. Leite da Silva, L.S. Rezende, L.A. da Fonseca Manso, L.C. de Resende, Reliability worth applied to transmission expansion planning based on ant colony system. Int. J. Electr. Power Energy Syst. **32**(10), 1077–1084 (2010)
18. S. Lumbreras, A.J.E.P.S.R. Ramos, The new challenges to transmission expansion planning. Survey of recent practice and literature review. Electr. Power Syst. Res. **134**, 19–29 (2016)
19. R. Mena, R. Escobar, Á. Lorca, M. Negrete-Pincetic, D.J.A.E. Olivares, The impact of concentrated solar power in electric power systems: A Chilean case study. Appl. Energy **235**, 258–283 (2019)
20. S.Z.J.I.J.O.E.P. Moghaddam, E. Systems, Generation and transmission expansion planning with high penetration of wind farms considering spatial distribution of wind speed. Int. J. Electr. Power Energy Syst. **106**, 232–241 (2019)
21. A. Rastgou, J. Moshtagh, Improved harmony search algorithm for transmission expansion planning with adequacy–security considerations in the deregulated power system. Int. J. Electr. Power Energy Syst. **60**, 153–164 (2014)
22. C. Rathore, R. Roy, Impact of wind uncertainty, plug-in-electric vehicles and demand response program on transmission network expansion planning. Int. J. Electr. Power Energy Syst. **75**, 59–73 (2016)
23. C. Rathore, R. Roy, U. Sharma, J. Patel, Artificial Bee Colony Algorithm based static transmission expansion planning, 2013 IEEE international conference on energy efficient Technologies for Sustainability (IEEE), Nagercoil-India. IEEE, USA, pp. 1126–1131 (2013). https://doi.org/10.1109/ICEETS.2013.6533544
24. G. Shrestha, P.J.I.T.O.P.S. Fonseka, Congestion-driven transmission expansion in competitive power markets. IEEE Trans Power Syst **19**(3), 1658–1665 (2004)
25. Ugranlı, F., Karatepe, E., 2015. Multi-objective transmission expansion planning considering minimization of curtailed wind energy. Int. J. Electr. Power Energy Syst. 65(\), 348–356
26. M. Wu, L. Kou, X. Hou, Y. Ji, B. Xu, H.J.I.J.O.E.P. Gao, E. Systems, A bi-level robust planning model for active distribution networks considering uncertainties of renewable energies. Int. J. Electr. Power Energy Syst. **105**, 814–822 (2019)
27. J. Zhan, W. Liu, C.J.I.T.O.P.S. Chung, Stochastic transmission expansion planning considering uncertain dynamic thermal rating of overhead lines. IEEE Trans Power Syst **34**(1), 432–443 (2019)
28. M. Zidar, P.S. Georgilakis, N.D. Hatziargyriou, T. Capuder, D. Škrlec, Review of energy storage allocation in power distribution networks: Applications, methods and future research, IET Gener. Transm. Distrib., IET Digital Library, UK. 10 (3), 645–652 (2016). https://doi.org/10.1049/iet-gtd.2015.0447

Chapter 11
Regulation of the Expansion of Electricity Transmission

Luis Olmos and Michel Rivier

11.1 Some Preliminary Considerations on the Organization of the Electricity Transmission Activity

The activities related to the electricity transmission activity include the planning of network investments, the construction of new assets, the operation planning of the network, the operation itself, the planning of the maintenance of network assets, and the maintenance itself. Investment planning is the process whereby the date of entry into service, location, capacity, and other characteristics of the reinforcements to a network are determined. The operation of the network concerns the online use of the network, making the proper decisions on generation redispatch, network topology changes, use of transformer taps, capacitor or reactance batteries, or any other network-related device to ensure technical network security limits are complied. The planning of the maintenance is the process whereby the time periods in which each line will be out of service are determined. In these periods, the transmission company carries out the repairs and tasks necessary to keep these facilities operational with a degree of adequate reliability. The construction of new assets and their maintenance are activities that can be carried out by specialized companies, not necessarily electricity companies.

Those activities that are directly related to the physical assets – construction, maintenance, and physical operation of the facilities – which are the transmission functions themselves, are carried out by transmission companies (or are subcontracted). On the other hand, the network planning activities and the operation planning of the facilities are typically in charge of the system's operating entity. Thus, the transmission companies are instructed by the system operator (SO) to

L. Olmos (✉) · M. Rivier
Comillas Pontifical University, Madrid, Spain
e-mail: luis.olmos@iit.comillas.edu; michel.rivier@iit.comillas.edu

S. Lumbreras et al. (eds.), *Transmission Expansion Planning: The Network Challenges of the Energy Transition*, https://doi.org/10.1007/978-3-030-49428-5_11

carry out certain actions on the network facilities. Besides, the SO is also in charge of ensuring the security and integrity of the electrical system.

The following considerations are relevant to assess the extent to which activities of transmission can be considered a natural monopoly [23]:

1. The duplication, or multiplicity, of competing transport or distribution networks for the provision of the same service is currently something unthinkable, in terms of cost, land use, and environmental impact. Both the common transmission network and the distribution network that is used to supply a specific end consumer must be unique.
2. The previous conclusion is enhanced by the fact that there are strong economies of scale in the construction of electrical networks. The unit cost of electric power transport (€ per unit of transported power) decreases strongly, ceteris paribus, with the total transmission capacity of a line. It is most economical, therefore, to have only one network with sufficient transmission capacity to meet the total demand of this service.
3. The transmission network must function and be operated as a whole. Power flows through all network facilities are instantaneously interlinked, and network failures and disturbances may have a very large geographical instant impact.

From the above, it can be concluded that there must be a single transmission network and that this network must be operated by a single entity: the SO. Different transmission facilities may be owned by different entities, as long as they all work in coordination under the mandate of the corresponding operator. There may, therefore, be a multiplicity of transmission companies.

The transmission companies have large market power, since any of them is essential for the electricity supply. Therefore, with certain exceptions, the remuneration of transmission facilities must be regulated, and they must be obliged to provide free access to their facilities without discrimination at a regulated transmission charge. Integration between network companies and the agents that operate the market can lead to a number of problems. First, if this occurs, cross subsidies may be paid by the regulated activity (transmission) to the competitive one, unduly favoring the latter against its competitors.

Likewise, the commercial activities of the integrated company that require access to the network, such as the purchase or sale of energy at the wholesale or retail level, could result in largely favorable conditions of access to the network facilities by this company at the expense of others. Therefore, access would cease to be free and non-discriminatory. Finally, the commercial activities within the integrated company could have privileged information about customers of the network activity. All this advises, from a regulatory point of view, that the electricity transmission activity is kept independent of the agents that operate in the market.

Apart from this, the operation of the electrical system, as well as the planning of the expansion and maintenance of the transmission network, largely affects the electricity market. Therefore, the independence of the system and market operators from any of the market agents is inexcusable.

11.1.1 *On the Relationship Between the System Operation and Transmission Activities*

The operational activities, i.e., the expansion planning, operation planning, and maintenance planning of the transmission network, affect the maintenance, operation, and construction of this network, to be carried out by transmission companies. However, the integration of both types of activities in the same company can lead to conflicts of interest that are detrimental to the efficiency of the operation of the system. Therefore, defining the relationship between the system operation activities and the transmission activities is a controversial issue discussed below.

There are several options when it comes to regulating the relationship between the transmission company and the SO:

- At one end of the spectrum of options, there is the complete separation between the entities responsible for both activities. Then, the property of both entities is different, i.e., the transmission and system operation activities are performed by companies with different shareholders.
- In the next step in the process of integrating both activities, the entities responsible for one and the other belong to the same business group, but each of them has an independent legal entity.
- Finally, authorities in a system may opt for a model whereby the same entity is responsible both for the operation of the system and for the physical transmission of electricity, although there may be an accounting separation of both activities.

As noted above, there are reasons for and against fully separating both activities. The reasons in favor have to do with the eradication of the conflicts that may arise between the interests of the transmission company and those of the SO. A list of the conflicts of interest that may arise follows:

- If the transmission company's remuneration is of the "cost plus service" type, increasing the volume of transmission assets results in an increase of its remuneration. Assuming that the interest rate applied by the regulator is attractive enough, if the SO is integrated with the transmission company, the former has an incentive to promote the construction of new transmission assets, even if these are not justified from the point of view of the system efficiency. Similarly, the operator could perform other functions to benefit its transmission subsidiary, instead of protecting the interests of the whole system or other entities. For example, if the connection of a new generator or demand increases the costs of the transmission company to a relevant extent, the SO could make obtaining the connection license more difficult.
- The SO is the entity with the technical knowledge to assess the causes and responsibilities of any problem of supply in the system. A conflict of interest raises whenever its transmission subsidiary may be one of the involved entities in the failure event.

- The SO can plan the maintenance of the transmission lines and other network assets according to the interests of its transmission subsidiary and not based on the interests of the whole system.
- If there is more than one transmission company, the SO may have an incentive to protect the interests of the transmission with which it is integrated at the expense of the rest of the companies of this kind. Thus, for example, the SO may try to award the construction and maintenance of a larger number of transmission facilities to its subsidiary. Going back to the example of the connection of new generators or demands, if this is providing large revenues, the SO may try to ensure that the largest possible number of generators and demands get connected to the network owned by its transmission subsidiary. Conversely, if the connection of new generators or demands implies an important cost for the transmission company, the operator will try to get them connected to networks that are owned by other companies.
- Moreover, if some of the network facilities owned by its transmission subsidiary are investments at risk, whose revenues correspond to the congestion rents they produce, the SO may try to increase these congestion rents, possibly at the expense of the system efficiency.

However, on the other hand, there are also advantages associated with the integration of both entities (transmission company and SO). The transaction costs should be lower within an integrated entity. At the same time, the level of coordination potentially achieved between the planning and maintenance activities within an integrated entity is higher. The transaction costs to be reduced include those associated with the exchange of information between both companies, the costs of monitoring the state of the network, and the costs of the economic and technical feasibility studies to be carried out for network investment projects, among others. Besides, the planning of the maintenance of lines and transformers and the construction and commissioning of new lines could be coordinated in a simpler and more efficient manner within an integrated company. All these aspects have certainly a direct impact on enhancing the security of the system which is a major issue.

Then, it is not entirely clear the degree of separation to be established between the operation and transmission activities. In the European Union, the vast majority of countries have opted for the creation of entities called "transmission system operators" (TSOs) carrying out both activities. These should be separated from generation, retailing, and distribution companies; see [9]. On the other hand, in the North American system, independent system operators, ISOs, and regional transmission operators, RTOs, which are pure system operators not owning any transmission assets, coexist with a multiplicity of transmission companies, or "Transcos," owning the transmission grid; see [11].

In the following paragraphs, we describe the main paradigms for the organization of the transmission activity and, specifically, the expansion of the grid and assess the merits of each option based on the previous considerations.

11.2 Regulatory Schemes for the Expansion of the Transmission Grid

With the introduction of competition in generation and retailing activities, the regulatory environment in which transmission network expansion planning takes place has changed substantially. Within most systems, the transmission expansion planning activity remains in the hands of a regulated monopoly, although some systems have established some competition mechanisms within the process of development of the grid. In other systems, the private initiative has some responsibilities in the promotion of the development of the grid. There are two basic options in this regard. The first one involves leaving the promotion of new transmission expansion projects, and sometimes even the payment of the cost of these projects, in the hands of coalitions of network users, subject to certain regulatory control. Alternatively, private investors may be in charge of promoting new projects and, afterward, commercially exploiting them: either buying energy at one end of the line, or project, that they have promoted and selling it at the other end or selling in advance the right to earn the congestion rents produced by the project or physically use its transmission capacity.

The incentives to undertake new transmission investments depend on the paradigm adopted for this activity. Ideally, agents that benefit from the construction of a transmission project should be the ones promoting it. However, identifying and assigning to specific agents, or network users, the benefits to be provided by a project tend to be difficult. Then, leaving the initiative in the hands of private promoters may not yield the desired results. This section discusses the main regulatory schemes that exist for the development of the transmission grid. The classification presented here is based on [21].

This description includes the institutional organization corresponding to each regulatory scheme (which entity performs each of the different functions involved in the network expansion process) and the remuneration of the owner of each line, which also depends on the adopted paradigm.

11.2.1 Centralized Network Expansion

11.2.1.1 Passive System Operator

In this regulatory approach, the SO is responsible for promoting the construction of new reinforcements to the network. The SO must propose a network expansion plan, which needs to be approved by the regulator. The SO may or may not be integrated with the transmission company. The responsibility for the construction of the network reinforcements finally approved may be assigned through a process of competitive bidding or, alternatively, may be carried out by the corresponding transmission company, which will charge the regulated costs corresponding to these investment projects.

Normally, before drafting and proposing a network expansion plan, the SO must consider and include, or reject, the reinforcement proposals made by the network users. In a deregulated context, the construction of new transmission projects may significantly affect the market benefits obtained by the network users. The regulator must decide on the approval of the proposed investment plan. If approved, the regulator must then decide on the individual approval of each of the proposed network reinforcements within the plan. In this case, the SO can be considered passive, because it proposes the undertaking of new projects but does not decide, ultimately, on their constructions. An example of this type of SO is the Spanish one, Red Eléctrica de España.

Both the SO, when proposing the undertaking of a project, and the regulator, when deciding on the approval of the corresponding project, must employ a set of rules to determine if the construction of the line is justified from a technical and economic point of view. This set of rules is called regulatory test and is described in a separate section.

The construction, operation, and maintenance of the new network facilities may be in the hands of the transmission company in charge of the area where these facilities are to be located (in many cases, the transmission company is the local SO). Alternatively, these tasks may be assigned through a competitive bidding mechanism. If this mechanism is implemented, the winner of the tenders is paid his/her bid, which is the basis for the computation of the regulated cost of the line.

Given that, normally, the operation and maintenance license for a transmission facility, or a set of them, has a limited duration; at the end of the license term, it is possible to put again on a tender these tasks for the next license period. It is also possible to set availability targets for the corresponding facilities and apply penalties or incentives according to the actual availability level achieved.

If the SO is the owner of the new transmission assets, his/her remuneration must be set according to some standard unit cost values defined in the regulation. Therefore, as aforementioned, the TSO may be encouraged to promote a larger number of new transmission projects than those justified from a technical or economic point of view. The larger the number of assets owned by the TSO, the larger his/her revenues, which will normally be set according to some standard unit costs high enough to make the undertaking of new regulated projects like these a profitable, low-risk business. Although the final decision on the construction of new transmission assets is made by the regulator, it will be difficult for this institution to reject an investment proposed by the TSO, since its absence could have a negative impact on the reliability of the system.

If the SO is not the owner of the network, his/her remuneration should not depend directly on its size. The operator should receive a predetermined amount of revenues that could undergo minor modifications related to the degree of efficiency achieved by this entity in the development of its functions. But, even in this case, both the regulator and the SO are encouraged to promote and accept the construction of a larger number of new transmission facilities than necessary, since both are considered responsible for the security of the system. Therefore, it is possible that the resulting network tends to be larger than the optimal one from an economic efficiency point of view.

Despite these potential problems, most authors who discuss this issue consider the traditional scheme of centralized planning of the network expansion as the most effective one to achieve a sufficient development of the network; see [12, 17, 20].

11.2.1.2 Active System Operator

Under this scheme, a private, for-profit company is responsible for determining the optimal network expansion plan and deploying it. This entity should have obtained the transmission service license in a competitive bidding process and for a limited period of time and be regulated as a monopoly. The licensee must follow certain network design guidelines (the so-called grid code) when planning the expansion of the grid and should be subject to efficiency incentives related to the achievement of network performance objectives. This scheme is similar to the one most commonly used to regulate the expansion of distribution networks.

The licensee of the transmission license can be referred to as the active SO, since this entity is ultimately responsible for the expansion of the network. This scheme has been implemented in the UK, where there is an active SO managing the expansion of the grid, National Grid Company, coexisting with several transmission companies. National Grid also owns the grid in England and Wales. Thus, it is one of the transmission companies active in the UK.

The remuneration of the transmission activity in this case should relate to the network that is optimally adapted to the system, not to the current network. The computation of the revenues corresponding to the whole grid in each area results from the application of an RPI-X type of scheme. The SO, which can also own all or part of the transmission network, as in the UK, provides the regulator with an expansion plan for the next regulatory period. The duration of this period must be set to maximize the efficiency and effectiveness of the scheme in achieving the construction of the required network investments in due time. According to this expansion plan and to the construction, operation, and maintenance cost forecasts also provided by the SO, the regulator determines the remuneration of this entity during this period. Finally, the regulator determines the values of the remuneration scheme parameters. These are, basically, the remuneration of the service corresponding to the first year and the annual adjustment factor X that allows to match the expected present value of the revenues and expenses of the transmission activity for the said period. Since investments in the new transmission lines can be assessed individually, the remuneration for a regulatory period should be computed explicitly taking into account the investments planned, the level of depreciation of the existing assets, the financial situation of the licensee, and the expected efficiency improvements.

Afterward, the transmission licensee decides unilaterally the reinforcements to be carried out. Incentives associated with the quality of service may be applied.

Obviously, in the following regulatory period, the regulator must take into account the investments that the licensee has actually undertaken and the quality of service achieved, when establishing the remuneration of this entity.

Since, under this remuneration scheme, the revenues of the licensee do not depend directly on the network investments undertaken, this company should reasonably invest in new facilities only to the extent necessary to comply with the minimum objectives determined in the network code. It should only undertake a larger amount of investments if the remuneration scheme adopted makes these extra investments attractive.

The compliance with the requirements defined in the network code can be guaranteed by applying severe administrative penalties to the breach of these requirements. It is unlikely that this mechanism of organization of the expansion of the network leads to the construction of lines that are not necessary. However, achieving an optimal development of the grid through this scheme seems unlikely as well. This depends on the efficiency of the incentives set for the active SO. When defining the efficiency incentives, it would be possible, as proposed in [16], to hold this company responsible for those redispatch costs above a certain level, as well as for the costs associated with a system collapse. However, these problems cannot be always attributed to the active SO. Therefore, according to this proposal, this entity would be exposed to very high risks that it cannot efficiently manage and are not commensurate with the revenues that the entity can obtain.

11.2.2 Investments by Coalitions of Network Users

As explained above, network users are the beneficiaries of the construction of new transmission projects. Therefore, those users who would benefit from a project have an incentive to cooperate to achieve its undertaking. In order for a project promoted by network users to be undertaken, the regulator must approve it. This is necessary to determine whether the corresponding network reinforcements are justified; see [18, 19]. Indeed, the physical laws that govern the distribution of flows through the network may drive some reinforcements that are beneficial for some users or for the promotor, to be, altogether, harmful for the system.

The construction of the network reinforcements within those projects that are approved can be assigned by means of a competitive bidding mechanism. Then, the regulated cost of these reinforcements can be determined based on the winning bid. If these projects are, in the end, owned by a regulated entity, their cost is recovered through network charges. Similarly to what occurs in other schemes, the authorities can set availability targets for the new assets and apply incentives or penalties associated with the fulfillment of these objectives.

Alternatively, the coalition of network users promoting a project may also be in charge of implementing and paying it, getting, in return, rights over the corresponding transmission capacity. Then, this coalition would own the project. A final alternative involves recovering part of the cost of the project through regulated tariffs and part through payments made by the members of the coalition promoting this project. Then, the promoters of the project should receive transmission rights over the fraction of the capacity of this project that they have paid. Therefore, there are different regulatory options for the implementation of this scheme; see [18, 19].

If the regulation allows market agents to participate in the development of the network, nodal energy pricing and transmission charges with geographical differentiation will provide the network users with incentives to promote the undertaking of those transmission projects that they would benefit from. Therefore, the success of this scheme largely depends on the economic signals related to the existence of the grid that are sent to the network users and the efficiency of these signals.

The authors in [2, 8] discuss the weak points of this scheme. This scheme is unlikely to result in the undertaking of those projects that, being socially efficient, produce benefits that are widely spread among a large number of network users (e.g., those projects improving the overall reliability of the electrical system). Most probably, no coalition of network users will form to promote this project, since other beneficiaries not being part of the coalition would probably free ride on it, while the benefits produced by this project for individual users will not be large enough to trigger them to build this coalition. The reader should keep in mind that, in systems whose network is much messed, there are many potential transmission projects whose benefits are widely distributed throughout the system.

Therefore, this network expansion scheme, when implemented, must be complemented by a centralized, back-up, planning mechanism that ensures that those socially efficient projects whose benefits are widely distributed are also undertaken.

11.2.3 Merchant Lines

The merchant lines are designed and managed to maximize the profits resulting from their commercial exploitation. The owner of a merchant line can use it to buy energy at one end of the line (the one where energy is cheaper) and sell it at the other end (where the energy is more expensive). Therefore, the revenues of a merchant line correspond, in principle, to the product of the difference in prices between both ends of the line and the amount of energy flowing through this line, i.e., the congestion rents produced by the line. Alternatively, the owner of this line may stabilize, or secure, these revenues in the long term selling financial transmission rights over the capacity of the line. Another possible way for a merchant entrepreneur to stabilize his/her revenues is selling the right to physically use the capacity of this line in return for being paid charges that are negotiated between the entrepreneur and the prospective line users.

Therefore, private developers may probably be only interested in building a merchant line in one of three situations:

1. If this line is expected to produce a relevant amount of congestion rents.
2. If this line is required to connect certain new network users to the rest of the system. However, the lines of this type can be expected to be built as regulated lines in most systems.
3. If this line is expected to largely benefit a reduced number of network users willing to pay specific charges to get it built.

The sale of financial, or physical, transmission rights over the capacity of a merchant line allows its private promoter, not only to get protected from the risk associated with the volatility of the congestion rents but also to finance its construction; see [3, 7, 16].

However, relying largely on the initiative of merchant promoters to drive the development of the grid is not advisable for a variety of reasons:

- First, the congestion rents produced by an efficiently developed network are not enough to recover the cost of this network [25]. Only heavily congested lines, those located in heavily congested corridors or those largely needed by a reduced number of network users, can be expected to be built as merchant lines [15].
- Second, the market revenues produced by merchant lines are highly uncertain, since these revenues may largely decrease as a result of the construction of new regulated lines easing the grid congestion that the former lines are addressing [6, 15]. As aforementioned, risks affecting the revenues of merchant lines may, in specific cases, be efficiently managed by selling in the long term transmission rights of any of several types over the capacity of these lines.
- Finally, one must take into account the discrete nature of the transmission network investments. As a result of this, building a merchant line whose capacity is smaller than the amount of new capacity required by the system could lead to an increase in the potential for the exercise of market power. Moreover, merchant line promoters have a natural incentive to build underinvested capacity network facilities – as compared to the socially efficient capacity to be built – to maximize the congestion rents they will earn. Once a small merchant line has been built, the centralized network expansion planner might not consider it socially efficient to build an additional line complementing the former.

In general terms, the weak points of any of the network expansion schemes here discussed are related to the differences between the interests of the promoters of new projects under these schemes and the interests of the whole system. Based on the analysis of these schemes carried out in this section, we now provide a blueprint of what could be an efficient scheme for organizing the expansion of the grid.

11.2.4 Defining an Efficient Scheme for the Expansion of the Grid

After discussing the different network expansion schemes, one should conclude that the centralized transmission expansion planning, either conducted by a passive or an active SO, should be the base, or default, mechanism for the expansion of the transmission grid in a system. Based on these schemes, regulated network investments would be undertaken to reinforce the grid. Centrally planning the expansion of the grid and promoting the construction of regulated investments are especially relevant in those systems where the network is highly meshed, as in most European countries and North American states.

When the SO is ultimately responsible for the construction of new lines, because this entity decides which investments to undertake (active SO), it is necessary to provide it with appropriate efficiency incentives that go beyond those strictly necessary to comply with the network code. However, when the SO is only responsible for the proposal of new investments (passive SO), while the regulator is responsible for authorizing these, the former will be inclined to promote more lines than necessary, since the operation of the system (for which the SO is directly responsible) will be safer, when there is excess capacity in the grid. Moreover, if the SO is also the owner of the transmission network, building more reinforcements will allow this entity to increase its revenues, and profits, as aforementioned. Then, the regulator should be aware of the overinvestment incentives that the SO is subject to and try to closely assess the investment proposals made by later making use of the appropriate tools. However, the regulator itself may also be interested in having a transmission capacity margin larger than the reasonable one. In this way, the probability that claims from network users and political authorities arise related to poor service quality levels would be lower. These claims may also negatively affect the position of the regulator, since this entity may be found partially responsible for service interruptions.

The reinforcements undertaken through regulated schemes, where the SO is in charge of their promotion of these, can be complemented with those other investments promoted by the private initiative. As indicated above, if an efficient energy pricing system and efficient network charges are in place, which should normally have some level of geographical differentiation, the network users will receive suitable incentives for the promotion of the construction of those transmission reinforcements whose beneficiaries are few and well defined, but not those others producing well-spread benefits. An example of the former can be those lines that connect a center of consumption or production of electricity with the rest of the system in radial networks. Those investments promoted and paid by network users are normally called investments at risk. This scheme can also be of application for the promotion of transmission projects in regional markets by those SOs in the region corresponding to the local, or national, systems significantly affected by these projects. In this case, the regulators in the region should normally approve the construction of these infrastructures and ensure that the revenues of the corresponding reinforcements are large enough.

The existence of congestion in the grid of a system can be an incentive for merchant investors to promote the construction of certain projects and then exploit them. However, in general, as previously discussed, these promoters will want to undertake projects whose size is smaller than the socially optimal one. A merchant project of a socially optimal size normally results in a relevant decrease of the congestion rents obtained by the promoter. Also, the revenues of these projects might be substantially reduced, with respect to the expectations of their private promoters, if new regulated investments take place decreasing the congestion in the corresponding area. For all this, merchant investments can only contribute to the expansion of the network in specific cases. These may play a relevant role in the expansion of radial networks or heavily congested interconnections among the systems in a region when there are some difficulties for these to be promoted as regulated investments. In any case, whenever a regulated option is also possible, this should be given priority over the merchant one.

The regulator should make sure that the size and design of merchant investments are optimal or, at least, reasonable, from a social point of view. In [24] some variants of the standard scheme for the promotion of merchant lines are discussed aimed at aligning the interests of the promoters with those of the system. According to these, a negotiation process among the promoters and the regulator could take place affecting the size of the project. Building a project of a larger size than that initially considered by the promoters could be possible if the promoters are guaranteed that during the first years of operation of the facilities making the project, only a fraction of the total capacity of the project will be made available for its use.

11.3 Driving the Proposal of Regulated Investments and their Assessment by the Authorities: The Regulatory Test

The "regulatory test" is the set of rules to apply to determine if the undertaking of a project is justified, based on the benefits it is expected to produce and its costs. These rules may vary depending on the entity that must carry out the test and the nature of the network reinforcement that is being analyzed. Therefore, different types of "regulatory test" need to be defined. First, a passive SO needs to carry out some tests to identify the reinforcements to propose, which should be part of the expansion plan. But some forms of test could also be used by an active SO to determine the benefits to be provided by a project or set of them. This should serve as an input in the negotiation process with the regulator about the definition of the projects that should be considered by the latter to compute the remuneration of the transmission activity in the next regulatory period. Finally, it is necessary to define the set of rules to be applied by the regulator to decide on the approval of a project proposed by some other entity. This proposal may have been made by the SO, by an association of network users, or by a merchant promoter.

The test to be applied by the SO must be used to identify the most efficient investments from the system's point of view (the ones that maximize the increase in social benefit) among a set of options (candidates) considered. In principle, in order to identify the following "n" reinforcements that should be carried out, this test should compute the optimal network expansion plan. However, in practice, the operator may analyze each of the candidates separately or in groups of them, making use of the knowledge about the functioning of the system that this entity has acquired over time, as well as any tool it has at its disposal.

To do this, the operator shall consider all the benefits and costs associated with each project (investment and operation costs, improvement of reliability, and environmental impact), either within a set of independent objectives to achieve or integrated in a single economic objective function. European and US legislation is supporting the deployment and use of cost-benefit analysis (CBA) methodologies for the decision makers to properly address investment decisions.

The test should consider the cost of opportunity corresponding to alternative reinforcements to that being analyzed at each time; see [1]. In other words, when a project is evaluated, the operator should consider the cost associated with the fact that, if this project is deployed, others will not. Potentially, this could bolt down to calculating the increase in the social benefit associated with each of the investment plans considered and choosing the one that results in the largest increase. However, together with the definition of the whole plan proposed, the operator should also come up with an estimate of the benefits and costs caused by each individual project making the plan. This is necessary because, most probably, some projects of the plan may be accepted while others are not or they are given a lower priority than the former. Thus, the information provided by the test should allow the SO and the regulator to consider each project separately.

On the other hand, the regulator should verify that the increase in the social benefit achieved through the implementation of the plan proposed by the operator is greater than that corresponding to an expansion plan taken as a reference. The regulator may also request the operator, or another entity, to analyze a set of investment alternatives to those proposed by the SO. In the case of investments proposed by private developers, the regulator has to verify that the construction of the proposed reinforcements is not harmful for the system, i.e., that their benefits are larger than their cost.

There is some controversy as to whether the negative benefits, or losses, borne by some network users as a result of the implementation of a project, or an expansion plan, should be taken into account along with the positive benefits that it creates, when deciding whether this project or plan is efficient and should be approved. If both the benefits and the losses caused by projects are considered, the rules to apply to decide on the undertaking of new projects would be coincident with those applied in a traditional, fully regulated context of minimization of system costs.

11.4 Allocating the Cost of Regulated Network Investment Projects

The allocation carried out of the cost of regulated network investment projects will condition the expansion of the transmission grid in two main regards. First, in a general context, efficiently allocating the cost of these investments should result in transmission charges applied on network users that allow them to internalize in the long-term decisions they make for their expected effect on the network development costs. If the network users are informed about the level of these charges before the aforementioned transmission projects have been undertaken, these charges could trigger changes in the long-term decision made by users that could result in the avoidance of projects that are not efficient socially speaking. Then, the development of demand and generation in the system could be coordinated with that of the transmission network.

Second, in a regional context, where several local systems exist within a region, allocating the costs of transmission projects affecting several of these systems in a way that they perceive as "fair" could avoid the opposition of these systems to the undertaking of these projects. Otherwise, those systems that have the impression that the fraction of the cost of regional projects they are made to pay is significantly larger than the fraction of the overall benefits created by these projects that they would receive could oppose and try to block the deployment of these projects, even if they are beneficial for the region as a whole. Then, appropriately allocating the cost of projects with a regional impact could have a relevant effect on achieving a sufficient expansion of the regional transmission grid; see [11].

Both to achieve an appropriate coordination of the expansion of generation, demand, and the network in a system and to achieve a sufficient development of the transmission grid a in a region, the cost of transmission projects should be allocated to individual network users, or local systems in the region, proportionally to the benefits that these individual users and local systems are expected to obtain from the transmission projects concerned. This is known as the beneficiary pays principle. The efficient network locational signals are those charging network users the network costs they cause; see [22]. According to the economic theory, these signals correspond to the long-term marginal impact of the consumption or production of electricity in each area or node of the system on the network development cost. However, long-term marginal costs cannot be computed in reality due to the lumpiness of network investments. Long-term incremental costs should be computed and applied instead, but they are difficult to compute as well.

However, the benefits that network users are expected to obtain of each expansion project are a good proxy of the true responsibility of these users in the undertaking of this project. This is due to the fact that network investments are made to produce some benefits for network users. Then, these benefits are the ultimate cause of the construction of network reinforcements. In other words, the larger the benefits to be earned by a network user from an investment project, the larger the responsibility this user could be deemed to have in the deployment of this project. At the same time, if the cost of projects is allocated proportionally to the benefits that network users, and local systems, are expected to obtain for the former, no one should oppose the undertaking of efficient projects. The benefits created by these projects should be larger than their cost. Then, according to the beneficiary pays principle, the cost of any of these projects paid by any user or system should be lower than the benefits it is expected to obtain from the project, i.e., socially efficient projects should be beneficial, in net terms, for all the stakeholders in the system.

Unfortunately, the computation of the benefits expected to be produced by projects is plagued with some difficulties that are briefly discussed next; see [22]:

1. Availability of the data required to compute the benefits of transmission projects: computing the benefits of projects requires determining the effect that these projects would have on the functioning (expansion and operation) of the system. These require computing the future evolution of the system both with and without the project concerned in place. The data required to compute the system evolution includes, among others, the investment and operation costs of power plants and network facilities, the utility that electricity has for individual consumers, the evolution of fuel and CO_2 (and other pollutant permits) prices, and the evolution of hydro inflows and primary intermittent energy resources (wind, solar irradiation) per area of the system. Some of these data are deemed confidential by the stakeholders having access to them. Others are difficult to determine because there is much uncertainty about their evolution.
2. Definition of the counterfactual: the counterfactual of the situation where a project is undertaken is the rational alternative situation to the former. Sometimes, assuming that the evolution of the rest of the system would remain unaltered by the absence of the project is reasonable. Others, the decision on whether to undertake a project will affect the development of generation, demand, or the rest of the network. This is evident when talking about a project used to connect a set of generators or consumers to the rest of the system. If this project were not built, these generators and consumers would not install in the same place, since they would be isolated. Then, an alternative to the deployment of this generation and demand must be defined for the "without project," or counterfactual, situation.
3. Computation of the benefits produced by individual projects: as discussed above, determining the benefits expected to be produced by individual projects is necessary, since final investment decisions are to be made on individual projects, not the expansion plan as a whole. But this is difficult because the benefits created by the individual projects within an expansion are, most of the times, intertwined, i.e., they are mutually dependent. Some benefits may only be realized if several of these projects are undertaken. They cannot be achieved by building any single project. On the other hand, there are other benefits that may be obtained when building any of several projects (these projects are substitutive in this regard). Then, the allocation of the benefits of the expansion plan to individual projects is complex and, to some extent, subject to relevant assumptions that may be deemed

disputable. For a description of relevant research related to the computation of the benefits of individual projects within a plan, consult [4, 5].

4. Consideration of the negative benefits produced by projects: projects may produce both positive and negative benefits for network users and local systems. However, the decision on whether and how to take into account the negative benefits produced by projects in the allocation of their cost is also controversial and subject to many considerations. Should those stakeholders harmed by the undertaking of a project be compensated for its implementation? Then, should those stakeholders obtaining positive benefits from the project pay just the cost of the projects or also some compensations to the stakeholders being negatively affected by the project? Probably, the negative benefits that are related to the increase in the level of competition in the system should not be compensated, while those negative benefits related to power system externalities (typically, the environmental ones) could give raise to compensations. However, in a regional context, countries, or local systems, negatively affected by a project may be able to block the undertaking of this projects, even if it is beneficial for the region as a whole. In this case, paying compensations for the negative benefits obtained by these systems may be justified. Otherwise, the whole region may be harmed by an insufficient development of the regional grid.

Even when the beneficiary pays principle may be deemed most appropriate to drive the allocation of the cost of projects, the difficulties involved in computing the project benefits have resulted in many systems and regions applying other approaches. A proxy to the economic benefits considered in many systems is the allocation of this cost based on the determination of the electrical usage that stakeholders are to be made of these projects. Then, their cost can be allocated proportionally to the usage made of them by the stakeholders. However, a multiplicity of usage-based cost allocation methods has been defined and implemented throughout the world; see, for instance, [23], producing widely different results. Thus, the allocation of the cost of projects based on the usage made of these projects is not free of controversy.

11.5 Regulation of the Expansion of the Grid in a Regional Context

Nowadays, local systems are coordinating their functioning within regional structures. Then, it is very valuable to assess the organization of the expansion of the grid at the regional level. In this last section, first, the desired features of the regulation of the expansion of grids at the regional level is discussed. Then, the approaches adopted in some of the most representative regions of the world regarding the management of the expansion of the transmission grid are briefly described.

In a regional context, the planning of the expansion of the grid should be carried out in a coordinated way as well. All the benefits that transmission projects are expected to render throughout the whole region should be jointly considered, together with the project costs, to assess the merits of each project and decide which ones to undertake. Besides, the interactions taking place among the projects to be undertaken in the whole region should be taken into account when computing the benefits produced by each project and how these benefits distribute among the several areas of the system, as discussed above. These are relevant to assess the investment proposals and to allocate the cost of the investment projects to the local systems in the region, respectively. Ideally, the planning of the expansion of the whole regional grid and the assessment of the proposals by the planning authorities should be centrally carried out by regional entities looking after the interest of all the stakeholders in the region, though this may not be possible in the short to medium term in many regions, as discussed next.

The financing of the regional transmission projects should take advantage of the financing instruments available within the region. These should, if possible, be tailored to the specific stage of the project development process considered. Public funds to be devoted to support new projects should consider the risk of the several stages. Thus, subsidies could be focused on the initial stages, where the risk of failure is high and the amount of funds to be mobilized is relatively small. Loans could be devoted to finance the later stages of the development process, mainly. As for the remuneration of investments, rate adders should be applied to standard rates of return reflecting the level of risk of projects. Large projects involving several countries, or local system, in the region and the use of less mature technologies should have access to larger rates of return.

Regulated investment projects centrally planned could be built, operated, and maintained by specialized transmission companies provided they have enough financial resources and technical knowledge to undertake projects of this complexity. TransCos would be assigned these tasks for new transmission projects in competitive auctions. TSOs could adopt this role when the previous conditions are not met, i.e., TransCos is not large and experienced enough, or the level of competition in the auction is low.

Next, the situation in several regions is described. Within the European Union, several national systems and sub-regional markets coordinate their functioning. Thus, the Internal Electricity Market of the EU is based on the cooperation among the local (national and sub-regional) systems and markets, which are deeply rooted in their territories. Market platforms have been merging into larger ones and may finally give raise to a single pan-European one. However, for the time being, national authorities keep large amounts of decision-making power regarding the management of the expansion and operation of the system.

The expansion of the European grid largely remains in the hands of national authorities. Investments to undertake must be approved by them. However, some coordination mechanism has already been put in place. Thus, the European Association of TSOs, together with the Agency for the Cooperation of Energy Regulators in Europe (ACER), is in charge of producing an indicative Europe-wide network expansion plan periodically (every 2 years). As aforementioned, most of the investments in this plan are indicative, i.e., only the regulators in each country have executive powers over the approval and construction of those of these reinforcements that lie within their territories. However, for those projects that are declared of high priority at the regional level and included in the list of Projects of Common Interest (PCI), regional authorities, represented by ACER, are the ones that have the last word to say regarding their approval and cost allocation. The allocation of the cost of regional projects is based on the negotiations established for this purpose among the countries affected by these projects. However, when an agreement on a PCI is not reached among countries, ACER is in charge of allocating its costs to the former. This, according to regulation, must be carried out based on the benefits that each country is expected to obtain from the project. A description of the regulation in Europe and recommendations for its adaptation to the new context of ever-increasing cooperation is provided in [11].

Within the USA, local systems have merged into regional markets, managed by the RTOs in many parts of the country. This undergoing process has been taking place for the last 20 years. RTO markets are fully integrated ones. Within them, the RTO is in charge of computing regional transmission expansion plans, i.e., there are regional expansion planning authorities in place in large parts of the country centrally planning the expansion of the grid within each region. Besides, RTOs are to cooperate among them to identify, with the help of market stakeholders and network users in general, network reinforcements that are relevant for several RTO markets or regions. The Federal Energy Regulatory Commission has authority over interstate trade. However, besides being approved at the federal level, the construction of new reinforcements must also be approved by local authorities, state and municipal ones. Reinforcements to regional transmission grids have been classified into those that are needed to increase the reliability of the system, those that are required to integrate larger amounts of RES generation and comply with policy objectives in this regard, and those that should increase the economic efficiency in the functioning of the system. The reliability reinforcements proposed by RTOs are normally approved without major opposition at any level. The construction of many reinforcements focused on the integration of RES generation has been opposed by the authorities of some states to be crossed by them. However, courts have normally ruled that they should be approved and have declared the arguments provided by states as not valid or irrelevant. On the other hand, the approval of the construction of the so-called economic reinforcements, focused on increasing the economic efficiency of the markets, has proved to be much more difficult.

Regarding the allocation of the cost of the regional and interregional grids eventually built, RTOs have applied fully integrated solutions within their territories. However, regions have cooperated to a small extent to define and implement generally applicable solutions for the allocation of the cost of interregional transmission projects. According to the regulation in place at the federal level, the cost of regulated investments should be allocated to network users based on the benefits that they are expected to obtain from the former. In reality, the cost allocation solutions implemented in many regions take some measure of the use of transmission facilities by stakeholders as a suitable proxy to the benefits they are obtaining from these facilities. The federal regulator (FERC) delivered in 2002 a proposal for the design of regional markets, the Standard Market Design (SMD) [14], where network development and cost allocation issues are discussed. Afterward, FERC Order 1000, [13], set the beneficiary pays principle as the one to be applied in network cost allocation matters.

Within Central America, there is a Regional Electricity Market (MER) which coexists with the six national markets in the region. A regional regulator, CRIE, for the Regional Commission for the Electricity Interconnection in Spanish, has been created. Together with this, a regional market and SO, EOR, and a regional body representing the six national governments have been implemented. These entities have executive powers over the planning of the expansion and the development of the regional transmission grid and the allocation of the cost of this grid in the form of regional transmission charges. However, the authorities within each country are in charge of allocating the transmission charges computed at the regional level to individual generators and loads in the country.

In practical terms, the only piece of transmission infrastructure that has been planned at the regional level and deployed, afterward, is the SIEPAC line crossing the entire region. Then, the adequate development of national grids is necessary to allow for regional power exchanges to take place. While, in some countries, the national grid has been sufficiently developed in this regard; in others, like Guatemala, this has not been the case, largely constraining the cross-border flows. The planning of the expansion of the grid at the regional level is focused on maximizing the social welfare resulting from the reinforcements computed.

As for the allocation of the cost of the SIEPAC lines and other future regional transmission facilities, the original design considered its allocation to countries and power injections and withdrawals within each country based on a combination of usage-based methods. However, this cost allocation mechanism was rendered very complex and replaced temporarily by another one socializing the costs of the facilities within the SIEPAC line, either across the whole region or within the corresponding country, depending on whether these facilities were classified as interconnectors or not, respectively. The rules of the regional market can be found in [10], the official documents for the regulation developed for the Regional Electricity Market in Central America.

One can see that the paradigm adopted for the expansion of the grid and the allocation of its cost in different regional markets is widely different. Probably, the level of success in the development of the grid in each region at a reasonable cost could provide some light on the features of a desirable regional framework.

11.6 Conclusions

Power systems in the world face the need to heavily decarbonize while undergoing their integration in many regions. In this situation, achieving a strong development of the transmission grid at the lowest possible cost is crucial. This will be needed for integrating large amounts of renewable generation located in remote areas and allowing increasing shares of power exchanges among local systems.

The selection of the transmission reinforcements to undertake and the allocation of their cost should be guided by the benefits expected to be produced by these reinforcements. Doing so will maximize the social welfare increase resulting from the expansion of the grid and should prevent the system stakeholders from opposing the undertaking of these network investments. Computing the benefits expected to be produced by individual transmission expansion projects poses a series of challenges. Then, the authorities should be ready to face them.

Out of the several regulatory paradigms that have been proposed and implemented for the expansion of the transmission grid, having this process managed by regulated entities looking after the interest of all the system stakeholders should, most probably, be the preferred approach. Only these institutions will take into account the whole social benefits to be produced by each project when deciding whether to promote and approve it. Investments promoted by this central regulated entities, the central planner and regulatory authorities, should be built, operated, and maintained by specialized transmission companies that have the financial muscle required to undertake the former, provided competition can be established among a sufficient number of them.

Other possible schemes for the expansion of the transmission grid, namely, merchant investments promoted by private promoters and those other investments promoted by associations of network users benefiting from them, should also be allowed. However, the regulated investments should have priority over alternative ones, which may complement the former when regulated investments are not an option to undertake some specific reinforcements that would benefit the system.

At the regional level and in line with what has been stated above, coordination among national, or local, systems should take place in the process of planning and approving the expansion of the transmission grid through regulated and private investments, as well as in the allocation of the cost of the regulated investments. Entities looking after the interest of network users in the whole region should be vested with the powers to implement such coordination schemes. However, at the same time, the features and history of the development of the power sector in each region should be taken into account when devising the solutions to apply in each case.

References

1. Australian Competition and Consumer Commission, Review of the Regulatory Test for Network Augmentations (2004). Available at: http://www.accc.gov.au/content/index.phtml/itemId/532002/fromItemId/54368
2. K.P. Anderson, A. McCarthy, Transmission pricing and expansion methodology: Lessons from Argentina. Util. Policy **8**(4), 199–211 (1999)
3. J.B. Bushnell, S.E. Stoft, Electric grid investment under a contract network regime. J. Regul. Econ. **10**, 61–79 (1996)

4. F. Bañez, L. Olmos, A. Ramos, J.M. Latorre, Estimating the benefits of transmission expansion projects: An Aumann-Shapley approach. Energy **118**, 1044–1054 (2017a)
5. F. Bañez, L. Olmos, A. Ramos, J.M. Latorre, Beneficiaries of transmission expansion projects of an expansion plan: An Aumann-Shapley approach. Appl. Energy **195**, 382–401 (2017b)
6. G. Brunekreeft, *Market-Based Investment in Electricity Transmission Networks: Controllable Flow*. CMI Electricity Project. Department of Applied Economics. Cambridge University (2003). Available at: http://www.econ.cam.ac.uk/electricity/publications/wp/index.htm
7. H.P. Chao, S. Peck, A market mechanism for electric power transmission. J. Regul. Econ. **10**, 25–29 (1996)
8. O.O. Chisari, P. Dal-Bó, C.A. Romero, Hight-tension electricity network expansions in Argentina: Decision mechanisms and willingness-to-pay revelation. Energy Econ. **23**, 697–715 (2001)
9. Council of European Energy Regulators-CEER, *Implementation of TSO and DSO Unbundling Provisions – Update and Clean Energy Package Outlook*. CEER Status Review. Ref: C18-LAC-02-08. 14 June 2019 (2019). Available at: https://www.ceer.eu/documents/104400/-/-/f69775aa-613c-78a5-4d96-8fd57e6b77d4
10. Comisión Regional de la Interconexión Eléctrica – CRIE, Reglamento del Mercado Eléctrico Regional. Libro III, De la Transmisión. Developed by the Comisión Regional de Interconexión Eléctrica. December 2005. p: 101 (2005). Available at: http://crie.org.gt/wp/wp-content/uploads/2019/02/RMER-010119.pdf (in Spanish)
11. De Clercq B., C. Ruiz Prada, M. Papon, B. Guzzi, S. Ibba, M. Pelliccioni, J. Sijm, A. Van Der Welle, K. De Vos, D. Huang, M. Rivier, L. Olmos, M. Golshani, G. Taylor, Y. Bhavanam, *Deliverable 5.1: Towards a Governance Model for the European Electricity Transmission Network in 2050*. A report within the European Commission 7FP project e-Highway 2050 (2015)
12. D.E. Dismukes, R.F. Cop III, D. Mesyanzhinov, Capacity and economies of scale in electric power transmission. Util. Policy **7**(3), 155–162 (1998)
13. Federal Energy Regulatory Commission-FERC, *Transmission Planning and Cost Allocation by Transmission Owning and Operating Public Utilities*, Docket No. RM10-23-001; Order No. 1000-A (2012). Available at http://www.ferc.gov/industries/electric/indus-act/trans-plan.asp
14. Federal Energy Regulatory Commission-FERC, *Standard Market Design Notice of Proposed Rulemaking*. Docket No. RM01-12-000 (2002)
15. P.L. Joskow, Patterns of transmission investment, in *Within Competitive Electricity Markets and Sustainability*, ed. by F. Lévêque, (Edward Elgar Publishing Limited, Cheltenham, 2006), pp. 131–187
16. P. L. Joskow, *Making Transmission Owners Accountable, presented in 'Investments in Transmission', within conference 'The Economics of Electricity Markets'*. Toulouse, France (2005)
17. P. Joskow, J. Tirole, *Transmission Investment: Alternative Institutional Frameworks*. Presented in 'Wholesale Markets for Electricity', Toulouse, France, November 22–23 (2002)
18. S. C. Littlechild, C.J. Skerk, *Regulation of Transmission Expansion in Argentina Part II: Developments Since the Fourth Line,* CMI Electricity Project. Department of Applied Economics. Cambridge University (2004a). Available at: http://www.econ.cam.ac.uk/electricity/publications/wp/
19. S.C. Littlechild, C.J. Skerk *Regulation of Transmission Expansion in Argentina Part I: State Ownership, Reform and the Fourth Line.,* CMI Electricity Project. Department of Applied Economics. Cambridge University (2004b). Available at: http://www.econ.cam.ac.uk/electricity/publications/w
20. D.M. Newbery, Network capacity auctions: Promise and problems. Util. Policy **11**(1), 27–32 (2003)
21. Pérez-Arriaga, I. J., L. Olmos (2006). Compatibility of investment signals in distribution, transmission and generation. In Competitive Electricity Markets and Sustainability. F. Lévêque, Edward Elgar Publishing Limited, Cheltenham p. 230–288
22. M. Rivier, L. Olmos, Cost allocation issues in transmission network investment, in Electricity Transmission Investment, ed. by J. Rosellón, I. Vogelsan, M. Hesamzadeh (Springer, 2019, to be published)

23. M. Rivier, I.J. Pérez-Arriaga, L. Olmos, Electricity transmission, in *Regulation of the Power Sector*, ed. by I. J. P. Arriaga, (Springer, London, 2013), pp. 2h51–2h340. https://doi.org/10.1007/978-1-4471-5034-3_6
24. C. Vázquez, J. Barquín, L. Olmos, M. Rivier, *Adecuación del Marco Regulatorio para la Integración Regional, Preparado para el Regulador del Mercado Peruano (OS INERG - GART)*, Instituto de Investigación Tecnológica, Universidad Pontificia Comillas p. 123 (in Spanish) (2005)
25. I.J. Pérez-Arriaga, F. Rubio-Odériz, J.F. Puerta Gutiérrez, J. Arcéluz Ogando, J. Marín, Marginal pricing of transmission services: An analysis of cost recovery. IEEE Transactions on Power Systems. **10**(1), 65–72 (1995)

Index

S. Lumbreras et al. (eds.), *Transmission Expansion Planning: The Network Challenges of the Energy Transition*, https://doi.org/10.1007/978-3-030-49428-5

T

The manufacturer's authorised representative in the EU is Springer Nature Customer Service Centre GmbH, Europaplatz 3, 69115 Heidelberg, Germany. If you have any concerns regarding our products, please contact ProductSafety@springernature.com

Printed and bound by CPI Group (UK) Ltd, Croydon, CR0 4YY
30/09/2026
02225695-0009